Biophysics for the Life Sciences

Series Editor: Norma Allewell

For further volumes:
http://www.springer.com/series/10230

Linda O. Narhi
Editor

Biophysics for Therapeutic Protein Development

Springer

Editor
Linda O. Narhi
Research and Development
Amgen, Inc.
Thousand Oaks, CA, USA

ISBN 978-1-4614-4315-5 ISBN 978-1-4614-4316-2 (eBook)
DOI 10.1007/978-1-4614-4316-2
Springer New York Heidelberg Dordrecht London

Library of Congress Control Number: 2013931098

Printed on acid-free paper

Springer is part of Springer Science+Business Media (www.springer.com)

Contents

Contributors

Fatma AlAzzam TechnoPharmaSphere LLC, Downingtown, PA, USA

Wasfi Al-Azzam Biopharmaceuticals Development, GlaxoSmithKline, King of Prussia, PA, USA

Nicolas Angell Process and Product Development, Amgen Inc., Thousand Oaks, CA, USA

Gerald W. Becker Drug Product Development, Amgen Inc., Seattle, WA, USA

Steven Bishop Formulation Sciences, Biopharmaceutical Development, MedImmune, Gaithersburg, MD, USA

Xiaolin Cao Department of Process and Product Development, Amgen Inc., Thousand Oaks, CA, USA

John F. Cipollo Division of Bacterial, Parasitic and Allergenic Products, Office of Vaccine Research and Review, CBER, FDA, Bethesda, MD, USA

Mary Cromwell Late Stage Pharmaceutical Development, Genentech, South San Francisco, CA, USA

Barthelemy Demeule Late Stage Pharmaceutical Development, Genentech, South San Francisco, CA, USA

John Gabrielson Analytical Sciences, Amgen Inc., Longmont, CO, USA

Feng He Drug Product Development, Amgen Inc., Seattle, WA, USA

Min Huang Pfizer Inc., Andover, MA, USA

Yijia Jiang Process and Product Development, Amgen Inc., Thousand Oaks, CA, USA

Sangeeta B. Joshi Department of Pharmaceutical Chemistry, Macromolecule and Vaccine Stabilization Center, University of Kansas, Lawrence, KS, USA

Chava Kimchi-Sarfaty Division of Hematology, Office of Blood Research and Review, Center for Biologics Evaluation and Research (CBER), Food and Drug Administration (FDA), Bethesda, MD, USA

James Kranz Biopharmaceuticals Development, GlaxoSmithKline, King of Prussia, PA, USA

Ozan S. Kumru Department of Pharmaceutical Chemistry, Macromolecule and Vaccine Stabilization Center, University of Kansas, Lawrence, KS, USA

Cynthia Li Process and Product Development, Amgen Inc., Thousand Oaks, CA, USA

Guiyang Li Department of Process and Product Development, Amgen Inc., Thousand Oaks, CA, USA

Jun Liu Late Stage Pharmaceutical Development, Genentech, South San Francisco, CA, USA

David Lowe MedImmune Ltd, Cambridgeshire, UK

Ewa Marszal FDA, Center for Biologics Evaluation and Research, Bethesda, MD, USA

C. Russell Middaugh Department of Pharmaceutical Chemistry, Macromolecule and Vaccine Stabilization Center, University of Kansas, Lawrence, KS, USA

Bilikallahalli Muralidhara MedImmune, Gaithersburg, MD, USA

Linda O. Narhi Process and Product Development, R&D, Amgen Inc., Thousand Oaks, CA, USA

Alla Polozova Biopharmaceutical Development, MedImmune, Gaithersburg, MD, USA

Jack A. Ragheb Division of Therapeutic Proteins, Office of Biotechnology Products, Center for Drug Evaluation and Research, FDA, Bethesda, MD, USA

Vladimir I. Razinkov Drug Product Development, Amgen Inc., Seattle, WA, USA

Atul Saluja Drug Product Science & Technology, Bristol-Myers Squibb, New Brunswick, NJ, USA

Hasige Sathish Formulation Sciences, Biopharmaceutical Development, MedImmune, One Medimmune Way, Gaithersburg, MD, USA

Zuben E. Sauna Division of Hematology, Office of Blood Research and Review, Center for Biologics Evaluation and Research (CBER), Food and Drug Administration (FDA), Bethesda, MD, USA

Ambarish Shah Formulation Sciences, Biopharmaceutical Development, MedImmune, Gaithersburg, MD, USA

Steven J. Shire Late Stage Pharmaceutical Development, Genetech, South San Francisco, CA, USA

Satish K. Singh Pfizer Inc., Chesterfield, MO, USA

Sreedhara Alavattam Late Stage Pharmaceutical Development, Genentech, South San Francisco, CA, USA

Evi B. Struble Division of Hematology, Office of Blood Research and Review, Center for Biologics Evaluation and Research (CBER), Food and Drug Administration (FDA), Bethesda, MD, USA

Juraj Svitel Process and Product Development, Amgen Inc., Thousand Oaks, CA, USA

Srivalli Telikepalli Department of Pharmaceutical Chemistry, Macromolecule and Vaccine Stabilization Center, University of Kansas, Lawrence, KS, USA

Gianni Torraca Department of Process and Product Development, Amgen Inc., Thousand Oaks, CA, USA

David B. Volkin Department of Pharmaceutical Chemistry, Macromolecule and Vaccine Stabilization Center, University of Kansas, Lawrence, KS, USA

Tingting Wang Department of Pharmaceutical Chemistry, Macromolecule and Vaccine Stabilization Center, University of Kansas, Lawrence, KS, USA

Ziping Wei Analytical Development, Novavax, Rockville, MD, USA

Zai-Qing Wen Department of Process and Product Development, Amgen Inc., Thousand Oaks, CA, USA

Sandeep Yadav Late Stage Pharmaceutical Development, Genentech, South San Francisco, CA, USA

Chanel K. Yee Department of Process and Product Development, Amgen Inc., Thousand Oaks, CA, USA

Qin Zou Pfizer Inc., Chesterfield, MO, USA

Chapter 1
Introduction

Linda O. Narhi

The last few decades have seen the evolution of protein therapeutics, from a few drugs isolated from natural sources (such as insulin) to numerous engineered molecules that are designed to target specific diseases; these biotherapeutics comprise an increasingly larger part of the commercial drug market (Walsh 2010; Pavlou and Belsey 2005). While more difficult to make than the traditional small molecule medications, proteins have the advantage of having a sustained half-life under physiological conditions, can be targeted to specific sites, and do not have the same concerns around metabolites generated during clearance in vivo, resulting in fewer side effects. Initial protein therapeutics were hormones and growth factors that were designed, using the new tools of molecular biology, to replace the native proteins that were missing or nonfunctional in certain disease states. As the field and industry matured, the focus shifted to designing molecules to act as antagonist or agonist targeted to specific components (such as receptors or ligands) on a particular cell type. Targeting to receptors commonly found on cancer cells has resulted in very effective oncology treatments with fewer of the debilitating side effects of the traditional chemotherapy treatments. Inflammation is another area that has seen the generation of effective biotherapeutics targeted to turn off the inflammatory response of the effected part of the body in autoimmune diseases (for instance the joints in rheumatoid arthritis). In order for proteins to be used as a drug, they must be stable, retaining both activity and structure, during isolation and storage for up to 2 years, often at high concentration. These are conditions that favor protein aggregation and denaturation.

Proteins are complex macromolecules that are polymers of the 20 amino acids connected through peptide bonds. This primary sequence is then folded into the

---

L.O. Narhi (✉)
Process and Product Development, R&D, Amgen Inc., One Amgen Center Drive,
Thousand Oaks, CA 91320, USA
e-mail: lnarhi@amgen.com

L.O. Narhi (ed.), *Biophysics for Therapeutic Protein Development*, Biophysics for the Life Sciences 4, DOI 10.1007/978-1-4614-4316-2_1,

secondary, tertiary, and quaternary structure required for function. The secondary structure consists of alpha helix, beta sheet, beta turns, or unstructured conformations which are defined by the configuration of the peptide backbone, while the tertiary structure is defined as the overall global protein fold or three-dimensional structure. Because of this higher order structure, a folded protein is stabilized by interactions between amino acids that are located far away from each other on the linear amino acid sequence. These can be hydrophobic, van der Waals or charge-charge interactions, hydrogen bonds, salt bridges, disulfide bonds, etc.

Proteins can be simple single domain globular proteins like the growth factors and cytokines that were the first biotech drugs or complex multi-domain molecules such as the antibodies, the active form of which is actually made of several different polypeptides held together by disulfide bonds between cysteines. Quaternary structure is the association of more than one polypeptide chain. This includes association necessary for the formation of a functioning molecule, like the antibodies and hemoglobin, association resulting in undesirable nonnative species such as heterogeneous partially unfolded protein aggregates that can be formed by stressing the native monomer, and association into the well-defined but nonnative structures characteristic of amyloids.

The conformation (overall global fold or three-dimensional structure) of a protein is important for maintaining biological activity and stability during long-term storage. It can also impact the safety profile and biological consequences of biotherapeutics (Dobsen 2004; Wang et al. 2007). Proteins can be unfolded by many different types of stress, including extremes of pH, high or low ionic strength and other solution conditions, heat, and mixing of the air-liquid interface in solution. All of these conditions can be encountered during the manufacturing, storage, and administration of the protein therapeutic. The amino acid residues comprising the molecule can also undergo chemical modifications which might or might not affect other molecular properties.

Proteins have an inherent affinity to self-associate under the appropriate solution conditions. This reaction in its simplest form is a concentration-dependent thermodynamic equilibrium, but can also proceed as a complex, multistep reaction that includes some irreversible steps. A very important irreversible reaction is the formation or shuffling of disulfide bonds between the monomeric subunits which can occur under appropriate red/ox conditions. These self-associated species, or aggregates, can range in size from dimer to species containing hundreds of monomers. Protein aggregates themselves are comprised of a heterogeneous population containing molecules of different sizes, morphologies, and chemical modifications, formed through different pathways and intermediates, all determined by the protein sequence and the conditions to which the protein has been exposed (Narhi et al. 2010).

Even subtle changes in conformation can result in changes in the safety and efficacy of the product, including decreased activity, inability to bind to receptors, changes in the pharmacokinetics and pharmacodynamic profiles, and the potential for immunogenicity. Biophysical methods are the principle techniques available to determine if a potential protein product has the appropriate higher order structure and can maintain the active conformation during manufacturing, storage, and delivery.

Thus, biophysical techniques play a key role during the development of protein therapeutics.

The development life cycle for pharmaceutical proteins begins with target identification and demonstration of the biological activity of an engineered protein. The next steps are identification of lead therapeutic candidates with this desired activity and selection of recombinant production cell lines. This is followed by process and formulation development and characterization and finally clinical trials and then, if successful, commercialization. The launch of a product changes the focus to product consistency and lot release assays, exploration of different delivery devices and therapeutic indications, comparability assessments, and support for product and process failure investigations.

Attributes that are important to screen during cell line development include the amount of monomer and aggregate of the target protein present in the cell culture media prior to purification. This information will contribute to selection of the final commercial cell line that delivers optimal yield and product quality. For selection of the product candidate itself, in addition to biological activity, the ideal molecule will need to be stable to process and storage conditions, including low pH for viral clearance if it is a mammalian cell line-derived molecule or refolding conditions if it is an *Escherichia coli*-derived protein in the form of inclusion bodies. The protein therapeutic also needs to withstand storage in solution at 4–8°C for 2 years, often at protein concentrations above 100 mg/mL (Mahler et al. 2009). Screening for this type of stability usually involves stressing the material and then assessing the impact of the stress on the integrity of the protein, with particular emphasis on protein aggregation and irreversible unfolding of the native three-dimensional structures.

The requirements for the biophysical assays used to characterize therapeutic proteins vary with stage of development. During the early stages of drug product development, there is very little of the therapeutic protein available and little time to perform the assessments; a relative ranking of the proteins or cell lines is often sufficient to select the candidate to move forward into development. Therefore, assays with minimal material requirements and high throughput are especially valuable, and categorizing candidates into pass or fail is an acceptable output.

Once the molecule to be developed as a therapeutic has been chosen, the focus becomes developing the production process and formulation to be used for the commercial product. At this stage material availability is no longer rate limiting, and more rigorous techniques comparing the higher order structure of the actual material obtained during the different processing steps can be used to ensure that the final product was not irreversibly damaged by the conditions being used for its manufacture.

During formulation development, the stability of the target protein is assessed in different buffer compositions, pH, storage conditions, delivery devices, etc. These studies typically involve the generation of many samples which need to be analyzed in order to arrive at the optimal formulation conditions. Many of the principles that apply during candidate selection apply here as well, and some of the same assays can even be used. Actual long-term stability studies are also performed to demonstrate that the protein drug in the formulation chosen has the desired shelf-life stability and stability to administration by the chosen delivery method. Determination of conformational integrity and amount of protein aggregation are important aspects of these studies, which usually involve more rigorous characterization and quantitative analyses.

As the product moves into production, the emphasis switches from developing and optimizing conditions to maintaining process/product control; biophysical techniques to follow the protein higher order structure are important parts of comparability studies and are required for obtaining licensure of the drug. In this case the methods must be shown to be fit for purpose, and the sensitivity of the assays to detect changes in the product needs to be determined. Another important aspect of preclinical and clinical development is the monitoring of stability samples, stored both under accelerated and recommended conditions, for comparability. Biophysical techniques are also used as tools to help ensure that changes in device, concentration, and formulation made as different indications or patient populations are added do not affect the conformation of the biotherapeutics. Techniques which can give reproducible and accurate results, and where the readout is understood, are most commonly used at this stage of development rather than the high throughput tests that were employed in the beginning of the product development life cycle. These analyses must be sensitive to changes in the protein conformation that can occur if the protein is exposed to slightly different process or storage conditions, as demonstrated by samples exposed to conditions outside the normal parameters.

Finally, during commercial production, batches occasionally fail the various lot release assays; biophysical techniques can be used o help identify root cause and contribute to the safety assessment of lots. In this case, very often, a single visible aggregate is being studied and so the methods must have the sensitivity to detect and analyze a very small amount of protein and provide a positive ID of the material if possible.

In this volume, organized to follow the product life cycle, the applications of biophysical analyses during the therapeutic protein development are presented. The initial chapters describe the underlying theory, strengths, and weaknesses of the different biophysical techniques commonly used during protein therapeutic development, while the later chapters present case studies to illustrate the application of biophysics to protein therapeutic development. Chapter 2 describes the strategy, development, and implementation of high throughput methods, while Chap. 3 provides an overview of the biophysical techniques commonly used for higher order structure analysis during the development of a biotherapeutic. One of the key product quality attributes that are be tracked during protein processing and storage is protein aggregation. The methods being developed and implemented for particle size distribution analysis and aggregate characterization are discussed in Chap. 4. And finally Chap. 5 describes the qualification of biophysical methods for the characterization of protein higher order structure and demonstration of fit for purpose of these methods. The sensitivity and limits of the methods must be demonstrated in order for them to be used in licensure filings, during characterization, and for comparability studies.

Chapters 6–10 present case studies organized to follow the product life cycle. Chapter 6 demonstrates the use of higher order structure analysis during discovery and candidate selection, Chap. 7 and 8 illustrates the use of biophysical techniques during process and product development. Chapter 9 focuses on characterization of protein aggregates, one of the key product attributes that have the potential to effect product safety. Chapter 10 contains several case studies demonstrating how biophysical techniques are used to identify particles and sources of product and process failure investigations. And finally Chap. 11 presents a regulatory perspective on higher order structure analysis across the life cycle. It is our hope that this volume will enhance the readers understanding and appreciation for the important role that biophysics plays in successful therapeutic protein development.

References

Dobsen CM (2004) Principles of protein folding, misfolding, and aggregation. Semin Cell Dev Biol 15:3–16

Mahler H-C, Friess W, Grauschopf U, Kiese S (2009) Protein aggregation: pathways, induction factors and analysis. J Pharm Sci 98:2909–2934

Narhi LO, Jiang Y, Deshpande R, Kang S, Shultz J (2010) Approaches to control protein aggregation during bulk production. In: Wang W, Roberts CJ (eds) Aggregation of therapeutic proteins. Wiley, Hoboken

Pavlou AK, Belsey MJ (2005) The therapeutic antibody market to 2008. Eur J Pharm Biopharm 59:389–396

Walsh G (2010) Biopharmaceutical benchmarks. Nat Biotechnol 28:917–924

Wang W, Singh S, Zeng DL, King K, Nema S (2007) Antibody structure, instability and formulation. J Pharm Sci 96:1–26

Chapter 2
High-Throughput Biophysical Approaches to Therapeutic Protein Development

Feng He, Vladimir I. Razinkov, C. Russell Middaugh, and Gerald W. Becker

2.1 Introduction

Protein therapeutic products typically experience many development cycles, in which decisions are empirically derived concerning the identity, manufacturing process, final product presentation, and administration methods. To obtain a quality therapeutic product, significant resources are spent on the search for appropriate development parameters throughout a product's life cycle. Due to the length and complexity of such development in the pharmaceutical industry, any process that can reduce the amount of time and resources while still reaching acceptable outcomes is highly preferred. One prominent area of such improvement is to increase the throughput capability of existing technologies. The focus of this chapter is on providing an introduction to the biophysical methods and techniques possessing high-throughput capabilities. We will especially focus on those techniques that are frequently used in the development of protein therapeutics. Though the definition of high throughput can be quite broad, the technologies and methodologies highlighted in this chapter all possess multi-sample throughput capability and/or automation-enabled measurements and analysis.

During the development cycle of protein therapeutics, it is critical to understand the physical integrity and other structural characteristics of the product. In the early stages of product development, preferred physical properties are often used as

F. He (✉) • V.I. Razinkov • G.W. Becker
Drug Product Development, Amgen Inc., 1201 Amgen Court West,
Seattle, WA 98119, USA
e-mail: fhe@amgen.com; razinkov@amgen.com; geraldb@amgen.com

C.R. Middaugh
Department of Pharmaceutical Chemistry, University of Kansas,
2030 Becker Dr., Lawrence, KS 66047, USA
e-mail: middaugh@ku.edu

L.O. Narhi (ed.), *Biophysics for Therapeutic Protein Development*, Biophysics for the Life Sciences 4, DOI 10.1007/978-1-4614-4316-2_2,

criteria to select protein candidates (described in Chap. 6). The selection philosophy can be based on an established relationship between a particular biophysical property and product quality, or simply follow the general assumption that better biophysical characteristics can lead to more stable products if other methods cannot differentiate the drug candidates. Typically, early stage development is conducted at small scale employing in vivo and in vitro systems because protein availability is limited. Thus, methods that consume less material but generate useful information are the most attractive. Once a protein drug candidate is selected, development is primarily focused on optimizing the parameters that effectively enable the manufacturing, packaging, storage, and delivery of the final product. Before these parameters can be finalized for commercial processes, however, numerous analyses need to be performed to define the space and limitations of physical conditions that best fit the product. High-throughput biophysical tools often play an important role in these efforts by providing a faster readout when the protein product is subjected to a range of experimental conditions.

Two areas during protein therapeutic development frequently employing high-throughput biophysical analyses are downstream and formulation development. The goal of downstream development is to find the most suitable purification process so that it can be properly scaled to deliver commercial product. Although the final optimization usually takes place at scales similar to the commercial settings, small-scale and high-throughput approaches are often utilized to predict protein behavior. For example, multiwell plate-based chromatographic techniques combined with high-throughput liquid handling instruments can provide a tremendous amount of information on protein–resin interaction and therefore help guide purification design with very affordable material input (Coffman et al. 2008). Although the ability of a process to perform adequately at a large scale remains to be derived empirically, small-scale development is critical because it provides opportunities to test the potential manufacturing steps while varying physical parameters. In fact, many regulatory authorities, when evaluating new drug product applications, require such small-scale data that can demonstrate the robustness of the processes as well as the edges of failure. Besides purification, the formulation development of protein therapeutics also requires a large number of experimental trials to select suitable conditions for final product presentation. The general aim is to find a formulation that will best retain the physical, chemical, and biological properties of the product while meeting the desired shelf life requirements. More recently, patient convenience and comfort have become major factors that influence formulation development. For instance, product formulations compatible with self-administration and above-freezing storage are preferred when treating chronic diseases. To derive suitable formulations, large arrays of factors are typically screened during the development phase for their ability to stabilize the drug product. Since it is well established that physical instability may negatively impact protein therapeutics, biophysical techniques offer relevant tools to monitor protein conformation in response to formulation and storage conditions. An important area of focus during formulation development is the evaluation of protein aggregation propensity. Protein aggregates are thought to often impose detrimental effects on the therapeutic potency and side effect profile of the drug which may even lead to significant clinical safety concerns (Jiskoot et al. 2012; Rosenberg 2006). As a result, protein aggregation is frequently used to differentiate formulation candidates. Over the last decade, it has been well documented in the literature that high-throughput biophysical methods are very capable of detecting the presence of protein aggregates over a wide size range (Mach and Arvinte 2011). The feasibility and performance of a protein therapeutic product can also be dependent on other key properties of the molecule. An emerging example is protein viscosity. To reach the desired bioavailability, protein drugs delivered via the subcutaneous route typically require high concentrations. This often leads to high solution viscosity that may cause significant difficulties during product manufacturing and administration of the protein (Shire et al. 2004). Traditional analytical methods have limitations in material consumption and throughput (Jezek et al. 2011), but newly developed biophysical techniques offer significant advantages, including a reduction of sample volume as well as increases in throughput capability (He et al. 2010a; Wagner et al. 2012).

In the following sections, the role of high-throughput biophysical analysis in the development of protein therapeutics is discussed. Technical background for selected biophysical methods is reviewed as well as their high-throughput utility. In addition, a general introduction to empirical phase diagram (EPD) is presented as an example of the application of high-throughput biophysical characterization and data interpretation in the development of protein therapeutics.

2.2 High-Throughput Biophysical Techniques That Can Be Applied to Protein Therapeutic Development

Due to increasing demands in sensitivity, diverse sample applicability and throughput, biophysical instrumentation has been transformed significantly over the past decade to implement multiwell measurements and automation modules. Such advancement has accelerated the research and development of protein-based therapeutic entities by decreasing the resource requirement and widening the experimental design space. In this section, a brief introduction is provided, concerning the background and utility of selected high-throughput techniques.

2.2.1 Surface Plasmon Resonance

Characterization of target binding is an essential component during the development of protein therapeutics. Since the majority of therapeutic targets bind to specific targets, the ability of protein therapeutics to interact with their respective sites of interaction is often used to differentiate among drug candidates. Many high-throughput biophysical tools are available to characterize the binding constant, K_d, and stoichiometry, (n), between protein product and therapeutic targets.

Protein–protein interactions can be characterized with the help of surface plasmon resonance (SPR) (Rich and Myszka 2000). This method is based on the measurement of changes in refractive index near a sensor surface caused by the binding or dissociation of a protein with its target (receptor). SPR instruments evaluate protein binding in real time without labeling, but surface immobilization of at least one component is required. SPR methods can be both qualitative and quantitative (Karlsson 2004). In the standard setup, measurements are performed sequentially with two flow cells: one that holds the sample and the second a reference solution. Recent microfluidic and automation technologies have created new opportunities for development of high-throughput instrumentation based on the SPR principle. The Biacore A100 biosensor can process multiple samples (1,000 samples/day) and allows high-quality data sampling. This instrument has been used to develop an in vitro high-throughput kinase assay (Takeda et al. 2006). Another SPR instrument, the Biacore Flexchip microarray device, has been used for rapid identification of high-affinity human antibodies from a phage display screen. Fab fragment analysis with surface plasmon resonance microarrays in a high-throughput format permits the determination of kinetic constants for 96 different Fab fragments in a single experiment (Wassaf et al. 2006). High-throughput antibody affinity characterization accelerates early discovery of lead candidates. The ProteOn™ XPR36 multiplexed SPR instrument from Bio-Rad Laboratories (Hercules, CA) (Bravman et al. 2006) employs microfluidics integrated into 6×6 interaction array. The ProteOn™ XPR36 has also been used for determination of antibody affinity (Bravman et al. 2006). Plexera® Bioscience (Woodinville, WA) offers a PlexArray™ HT instrument which can test thousands of protein interactions in only 30 min. This system is based on a high-density array with the capacity to evaluate more than 1,000 spots with a spot size as small as 100 μm. Microarray technology has been widely applied in SPR-based settings to miniaturize spot size and sample volume to increase throughput (Otsuki and Ishikawa 2010). A wide variety of other instruments based on SPR and related phenomena are available as well.

2.2.2 Liquid Chromatography

Liquid chromatography (LC) is perhaps the most heavily used tool in the field of biotechnology. LC methods are the main components for the purification and analysis of protein therapeutics during their development cycle (Ahrer and Jungbauer 2006; Andrew and Titus 2001). Understandably, most LC technologies have adopted the high-throughput scheme by automating and streamlining the sampling mechanism. The principle of protein LC is based on the nature of protein–resin interactions in a given liquid mobile phase. An initial major step towards high-throughput LC technology is the implementation of high-performance liquid chromatography (HPLC) (Swadesh 2001). This technique uses a fast flow rate under high enclosed pressure, instead of gravity force, to drive the mobile phase through the column. This permits faster sampling throughput and improves the resolution of sample partition. A recent advancement has been categorized as ultra-performance liquid chromatography (UPLC) where smaller diameter particles and subsequently higher pressures enable even faster separation times (Xiang et al. 2006). UPLC technology has clearly shown its advantage in reversed phase (RP) chromatography and has become the primary separation method for mass spectrometry analysis (Stackhouse et al. 2011; Szapacs et al. 2010).

Another application of high-throughput LC technology is the development of protein purification methods. The goal is to obtain the largest amount of the therapeutic protein in its active form while reaching the highest purity possible. The range of such methods is currently extremely wide, and the specific purification steps often need to be derived empirically. Primary methods include affinity, size-exclusion, ion-exchange, and hydrophobic interaction chromatographies. Multiwell plates or miniaturized columns are frequently used to screen a large number of parameters including type of resin, protein loading and elution conditions, as well as efficiency for the product of interest (Fahrner et al. 2001). The quantity of column resin and protein material needed is often measured in microliters, and the experimental steps are simply executed by gravity, centrifugal force, or a pump. This type of high-throughput approach offers an opportunity to evaluate as many variables as possible early in development with a minimum of time and protein. The outcome of the screening can often help to select product-specific processes for further development and may even allow companies to bypass intellectual property restricted common practices. The latest improvement in this field highlights the use of automated liquid handling systems coupled to positive displacement liquid transfer technology (Susanto et al. 2008; Wiendahl et al. 2008). Such a system has greater similarity to large-scale chromatographic instruments used in manufacturing by generating a pressurized liquid flow through the microcolumns. The results obtained using such high-throughput techniques are believed to be more representative of a protein's behavior during large batch purification.

2.2.3 Light Absorption Spectroscopy

Protein concentration measurements are essential during purification and formulation development, and high-throughput compatible UV spectroscopy is a well-utilized tool (Zhao et al. 2010). Besides protein concentration measurements, second derivative UV spectroscopy has been frequently used to characterize protein tertiary structure (Kueltzo et al. 2003; Mach and Middaugh 2011). Many absorption spectrometers are compatible with a multiwell plate format and/or automation modules. Typically, light passes through the plate vertically while the light source and detector move from sample to sample. Alternatively, the plate itself can be moved. Since the signal from proteins is typically strong, UV detection and absorbance integration are generally quick, allowing fast analysis of a large number of samples. The disadvantages of this technique are also very well understood.

The path length of the vertical light absorbance is poorly controlled and highly dependent on the amount of sample present in the well. In addition, proteins with high extinction coefficients, such as monoclonal antibodies, can easily saturate the light detector. New adjustable path length spectrophotometers have recently become available, although their throughput is not as high and light scattering frequently complicates the use of this method.

Compared to UV absorbance, high-throughput optical density (OD), or turbidity, assessment is more widely applied to the protein therapeutic development. Turbidity usually refers to the obscuration of a sample at wavelengths near the visible light range where proteins in solution do not manifest significant absorbance. The most widely employed wavelength range for this purpose is 350–400 nm, which avoids any specific absorbance arising from amino acid side chains or common color pigments. Turbidity is proportional to the amount of light blocked or scattered by the solution components. In a protein sample, aggregates and precipitates are known to give rise to solution turbidity as measured by OD, making it a quick method to assess the quality of protein samples with respect to the presence of protein aggregates. High-throughput turbidity assessment is frequently used during protein formulation development to evaluate a large number of samples that are put on storage or under environmental stresses (Zhao et al. 2010; Capelle et al. 2009). Though quantitative determination of protein degradation is not usually possible with turbidity measurements, the information obtained is sufficient to discriminate or rank order formulations. Turbidity assessment is generally noninvasive and can be applied using a variety of spectroscopic instrumentation and sample cells, including microtiter plate readers and pharmaceutically relevant containers. The latter provide a unique opportunity to assess sample quality of protein therapeutics in their actual storage units, and permit a real-time monitoring of aggregation during the manufacturing and distribution of a protein commercial product.

Another common tool for protein analysis is circular dichroism (CD). CD is a technique that measures the difference in absorbance of left- and right-handed circularly polarized light. CD is generally divided into near-UV (250–350 nm) and far-UV (190–250 nm) measurements (Li et al. 2011). Near-UV CD is sensitive to the tertiary structure of proteins, due to the presence of optically active chromophores including the aromatic amino acid side chains and disulfide bonds. Far-UV CD, on the other hand, is used to study the secondary structure of proteins. Alpha helix, beta sheet, turns, and disordered structure all display unique CD spectra. Expanding the high-throughput capabilities of CD instruments is achieved either by increasing the number of cuvettes that an instrument can employ or the use of autosamplers. In addition, it has recently become possible to obtain data from the near- and far-UV regions in a single scan. CD is typically more time-consuming than other light absorbance measurements and can be significantly affected by both light scattering and absorption flattening phenomena. In order to minimize the interference, high-concentration protein samples are often measured via the use of short path length (μm range) cells (Harn et al. 2007).

2.2.4 Vibrational Spectroscopy

Vibrational spectroscopy includes Raman, Fourier transform infrared (FTIR), and near-infrared (NIR) spectroscopy. With respect to protein therapeutics development, various Raman techniques and FTIR are commonly used to analyze protein secondary structure, while NIR is often applied to the analysis of the sample components other than protein, such as organic compounds (Siebert and Hildebrandt 2008). Because most chemical materials have identifiable spectral patterns, these techniques are also frequently used for raw material and forensics analysis (described in Chap. 10). Another common advantage of these techniques with protein therapeutics development is that they can be used to analyze lyophilized protein samples directly, a task at which many other biophysical techniques fail. Near-IR instrumentation has successfully incorporated an automated sample presentation unit that permits higher throughput analysis. The high-throughput development of FTIR, however, is challenging due to geometrical problems. The only currently available high-throughput option for FTIR involves the drying of liquid samples, which might produce structural changes in the protein. Novel solutions to this problem are known to be under investigation. In contrast, incorporation of multiwell plate and microarray technologies instrumentation has been demonstrated with Raman spectroscopy (Anquetil et al. 2003). The advantage of Raman-based technologies is that they are typically compatible with a wide range of transparent container types and suffer less from interference by water. The major disadvantage of Raman spectroscopy is that signals from proteins are normally weak. Resonance and surface-enhanced Raman technologies can partially overcome this problem but suffer from limited applicability for a variety of reasons. In summary, vibrational spectroscopic techniques possess unique abilities to provide quality and conformational information on protein products, including lyophilized samples. High-throughput applications of these techniques, however, clearly require further development.

2.2.5 Fluorescence Spectroscopy

The aromatic side chains in proteins serve as fluorophores which emit photons at higher wavelengths once excited with UV range light. The excitation and emission profile is highly dependent on the polarity of the side chain environment and, therefore, is sensitive to the local tertiary structure of the protein (Lakowicz 2006). This fluorescent property is referred to as protein intrinsic fluorescence. Since most large proteins have widely distributed aromatic amino acids, the overall intrinsic fluorescence provides a good measure of protein folding. Upon conformational changes, the intrinsic fluorescence emission peak generally shifts in wavelength. If present, tryptophan (Trp) emission dominates the fluorescence of proteins with tyrosine (Tyr) contributing indirectly. This method is often applied to protein formulation studies to detect changes in protein conformation as a result of stresses such as temperature, pH, and solute.

The high-throughput capability of fluorescence techniques also typically relies on a multi-sample cuvette holder, a multiwell plate compartment, or an autosampler. Generally, fluorescence intensity is detected at a 90° angle to the light source to minimize scattering. In the case of multiwell plates, fluorescence signal is usually acquired via a top or bottom reading method. Because of the noninvasive nature of the intrinsic fluorescence measurements, it can be applied to analyze protein conformation in samples pulled from long-term studies. The utility of intrinsic fluorescence can be extended by fluorescence lifetime and anisotropy measurements, through which further information concerning protein conformation is obtained (Fowler et al. 2002; Owicki 2000). Use of front surface geometry permits the fluorescence from highly scattering samples to also be obtained. Recently, a high-throughput, microtiter plate-based fluorometer (the Avacta Optim 1000) has been developed, which measures fluorescence spectra and light scattering simultaneously. This instrument is capable of performing rapid thermal melts with a resultant overall increase in throughput of better than a factor of 10.

Useful fluorescence signals can also be obtained from extrinsic fluorescent probes such as small molecule dyes. There is a wide variety of fluorescent dyes commercially available for protein research and development (Hawe et al. 2008a). Different dyes can be employed to reveal different properties of protein solutions. For example, hydrophobic dyes are typically used to probe the presence of apolar sites on a protein. Anilinonaphthalene sulfate (ANS)-based dyes are the most common hydrophobic probe choices. An increase in surface hydrophobicity is believed to be an indication of protein unfolding. This usually results in enhancement of fluorescence intensity by the associated dye. Another widely used dye is SYPRO Orange, a probe originally developed as a gel stain. Recently, SYPRO Orange has been used to detect protein aggregation in monoclonal antibody product samples and has been shown to display specificity to structurally perturbed protein aggregates (He et al. 2010b; Mach et al. 2011). Similarly, a number of other dyes have been reported to be sensitive to the aggregated protein species (Hawe et al. 2010a; Hawe et al. 2010b). SYPRO Orange has also been employed to study protein unfolding (described in Sect. 2.7). Other extrinsic probes have been employed to measure solution viscosity (Schäfer and Schmidt 2006; Haidekker et al. 2005; Kung and Reed 1989). Though the application of these viscosity-sensitive dyes has been extensively demonstrated in biologically relevant samples, their utility in probing therapeutic protein viscosity remains to be fully tested. One major concern is that these dyes can interact with proteins in high-viscosity samples, which typically correlates with high protein concentration (Hawe et al. 2010a). In addition to probing the protein itself, other extrinsic fluorescent tools are available to study formulation buffers. For instance, 1-N-phenylnaphthylamine (NPN) and 9-anthryldiazomethane (ADAM) have been shown to detect polysorbate, a common surfactant used in protein formulation, and its degraded form in solution (Khossravi et al. 2002). Fluorescence spectroscopy is also compatible with HPLC instrumentation as an on-line detection method to monitor elution (Hawe et al. 2008b), offering yet another high-throughput analytical option. The combination of chromatography and fluorescence is particularly attractive, since real-time information can be obtained on specific protein species as they fractionate and elute from the column.

2.2.6 Isothermal Titration Calorimetry

Isothermal titration calorimetry (ITC) is another sensitive method to evaluate interactions in solution (Torres et al. 2010). Understanding how proteins interact with other solution components, such as surfactants and excipients, could be informative and beneficial for the development of protein therapeutics. Unlike SPR, ITC does not require immobilization or chemical modifications of the interacting compounds (Pierce et al. 1999). This enables ITC to be applied to more types of interactions that occur with proteins. In its standard format, this technique requires a significant amount of protein and the experiment includes long-duration titrations. The high-throughput development of ITC has been aimed at reducing the sample consumption as well as improving the measurement speed. Recently ITC instruments with reduced sample volumes have become commercially available, including iTC200 and Auto-iTC200 calorimeters from MicroCal (GE Healthcare, Waukesha, WI), as well as Nano ITC instrument (TA Instruments, New Castle, DE) which offers higher sensitivity and titration speed and employs a significantly smaller sample cell compared to the two-cell MicroCal VP-ITC instrument (Peters et al. 2009). The Auto-iTC200 can run up to 384 titrations automatically through the use of a temperature-controlled autosampler. A miniaturized ITC microcalorimeter can reduce the protein requirement up to seven times (Verhaegen et al. 2000). Forty-eight samples in each array can be measured with a reaction volume of 10–20 μL. In addition, closed-chamber microfluidic calorimeters with thermopile heat sensors have also been described in the literature (Lee et al. 2009). These calorimeters achieve enhanced sensitivity by surrounding the measurement chamber with a vacuum. Flow microfluidic devices can use as little as 10–20 μL of sample volume (Lerchner et al. 2008). In addition to the single-chamber settings, further advancements utilizing array technologies might be possible.

2.2.7 Differential Scanning Calorimetry/Differential Scanning Fluorometry

Stability of protein therapeutics under thermal stress is believed to be critical and often assessed during development (Bruylants et al. 2005). A low temperature of unfolding can lead not only to protein instability but also a decreased energy barrier for unfolding events caused by protein interactions such as protein–surface (Chang et al. 1996), protein–solvent, (Schiffer and Dotsch 1996) or other interactions. Screening approaches focused on protein thermostability can reveal undesired protein therapeutic candidates or manufacturing and storage conditions early during the development. Differential scanning calorimetry (DSC) is a quantitative method widely used for determination of protein thermal stability (Privalov and Privalov 2000). However, a low-throughput DSC takes a significant amount of sample and time to complete a typical experiment. Automated autosampling instruments such as the VP-Capillary DSC Platform from GE Healthcare (Piscataway, NJ), the PYRIS Diamond™ DSC Autosampler from PerkinElmer (Waltham, MA), or the Nano DSC Autosampler System™ from TA Instruments (New Castle, DE) significantly improve sample preparation, but analysis time is still lengthy and the significant amounts of protein required make analysis difficult for the early screening of hundreds of drug candidates and formulation excipients. Improvements in microfluidic technology, creation of array-based calorimetry microchips, and the use of microplates will enable higher-throughput calorimetry (Vermeir et al. 2007; Lerchner et al. 2006). Microelectromechanical system (MEMS)-based calorimetry requires only 1.2 μL of solution, and the sensitivity is about 5 mg/mL of protein concentration (Wang et al. 2008). Since the dimensions of this device are approximately 5×5 mm, there are opportunities for microarray fabrication.

DSC can detect the unfolding transitions of several domains in a multidomain protein such as an immunoglobulin. It is, however, often sufficient to screen the thermal stability of proteins based only on their lowest melting temperature. Single-domain proteins can, of course, also be characterized by determination of their single transition. In cases such as this, a high-throughput method, known as differential scanning fluorometry (DSF), based on the extrinsic fluorescence of probes sensitive to the polarity of their environment can be used (Pantoliano et al. 2001a). Unfolding of proteins usually exposes hydrophobic regions to the solution resulting in a significant increase in the fluorescence of these probes when bound. The method was originally used for screening of small molecule interactions with proteins and can therefore be used to identify some types of excipients (Pantoliano et al. 2001b). The unfolding temperature of a protein shifts after binding of a ligand and binding parameters can be determined by this shift in the melting temperature. In other studies this method has been applied to the evaluation of crystallization and general stability (Malawski et al. 2006; Ericsson et al. 2006). Because of low background fluorescence in the presence of native antibodies, DSF has been successfully used for mAb formulation development (He et al. 2010c). Examples of DSF scans obtained during formulation screening for a therapeutic protein are shown in Fig. 2.1. A 96-well plate with different formulations was screened to determine the most thermally stable formulation. The midpoint of the transition determined by the increase in fluorescence intensity was found to correlate well with the unfolding transition measured by DSC. One significant disadvantage of DSF in applications for protein formulation development is the high-fluorescence background in the presence of detergents commonly used in protein formulations (see high-fluorescence background in Fig. 2.1). New probes, such as the thiol-specific fluorochrome N-[4-(7-diethylamino-4-methyl-3-coumarinyl) phenyl] maleimide (CPM), seem to overcome this detergent problem and have been used for stability profiling of membrane proteins under different solution and ligand conditions (Alexandrov et al. 2008). Fluorescent plate readers with a thermostat are necessary to make these measurements. The maximal temperature of a typical scan is high and standard plate-based fluorometers usually cannot provide the necessary temperature range. Instead real-time polymerase chain reaction (RT-PCR) instruments equipped with 96, 384 and even 1024-well plates are well suited for DSF measurements.

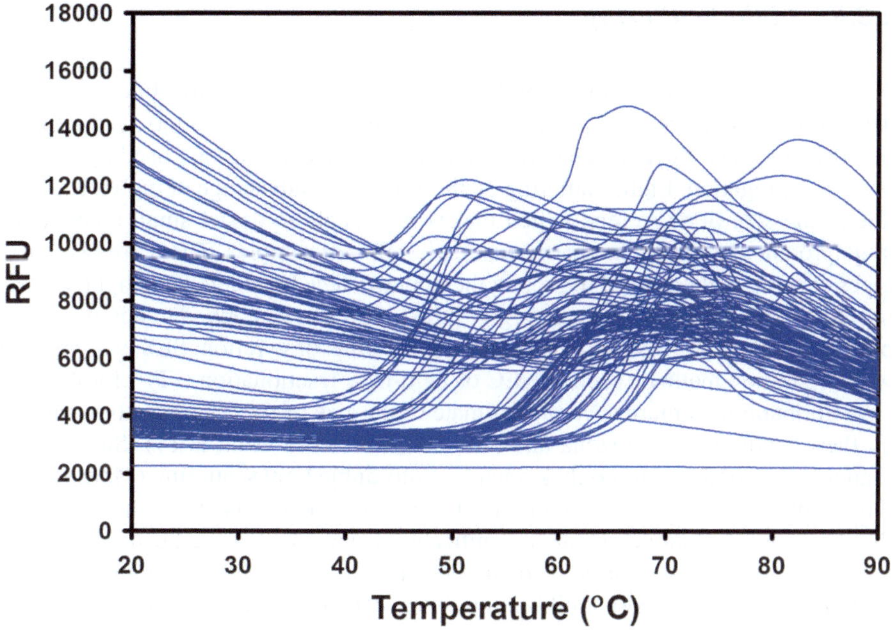

Fig. 2.1 Representative differential scanning fluorometry spectra obtained with a therapeutic protein on 96-well plate filled with different formulations for each well. Fluorescence intensity in relative fluorescence units (RFU) is shown as a function of temperature. Protein concentration is 1 g/L. The sample volume is 30 μL. Data were obtained using the Bio-Rad CFX96 RT-PCR plate reader

Many companies that manufacture these instruments have realized the popularity of this application and have included software options to obtain and analyze DSF data. Another advantage of DSF is the wide range of protein concentration that can be used. Depending on protein properties, transitions can be detected at as low as 0.05 mg/mL or as high as at 100 mg/mL. Another high-throughput method, the ProteoStat™ assay from Enzo Life Sciences (Farmingdale, NY), provides an improved thermal shift approach based on extrinsic fluorescence for assessment of protein stability through monitoring protein aggregation, rather than protein unfolding. These methods, which employ high-throughput technologies, further expand the application of thermal analysis to modern pharmaceutics. While it is not guaranteed that thermal stability correlates with a protein's physical stability during storage, better thermal stability usually indicates a greater energy requirement to unfold the protein. When executed under similar experimental conditions, high-throughput thermal analysis offers useful information that can be used to rank protein or formulation candidates. It seems safe to state that when all other properties are comparable, the protein constructs or formulations which lead to better thermal stability will always be more desirable.

2.2.8 Light Scattering

Aggregation is one of the major problems in protein pharmaceutical development. The presence of aggregated protein can compromise the purity, safety, and efficacy of a drug product. Several separation and detection techniques are used to monitor protein aggregation. Light scattering methods have the advantage of high sensitivity due to the size of scatterers, which makes it possible to detect small amounts of large protein aggregates in pharmaceutically relevant samples. There are two basic types of such methods used in protein therapeutics development: static light scattering (SLS) and dynamic light scattering (DLS). SLS can be applied to determine the protein's molecular mass and the mean square radius of gyration. SLS is often used with separation methods such as SEC or field-flow fractionation (FFF) for the purpose of obtaining a more accurate estimate of the size of the components separated by these techniques (Tarazona and Saiz 2003; McEvoy et al. 2011). Such online scattering method is commonly known as multi-angle light scattering (MALS). The throughput of SLS analysis used in this fashion often depends on the throughput of the separation technique. Though significant advancements have been achieved in numerous types of separation technologies, a variety of physical parameters, such as high pressure and short equilibration times, can be problematic for the coupling of light scattering detectors. Light scattering can also be coupled to plate readers. Simple monitoring of the scattered light at fixed wavelengths and angles can provide sensitive detection of aggregates formation. The Stargazer-384™ system (Harbinger Biotechnology and Engineering Corporation, Toronto, Canada), a 384-well microplate reader, was used in a study to monitor colloidal stability of mAbs at elevated temperatures (Goldberg et al. 2010). As previously discussed, the high-throughput Avacta Optim® 1000 (Pall Corporation, Port Washington, NY) is available for light scattering as well as fluorescence measurements with a volume as low as 1 μL. In these instruments the light scattering signal is used to monitor aggregation by detecting the increased intensity, similar to turbidity measurements but with higher sensitivity. It should be noted, however, that if the aggregates formed are less dense than the monomeric protein, decreases in scattering can also be seen.

Dynamic light scattering is based on the measurement of the fluctuations in intensity of scattered light. An autocorrelation function, derived from fluctuation analysis, can reveal the distribution of the hydrodynamic radii of protein molecules present in solution (Schmidt 2010). Separation is not necessary, but resolution of species depends on their size difference and concentration. Because of the exponential form of the autocorrelation function, no more than two to four components can be comfortably resolved in the same solution. DLS plate readers have become very popular in the biopharmaceutical industry for protein and vaccine characterization (Vincentelli et al. 2004). Multiwell plate formats, small volumes, and automated procedures for measurement and data analysis make the DLS method high throughput and easy to apply. Antibody self-association has been studied with the help of gold nanoparticles and their characterization by dynamic and static light scattering. Nanoparticle–antibody conjugates displayed complex aggregation behavior dependent on pH and

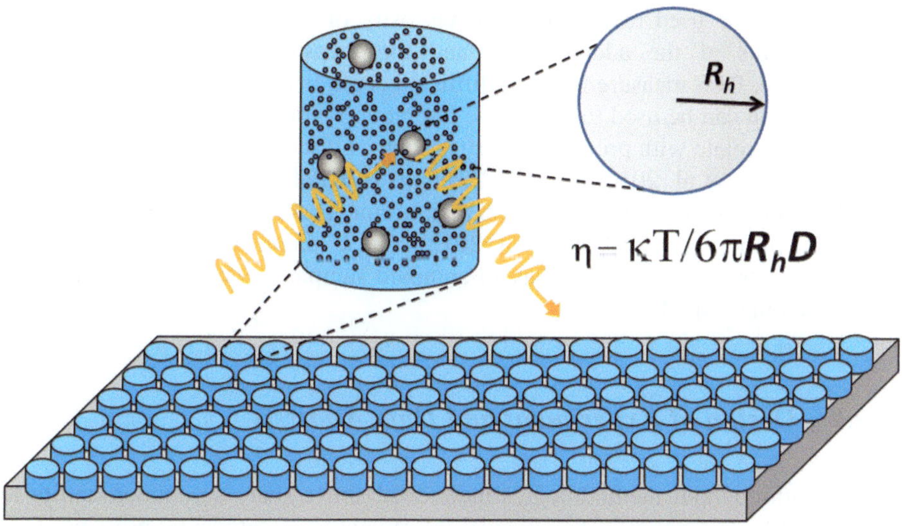

Fig. 2.2 High-throughput method for viscosity measurements based on dynamic light scattering determination of the diffusion coefficient of polystyrene beads externally added into a protein solution

ionic strength of the solution. Use of a DLS plate reader was a significant part of the high-throughput analytical development (Sule et al. 2011).

In biopharmaceutical drug development DLS has been used not only for direct detection and characterization of aggregation but also for the study of large colloidal structures. Certain large colloid-like aggregates have been shown to inhibit enzymes leading to false-positive HTS leads. These so-called promiscuous inhibitors were detected and screened by DLS using a plate reader (Feng et al. 2005). Results from such high-throughput assays for promiscuous inhibitory aggregates have been used to develop new computational models of this phenomenon. A method for quantitative characterization of macromolecular interactions using DLS has been introduced in a temperature-controlled plate reader format (Hanlon et al. 2010). This technique enabled determination of equilibrium dissociation constants and thermodynamic parameters. The low volume of plate-based DLS reduced the sample amount to a few microliters per experiment, with detection limits in the femtomolar range.

Biopharmaceutical products are often formulated at high concentrations to maximize delivery dosage and efficiency, and solutions of some proteins become very viscous at high concentrations (Yadav et al. 2010), creating significant problems for processes like purification, filtration, and injection through syringes. Standard methods for viscosity measurements have low throughput and require large quantities of protein. Thus, there is increasing demand for higher-throughput viscosity screening. A DLS assay based on measurement of the diffusion coefficient of beads added directly to the protein solution is high throughput and run in a multiwell plate format (He et al. 2010a). As shown in Fig. 2.2 the Stokes–Einstein equation can be used to calculate the viscosity of a protein solution using the known radius of the added beads and the measured diffusion coefficient. Furthermore, DLS measurements of diffusion coefficients as a function of protein concentration can be used to derive the interaction parameter, k_D, which has been shown to correlate with protein properties such as viscosity and particulation propensity (Yadav et al. 2010; He et al. 2011). It is widely accepted that the second virial coefficient, B_{22}, obtained by SLS measurements contains information on protein–protein interaction (Printz et al. 2012). The k_D parameter derived from DLS measurements offers a simple way to compare samples under similar conditions (discussed in Chap. 3).

2.2.9 Design of Experiment and Data Analysis

All high-throughput methodologies mentioned above share a common ability to generate a large amount of biophysical information on therapeutic proteins. This enables more complex experimental designs at various stages of pharmaceutical development. The Quality by Design (QbD) concept has gained popularity in recent years among the biopharmaceutical industry and regulatory agencies (Rathore and Winkle 2009; Rathore and Devine 2008). Design of product quality is built on comprehensive understanding of a well-defined process and product space where the protein therapeutic is in its most desired form. The principle of design of experiment (DOE) is often applied to systematically evaluate the protein of interest under a variety of conditions, which are often selected based on the types of stresses that a protein therapeutic is subjected to during manufacturing, storage, and administration. The degree of these stresses often exceeds reality, and the experimental results can be used to predict the protein behavior when failures occur during the product life cycle. The combination of high-throughput measurement and DOE can also help enhance the statistical power when interpreting the results. More importantly, the implementation of high-throughput methodology offers opportunities to simultaneously assess a large number of samples. This is particularly valuable when employing methods that are only used to qualitatively rank order protein samples.

Application of high-throughput techniques results in large data sets, often requiring mathematical tools for rigorous analysis. Statistics helps to establish correlations among measured properties of a molecule. Commonly used statistical analyses that are frequently applied to the development of protein therapeutics include Gaussian modeling, analyses of variance (ANOVA), and the *t*-test. In addition, biophysical characterization often involves spectroscopic methods such as CD, FTIR, and fluorescence, and the results usually include measured values as a function of wavelength of the optic source. Such data can be analyzed by methods of chemometrics including the singular value decomposition technique which has been used to determine the maximum changes in protein properties caused by particular factors like pH or ionic strength. This methodology has been applied to the interpretation of CD and FTIR spectra obtained during the production of antibodies (Greenfield 2006; Sellick et al. 2010). Another useful mathematical tool is polynomial-based data fitting. This approach involves fitting an arbitrary polynomial model to a limited data set and then using the same model to predict protein behavior outside of the tested range. For example, discrete data at pH 5, 5.5, and 6 can be used to generate a polynomial model that best fit the experimental results. The continuous mathematical model can then be used to predict results at pH 5.8, and even at pH values outside of 5–6, if the assumption holds. The polynomial method is especially effective when the source data set is large. Even more predictive information can be obtained while considering multiple variables simultaneously (Sall et al. 2007).

2.3 Empirical Phase Diagrams as Tools to Interpret Results from High-Throughput Biophysical Approaches

2.3.1 Combination of Biophysical Techniques and Data Analysis

As discussed above and in other chapters in this text, high-throughput screening (HTS) is usually performed with only one or two low-resolution techniques of the type previously considered. The selection of the particular technique is typically based on what is known about degradation pathways of the target, convenience, and availability of appropriate instrumentation as well as speed. It has recently become possible to combine the results from multiple techniques with the goal of providing a more comprehensive picture of a protein's structure and its response to various environmental perturbations. This can be used to select optimal methods for HTS as well as for various forms of comparative analysis. A number of methods are available for this purpose. We will consider here only the one that has been most thoroughly described in the literature [reviewed in Maddux et al. (2011)], but other approaches such as Chernoff faces and star charts (Yau 2011) in which information is encoded in facial features or abstract geometric shapes are under consideration. The former method is known as the EPD. The word empirical is inserted in front of the phrase "phase diagram" to differentiate it from the well-known thermodynamic or equilibrium phase diagram since equilibrium conditions are not implied in the former.

The basic idea behind the EPD is to represent the protein (application has also been made to peptides, nucleic acids, virus-like particles, viruses, and bacteria cells) as a vector in which the components of the vector are experimental values obtained from the various methods employed as a function of solution variables. Preparation of EPDs generally involves buffer subtraction of the data, peak selection for data analysis (entire spectra can also be used), averaging of multiple data acquisitions, normalization, input matrix synthesis, singular value decomposition, and finally color mapping of the most significant data using an RGB color scheme. A detailed description of the method including the mathematics involved is presented in Maddux et al. (2011). In general, far-UV circular dichroism (CD) is used to monitor secondary structure although FTIR and Raman spectroscopies can also be employed. Tertiary structure is most commonly analyzed with intrinsic fluorescence, near-UV CD, or high-resolution derivative absorption spectroscopy. Dye binding using compounds such as 8-anilino naphthalene sulfonic acid (ANS) is frequently used to probe the exposure of apolar regions in the protein. Dissociation and association (including aggregation) are typically probed with static and/or DLS (although see below). Overall thermal stability is often studied with differential scanning calorimetry. The most common independent variables (i.e., forms of stress that have previously been employed) are temperature and pH although a wide variety of other variables have been used as described below. Perhaps the major limitation of the EPD approach has been the time and instrumentation necessary to prepare an EPD. This has recently changed with the advent of equipment capable of performing multiple different types of measurements simultaneously. Originally an EPD typically required a fluorometer, CD spectropolarimeter, light scattering system, and perhaps a DSC or FTIR spectrometer. Several newly developed instruments now at least partially overcome this limitation. For example, recent improvements in CD instruments now permit both near- and far-UV spectra, near-UV absorption spectra, fluorescence, and SLS (turbidity or scattering at the fluorescence emission wavelength) to all be acquired simultaneously in a four-position sample chamber under variable temperature conditions (Hu et al. 2011). This permits EPDs to be generated in less than a day. A similar "protein machine" with a six-position sample chamber has also been recently described (Maddux et al. 2012). Perhaps the simplest version of a system with rapid EPD generation capability is a UV absorption spectrometer, typically of the diode array variety, to permit sufficient resolution of derivative peaks (Kueltzo et al. 2003). At high resolution (usually second), the derivative spectrum of a protein will usually manifest distinct peaks for phenylalanine, tyrosine, and tryptophan (if present). Since the residues are usually buried, Tyr is often interfacial, and Trp present in highly variable environments, temperature, and pH-induced peaks shifts can frequently provide a fairly detailed picture of a protein's structural response to various perturbations. Such data can easily be represented in the form of an EPD (Kueltzo et al. 2003). Similarly, fluorescence microtiter plate-based fluorometers have been developed which employ multiple fluorescence and light scattering measurements as a function of temperature, as also described in the fluorescence section. The latter is of especially high throughput, permitting the generation of many EPDs in a single day. EPDs for four model proteins obtained from multiple instruments, two CD-based spectrometers, and a high-throughput fluorometer are shown in Fig. 2.3, where it can be seen that all produce similar EPDs although small differences are apparent due to the various types of measurements used to construct each EPD.

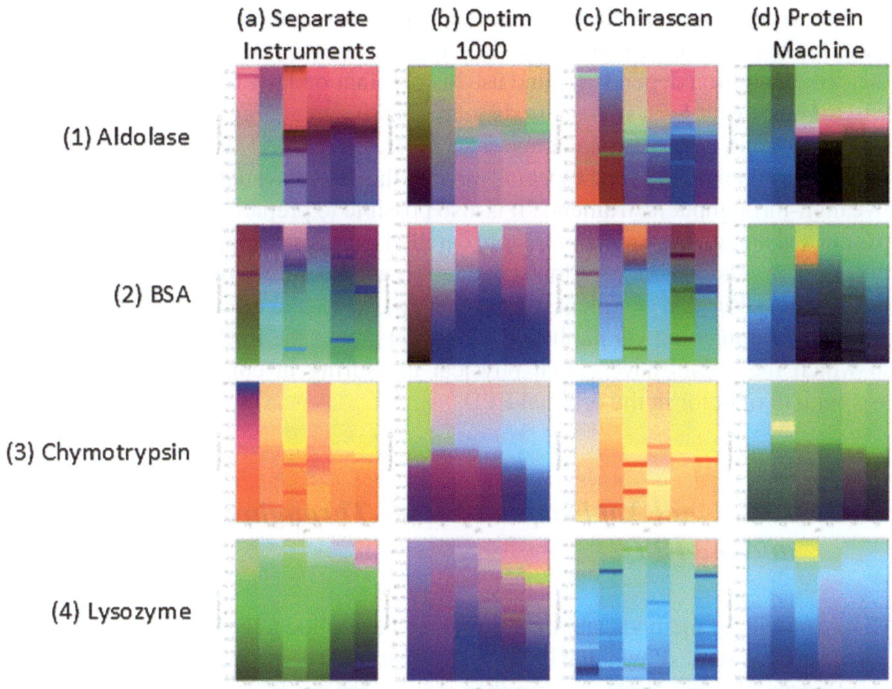

Fig. 2.3 Empirical phase diagrams (EPDs) of four model proteins (*1*) aldolase, (*2*) BSA, (*3*) chymotrypsin, and (*4*) lysozyme constructed using data collected from various instruments: (*a*) intrinsic fluorescence (FL) and static light scattering (SLS) data from a Photon Technology International (PTI) fluorometer, and circular dichroism (CD) from an Applied Photophysics Chirascan, (*b*) FL and SLS by an Avacta Optim 1000 microtiter plate fluorometer, (*c*) FL and CD Applied Photophysics Chirascan, and (*d*) FL, SLS, CD, and UV absorbance by an Olis Protein Machine

2.3.2 *High-Throughput Characterization and Preformulation Development*

The EPD method provides a comprehensive overview of how a protein responds to environmental alteration in the form of a colored diagram in which regions of different color correspond to different structural states of the target molecules. By reference to the original data, native partially folded and molten globule, extensively unfolded, dissociated, oligomerized, and various aggregated states can all be identified. This provides the scientist with clues to trouble spots in a protein and a basis with which to select assays with which to screen for potential stabilizers. For example, if aggregation or a particular structural change occurs under moderate temperature and/or pH conditions, one can select a less stable condition and one or more techniques sensitive to selected degradation events for screening purposes. Typically, a supplemental GRAS (generally regarded as safe) library containing a selection of buffers, sugars, sugar alcohols, amino acids, polymers, detergents, and osmolytes is used. In the initial screen relatively high concentrations of compounds are used with their concentration dependence and use in combination later optimized. It is usually wise to employ at least two methods: one sensitive to aggregation (e.g., light scattering) and one to structural change (e.g., fluorescence, CD) for this purpose. DSC is also commonly employed especially due to the recent availability of highly sensitive high-throughput instruments. It is also possible to prepare EPDs in the presence of selected stabilizers to permit a more detailed analysis/comparison of their effects on a protein. The information thus obtained by a temperature/pH EPD thus provides a basis for buffer and excipient selection at an early stage of pharmaceutical development. Although not yet published, a new version of the EPD has been developed in which the colors have actual physical measuring in contrast to the arbitrary assignment of color in the original EPD.

2.3.3 Additional Application of High-Throughput Methods and the EPD

High-throughput methods and EPDs can also be applied to a wide variety of different situations, some of which will be briefly described here. Two commonly encountered forms of stress in the protein therapeutic area are freeze/thaw and shear. It is often necessary to freeze and then thaw both during development and manufacturing situations. An EPD can be created using the number of freeze/thaw cycles under defined conditions as an independent variable accompanied by temperature and pH stress. All three variables (temperature, pH, freeze/thaw cycles) can be combined into a three-dimensional representation in which the EPD is presented as a colored surface. Shear stress is also often encountered in the development, manufacture (especially filling), and shipping of protein pharmaceuticals. To explore this potential degrading stress, the intensity of the shear can be varied by a mechanical process such as stirring, shaking, or some other forms of agitation, and this is used as a variable in EPD production.

Protein concentration is another important variable that has assumed increasing importance with the use of high-concentration formulations. This variable can be typically evaluated over the range of 0.05–300 g/L depending on the solubility of the protein and the methods employed in the analysis. Proteins usually alter their structure to little or no extent as a function of protein concentration, but aggregation and surface adsorption are both highly concentration dependent. Thus, aggregation-sensitive techniques such as light scattering are often of special importance in protein concentration-dependent studies. Another common variable of particular importance is ionic strength. In the Debye–Huckel charge shielding regime (0–0.15 ionic strength), a number of intermediate concentrations should be evaluated to probe electrostatic effects. At higher salt concentrations both preferential hydration and binding effects usually dominate with salt concentration into the molar range appropriately examined.

It is also possible to create EPDs based on phenomena such as aggregation. A variety of different types of aggregates have been identified in protein solutions based on their relative size and the nature of a protein's conformation (altered or native) within the aggregate. A number of methods are available that are sensitive to these features (see Chap. 9), permitting an aggregation-based EPD to be created. Again, using variables like temperature, pH, ionic strength, freeze/thaw, and shear, soluble protein aggregates can be detected by methods such as size-exclusion HPLC, sedimentation velocity analytical ultracentrifugation, FFF, and DLS. Complimentary structural data can be obtained by the methods described above with FTIR and Raman spectroscopy especially useful because of the particulate nature of such samples. Larger (i.e., submicron) aggregates can be characterized by DLS including single-particle microscope-based approaches (nanoparticle tracking analysis) and classic microscopy-based techniques (atomic force microscopy, scanning electron microscopy, and transmission electron microscopy) although they are difficult to quantitate and new methods such as quartz crystal microbalances and nanomechanical resonators are seeing increasing use. Subvisible and visible particles can be analyzed by methods such as coulter counting, light obscuration, microflow digital imaging, and various visual procedures. Using parameters such as size, composition, structure, and particle number as dependent variables, an EPD can be generated that provides a comprehensive picture of the nature of protein aggregates that form under a wide variety of stress conditions.

Although not yet described in the scientific literature, chemical degradation can also be analyzed in various high-throughput modes and be summarized in EPDs. The application of EPDs described above has all been to various physical processes in which covalent bonds are not broken. Of equal importance to protein degradation, however, are chemical changes such as oxidation and deamidation events. Chemical changes are usually quantitatively determined by peptide mapping combined with mass spectrometry (MS) in the form of HPLC-MS experiments which permit both the amount and location of a residue modification within a protein's amino acid sequence to be determined. To increase throughput, once the nature of any changes has been identified, HPLC-MS analysis may be replaced by methods such as RP-HPLC and capillary electrophoresis or isoelectric focusing. A convenient way to present and analyze such data is in the form of rate constants for the individual, for example, deamidation and oxidation events. This of course requires time-dependent measurements. The rate constants can be used as the dependent variables in an EPD since they are typically determined as function-independent variables such as pH and temperature. Of special interest is the comparative use of physical and chemical EPDs which permits an exploration of the relationship between structural changes and chemical events (and vice versa).

As a final example, EPDs can be used in a strictly comparative mode. Structural comparisons between both similar and assumed identical proteins are often a very important element of pharmaceutical analysis. For example, when manufacturing changes are made, during the development of biosimilars or when investigating mutant proteins, a detailed comparison of the various species is a critical part of the analysis. A direct comparison of EPDs of the target molecules provides a convenient and sensitive way to see if structural identity has been obtained. As an example, second-generation functional mutants of fibroblast growth factor one (FGF-1) have been compared using EPDs and used to select molecules that are not dependent on heparin for their activity (Alsenaidy et al. 2012). Such EPD-dependent comparisons have even been performed with different rotavirus serotypes despite their individual complexity (Esfandiary et al. 2010). Various mathematical (difference) methods exist to facilitate such comparisons.

2.4 Advantages and Challenges of Implementing High-Throughput Technology in Therapeutic Protein Development: An Industry Perspective

High-throughput technologies can be applied across all aspects of biopharmaceutical discovery and development from the early stages of candidate screening and selection to the later stages of formulation development. The benefits of a successful high-throughput screening strategy in this environment are many. The most obvious one is speed. Faster assays mean that sufficient data to drive a decision can be collected within a shorter timeframe leading to more rapid decisions and ultimately an accelerated development timeline. In the pharmaceutical industry where product development is a protracted process, any acceleration of the timeline can mean getting a promising drug candidate into clinical testing sooner and ultimately to market faster, potentially providing a competitive advantage and leading to an earlier revenue stream. A second advantage is that high-throughput assays typically require smaller volumes and fewer samples than standard assay formats. This sample sparing feature is especially important early in development in which the purification process is an early stage of development and consequently the quantities of the candidates being tested may be in short supply. Another advantage is that high-throughput procedures permit many more drug candidates to be tested than would be possible using a standard approach. In the early stages of discovery research where many candidates are being evaluated based on screening assays, a good high-throughput screen for binding permits many more candidates to be tested and allows this to be accomplished within a shorter period of time. Finally, high-throughput assays permit additional molecular features to be evaluated. A good example of this advantage is found in formulation development where the influence of a variety of solution conditions and potential excipients will need to be tested. The solution pH, buffer salts used, excipients, surfactants, and the presence or absence of salt all need to be evaluated, and this can be accomplished much more easily using high-throughput approaches.

High-throughput assays have been a standard approach to the discovery of novel small molecule drugs for several years. These assays were typically designed using an isolated drug target, often the soluble domain of a membrane bound receptor, and a library of small molecules, often containing tens of thousands of compounds, would be screened with a readout indicating a binding event or inhibition of binding. With the birth of the biotechnology industry approximately three decades ago and the beginning of the development of recombinant proteins as therapeutic entities, the need for high-throughput approaches for biotherapeutic drug discovery and development has grown. The greater chemical complexity of proteins and the need to maintain native three-dimensional structure during both processing and storage of the drug substance and drug product have contributed to the application of the various biophysical methods discussed in this chapter. Another property of many proteins, the tendency to self-associate to form dimers and higher oligomers, in some cases an undesirable property, has also demanded the use of highly selective biophysical techniques. The development of high-throughput approaches to the application of these biophysical methods has lagged behind other methods, but the burgeoning biotechnology industry with its unique analytical requirements has helped to recently accelerate this development.

The implementation of a high-throughput screen, however, is not a simple matter of just miniaturizing an assay and running it in a 96-well plate. High-throughput assays must be carefully developed and validated before use in drug development. Special equipment is often necessary to successfully perform a high-throughput assay. Depending on the readout of the assay, a special instrument with the capability to measure signals from a plate may be needed. Special devices for heating or cooling a plate are often required to maintain the temperature required for the assay, and, in some cases, liquid handling equipment is necessary for setting up the multiple conditions to be examined. One of the biggest challenges to implement a high-throughput strategy is the handling of the data produced. High-throughput methods have the potential to generate vast quantities of data which not only must be collected, stored, and archived but also must be analyzed and then presented in a form that can be understood by others. The generation of these very large data sets has perhaps had an unintended consequence. As described above, statistical methods are an absolute requirement for the analysis of these data. This has led to a much greater understanding of the relationships among the variables in an experiment. At the extremes of an experiment are experimental design and presentation of results. DOE has been critical to the success of high-throughput experimental design. When working with multiple molecules or a large number of variables, it is a physical impossibility to design and execute experiments that examine all possibilities in a meaningful way, even with the use of high-throughput methods. DOE provides a way of designing experiments that provide the necessary data while significantly decreasing the number of experiments. This approach is expected to grow within the biopharmaceutical industry as more high-throughput assays are brought on line. Equally demanding is the presentation of the processed data from high-throughput experiments. One approach described above which draws on the statistical analysis of large multiple data sets is the EPD in which a protein is represented as a vector in which the components of the vector are experimental values obtained from the various methods employed as a function of solution variables. The concept and preparation of EPDs have been discussed earlier in this chapter, but the outcome is a presentation using color mapping to denote the most significant data.

This approach as well as others allows a very complex data set to be presented in a way that is readily understandable.

Although high-throughput biophysical methods become increasingly important in the development of therapeutic proteins, there are still biophysical methods that have not been adapted to a high-throughput format. Examples include circular dichroism and mass spectrometry which are still in their infancy but which, no doubt, will have high-throughput adaptations in the near future. It is clear that there is a need for new and better high-throughput biophysical assays and that the needs of the biopharmaceutical industry will play a role in their development.

References

Ahrer K, Jungbauer A (2006) Chromatographic and electrophoretic characterization of protein variants. J Chromatogr B Analyt Technol Biomed Life Sci 841(1–2):110–122

Alexandrov AI, Mileni M, Chien EY, Hanson MA, Stevens RC (2008) Microscale fluorescent thermal stability assay for membrane proteins. Structure 16(3):351–359

Alsenaidy MA, Wang T, Kim JH et al (2012) An empirical phase diagram approach to investigate conformational stability of “second-generation” functional mutants of acidic fibroblast growth factor-1. Protein Sci 21(3):418–432

Andrew SM, Titus JA (2001) Purification of immunoglobulin G. Curr Protoc Immunol. Chapter 2:Unit 2.7

Anquetil PA, Brenan CJ, Marcolli C, Hunter IW (2003) Laser Raman spectroscopic analysis of polymorphic forms in microliter fluid volumes. J Pharm Sci 92(1):149–160

Bravman T, Bronner V, Lavie K, Notcovich A, Papalia GA, Myszka DG (2006) Exploring “one-shot” kinetics and small molecule analysis using the ProteOn XPR36 array biosensor. Anal Biochem 358(2):281–288

Bruylants G, Wouters J, Michaux C (2005) Differential scanning calorimetry in life sciences: thermodynamics, stability, molecular recognition and applications in drug design. Curr Med Chem 12:2011–2020

Capelle MA, Gurny R, Arvinte T (2009) A high throughput protein formulation platform: case study of salmon calcitonin. Pharm Res 26(1):118–128

Chang BS, Kendrick BS, Carpenter JF (1996) Surface-induced denaturation of proteins during freezing and its inhibition by surfactants. J Pharm Sci 85(12):1325–1330

Coffman JL, Kramarczyk JF, Kelley BD (2008) High-throughput screening of chromatographic separations: I. Method development and column modeling. Biotechnol Bioeng 100(4):605–618

Ericsson UB, Hallberg BM, Detitta GT, Dekker N, Nordlund P (2006) Thermofluor-based high-throughput stability optimization of proteins for structural studies. Anal Biochem 357(2):289–298

Esfandiary R, Yee L, Ohtake S et al (2010) Biophysical characterization of rotavirus serotypes G1, G3 and G4. Hum Vaccin 6(5):390–398

Fahrner RL, Knudsen HL, Basey CD et al (2001) Industrial purification of pharmaceutical antibodies: development, operation, and validation of chromatography processes. Biotechnol Genet Eng Rev 18:301–327

Feng BY, Shelat A, Doman TN, Guy RK, Shoichet BK (2005) High-throughput assays for promiscuous inhibitors. Nat Chem Biol 1(3):146–148

Fowler A, Swift D, Longman E et al (2002) An evaluation of fluorescence polarization and lifetime discriminated polarization for high throughput screening of serine/threonine kinases. Anal Biochem 308(2):223–231

Goldberg DS, Bishop SM, Shah AU, Sathish HA (2010) Formulation development of therapeutic monoclonal antibodies using high-throughput fluorescence and static light scattering techniques: role of conformational and colloidal stability. J Pharm Sci 100(4):1306–1315
Greenfield NJ (2006) Using circular dichroism spectra to estimate protein secondary structure. Nat Protoc 1(6):2876–2890
Haidekker MA, Brady TP, Lichlyter D, Theodorakis EA (2005) Effects of solvent polarity and solvent viscosity on the fluorescent properties of molecular rotors and related probes. Bioorg Chem 33(6):415–425
Hanlon AD, Larkin MI, Reddick RM (2010) Free-solution, label-free protein-protein interactions characterized by dynamic light scattering. Biophys J 98(2):297–304
Harn N, Allan C, Oliver C, Middaugh CR (2007) Highly concentrated monoclonal antibody solutions: direct analysis of physical structure and thermal stability. J Pharm Sci 96(3):532–546
Hawe A, Sutter M, Jiskoot W (2008a) Extrinsic fluorescent dyes as tools for protein characterization. Pharm Res 25(7):1487–1499
Hawe A, Friess W, Sutter M, Jiskoot W (2008b) Online fluorescent dye detection method for the characterization of immunoglobulin G aggregation by size exclusion chromatography and asymmetrical flow field flow fractionation. Anal Biochem 378(2):115–122
Hawe A, Filipe V, Jiskoot W (2010a) Fluorescent molecular rotors as dyes to characterize polysorbate-containing IgG formulations. Pharm Res 27(2):314–326
Hawe A, Rispens T, Herron JN, Jiskoot W (2010b) Probing bis-ANS binding sites of different affinity on aggregated IgG by steady-state fluorescence, time-resolved fluorescence and isothermal titration calorimetry. J Pharm Sci 100(4):1294–1305
He F, Becker GW, Litowski JR, Narhi LO, Brems DN, Razinkov VI (2010a) High-throughput dynamic light scattering method for measuring viscosity of concentrated protein solutions. Anal Biochem 399(1):141–143
He F, Hogan S, Latypov RF, Narhi LO, Razinkov VI (2010b) High throughput thermostability screening of monoclonal antibody formulations. J Pharm Sci 99(4):1707–1720
He F, Phan DH, Hogan S et al (2010c) Detection of IgG aggregation by a high throughput method based on extrinsic fluorescence. J Pharm Sci 99(6):2598–2608
He F, Woods CE, Becker GW, Narhi LO, Razinkov VI (2011) High-throughput assessment of thermal and colloidal stability parameters for monoclonal antibody formulations. J Pharm Sci 100(12):5126–5141
Hu L, Olsen C, Maddux N, Joshi SB, Volkin DB, Middaugh CR (2011) Investigation of protein conformational stability employing a multimodal spectrometer. Anal Chem 83(24):9399–9405
Jezek J, Rides M, Derham B et al (2011) Viscosity of concentrated therapeutic protein compositions. Adv Drug Deliv Rev 63(13):1107–1117
Jiskoot W, Randolph TW, Volkin DB et al (2012) Protein instability and immunogenicity: roadblocks to clinical application of injectable protein delivery systems for sustained release. J Pharm Sci 101(3):946–954
Karlsson R (2004) SPR for molecular interaction analysis: a review of emerging application areas. J Mol Recognit 17(3):151–161
Khossravi M, Kao YH, Mrsny RJ, Sweeney TD (2002) Analysis methods of polysorbate 20: a new method to assess the stability of polysorbate 20 and established methods that may overlook degraded polysorbate 20. Pharm Res 19(5):634–639
Kueltzo LA, Ersoy B, Ralston JP, Middaugh CR (2003) Derivative absorbance spectroscopy and protein phase diagrams as tools for comprehensive protein characterization: a bGCSF case study. J Pharm Sci 92(9):1805–1820
Kung CE, Reed JK (1989) Fluorescent molecular rotors: a new class of probes for tubulin structure and assembly. Biochemistry 28(16):6678–6686
Lakowicz JR (2006) Principles of fluorescence spectroscopy, 3rd edn. Springer, New York
Lee W, Fon W, Axelrod BW, Roukes ML (2009) High-sensitivity microfluidic calorimeters for biological and chemical applications. Proc Natl Acad Sci USA 106(36):15225–15230

Lerchner J, Wolf A, Wolf G et al (2006) A new micro-fluid chip calorimeter for biochemical applications. Thermochim Acta 445:144–150

Lerchner J, Maskow T, Wolf G (2008) A new micro-fluid chip calorimeter for biochemical applications. Thermochim Acta 47:991–999

Li CH, Nguyen X, Narhi L et al (2011) Applications of circular dichroism (CD) for structural analysis of proteins: qualification of near- and far-UV CD for protein higher order structural analysis. J Pharm Sci 100(11):4642–4654

Mach H, Arvinte T (2011) Addressing new analytical challenges in protein formulation development. Eur J Pharm Biopharm 78(2):196–207

Mach H, Middaugh CR (2011) Ultraviolet spectroscopy as a tool in therapeutic protein development. J Pharm Sci 100(4):1214–1227

Mach H, Bhambhani A, Meyer BK et al (2011) The use of flow cytometry for the detection of subvisible particles in therapeutic protein formulations. J Pharm Sci 100(5):1671–1678

Maddux NR, Joshi SB, Volkin DB, Ralston JP, Middaugh CR (2011) Multidimensional methods for the formulation of biopharmaceuticals and vaccines. J Pharm Sci 100(10):4171–4197

Maddux NR, Rosen IT, Hu L, Olsen CM, Volkin DB, Middaugh CR (2012) An improved methodology for multidimensional high-throughput preformulation characterization of protein conformational stability. J Pharm Sci 101(6):2017–2024

Malawski GA, Hillig RC, Monteclaro F et al (2006) Identifying protein construct variants with increased crystallization propensity – a case study. Protein Sci 15(12):2718–2728

McEvoy M, Razinkov V, Wei Z, Casas-Finet JR, Tous GI, Schenerman MA (2011) Improved particle counting and size distribution determination of aggregated virus populations by asymmetric flow field-flow fractionation and multiangle light scattering techniques. Biotechnol Prog 27(2):547–554

Otsuki S, Ishikawa M (2010) Wavelength-scanning surface plasmon resonance imaging for label-free multiplexed protein microarray assay. Biosens Bioelectron 26(1):202–206

Owicki JC (2000) Fluorescence polarization and anisotropy in high throughput screening: perspectives and primer. J Biomol Screen 5(5):297–306

Pantoliano MW, Petrella EC, Kwasnoski JD et al (2001) High-density miniaturized thermal shift assays as a general strategy for drug discovery. J Biomol Screen 6:429–440

Peters WB, Frasca V, Brown RK (2009) Recent developments in isothermal titration calorimetry label free screening. Comb Chem High Throughput Screen 12(8):772–790

Pierce MM, Raman CS, Nall BT (1999) Isothermal titration calorimetry of protein–protein interactions. Methods 19(2):213–221

Printz M, Kalonia DS, Friess W (2012) Individual second virial coefficient determination of monomer and oligomers in heat-stressed protein samples using size-exclusion chromatography-light scattering. J Pharm Sci 101:363–372

Privalov GP, Privalov PL (2000) Problems and perspectives in microcalorimetry of biological macromolecules. Methods Enzymol 323:31–62

Rathore AS, Devine R (2008) PDA workshop on "Quality by Design for Biopharmaceuticals: Concepts and Implementation", May 21–22, 2007, Bethesda, Maryland. PDA J Pharm Sci Technol 62(5):380–390

Rathore AS, Winkle H (2009) Quality by design for biopharmaceuticals. Nat Biotechnol 27(1):26–34

Rich RL, Myszka DG (2000) Advances in surface plasmon resonance biosensor analysis. Curr Opin Biotechnol 11(1):54–61

Rosenberg AS (2006) Effects of protein aggregates: an immunologic perspective. AAPS J 8(3):E501–E507

Sall J, Creighton L, Lehman A (2007) JMP start statistics: a guide to statistics and data analysis using JMP, 4th edn. SAS Press, Cary

Schäfer G, Schmidt H (2006) High-throughput spectroscopic viscosity measurement of nanocomposite sols with ETC-effect. J Sol–Gel Sci Technol 38(3):241–244

Schiffer CA, Dotsch V (1996) The role of protein-solvent interactions in protein unfolding. Curr Opin Biotechnol 7(4):428–432

Schmidt R (2010) Dynamic light scattering for protein characterization. Encyclopedia of Analytical Chemistry, 20–24

Sellick CA, Hansen R, Jarvis RM et al (2010) Rapid monitoring of recombinant antibody production by mammalian cell cultures using Fourier transform infrared spectroscopy and chemometrics. Biotechnol Bioeng 106(3):432–442

Shire SJ, Shahrokh Z, Liu J (2004) Challenges in the development of high protein concentration formulations. J Pharm Sci 93(6):1390–1402

Siebert F, Hildebrandt P (2008) Vibrational spectroscopy in life science. Wiley-VCH, Weinheim

Stackhouse N, Miller AK, Gadgil HS (2011) A high-throughput UPLC method for the characterization of chemical modifications in monoclonal antibody molecules. J Pharm Sci 100(12):5115–5125

Sule SV, Sukumar M, Weiss WF, Marcelino-Cruz AM, Sample T, Tessier PM (2011) High-throughput analysis of concentration-dependent antibody self-association. Biophys J 101(7):1749–1757

Susanto A, Knieps-Grünhagen E, von Lieres E, Hubbuch J (2008) High throughput screening for the design and optimization of chromatographic processes: assessment of model parameter determination from high throughput compatible data. Chem Eng Technol 31(12):1846–1855

Swadesh J (2001) HPLC: practical and industrial applications, 2nd edn. CRC Press, Boca Raton

Szapacs ME, Urbanski JJ, Kehler JR et al (2010) Absolute quantification of a therapeutic domain antibody using ultra-performance liquid chromatography-mass spectrometry and immunoassay. Bioanalysis 2(9):1597–1608

Takeda H, Fukumoto A, Miura A, Goshima N, Nomura N (2006) High-throughput kinase assay based on surface plasmon resonance suitable for native protein substrates. Anal Biochem 357(2):262–271

Tarazona MP, Saiz E (2003) Combination of SEC/MALS experimental procedures and theoretical analysis for studying the solution properties of macromolecules. J Biochem Biophys Methods 56(1–3):95–116

Torres FE, Recht MI, Coyle JE, Bruce RH, Williams G (2010) Higher throughput calorimetry: opportunities, approaches and challenges. Curr Opin Struct Biol 20(5):598–605

Verhaegen K, Baert K, Simaels J, Van Driessche W (2000) A high-throughput silicon microphysiometer. Sens Actuators A Phys 82(1):186–190

Vermeir S, Nicolai BM, Verboven P et al (2007) Microplate differential calorimetric biosensor for ascorbic acid analysis in food and pharmaceuticals. Anal Chem 79(16):6119–6127

Vincentelli R, Canaan S, Campanacci V et al (2004) High-throughput automated refolding screening of inclusion bodies. Protein Sci 13(10):2782–2792

Wagner M, Reiche K, Blume A, Garidel P (2012) Viscosity measurements of antibody solutions by photon correlation spectroscopy: an indirect approach – limitations and applicability for high-concentration liquid protein solutions. Pharm Dev Technol. doi:10.3109/10837450.2011.649851

Wang L, Wang B, Lin Q (2008) Demonstration of MEMS-based differential scanning calorimetry for determining thermodynamic properties of biomolecules. Sens Actuators B Chem 134:953–958

Wassaf D, Kuang G, Kopacz K et al (2006) High-throughput affinity ranking of antibodies using surface plasmon resonance microarrays. Anal Biochem 351(2):241–253

Wiendahl M, Schulze Wierling P, Nielsen J et al (2008) High throughput screening for the design and optimization of chromatographic processes – miniaturization, automation and parallelization of breakthrough and elution studies. Chem Eng Technol 31(6):893–903

Xiang Y, Liu Y, Lee ML (2006) Ultrahigh pressure liquid chromatography using elevated temperature. J Chromatogr A 1104(1–2):198–202

Yadav S, Liu J, Shire SJ, Kalonia DS (2010) Specific interactions in high concentration antibody solutions resulting in high viscosity. J Pharm Sci 99(3):1152–1168

Yau N (2011) Visualize this: the FlowingData guide to design, visualization, and statistics. Wiley, Indianapolis

Zhao H, Graf O, Milovic N et al (2010) Formulation development of antibodies using robotic system and high-throughput laboratory (HTL). J Pharm Sci 99(5):2279–2294

Chapter 3
Techniques for Higher-Order Structure Determination

James Kranz, Fatma AlAzzam, Atul Saluja, Juraj Svitel, and Wasfi Al-Azzam

3.1 Introduction: Protein Structure

Proteins are the building blocks for major parts of living systems that are responsible for all critical cellular functions. Their roles vary from catalyzing reactions (enzymes) to facilitating movement (cytoskeletal and motor proteins) and acting as messengers for signal transduction, to name a few. The ability of proteins to carry out such broad structural and functional roles is due to their unique three-dimensional structures, as well as thermodynamic characteristics that allow the functional form of a protein to retain stability under many different environmental conditions, such as changing pH or temperature. Understanding protein structure is a prerequisite for the development of protein therapeutics, from the identification of potential targets for therapeutic agents, being developed and administered as the therapeutic agent or being used in more novel applications as carriers of therapeutic agents and sustained released devices, etc. Control, optimization, and quantitation of protein structure are critical to the success of protein pharmaceuticals.

J. Kranz • W. Al-Azzam (✉)
Biopharmaceuticals Development, GlaxoSmithKline, 709 Swedeland Road, UW2108, King of Prussia, PA 19406, USA
e-mail: james.k.kranz@gsk.com; wasfi.a.al-azzam@gsk.com

F. AlAzzam
TechnoPharmaSphere LLC, 822 Geddy Lane, Downingtown, PA 19335, USA
e-mail: fatma.alazzam@technopharmasphere.com

A. Saluja
Drug Product Science & Technology, Bristol-Myers Squibb, 1 Squibb Drive, New Brunswick, NJ 08903, USA
e-mail: Atul.Saluja@bms.com

J. Svitel
Process and Product Development, Amgen Inc., One Amgen Center Drive, Thousand Oaks, CA 91320, USA
e-mail: jsvitel@amgen.com

L.O. Narhi (ed.), *Biophysics for Therapeutic Protein Development*, Biophysics for the Life Sciences 4, DOI 10.1007/978-1-4614-4316-2_3,

Protein structure starts with the linear polymer composed of 20 different amino acids covalently linked through peptide bonds; this is the primary structure or sequence. The peptide bonds are also responsible for the formation of intraresidue hydrogen bonding that gives rise to secondary structures. Each of the 20 different amino acids has a specific side chain which determines the chemical and functional properties of the residue; they are classified as nonpolar, polar, acidic, basic, or aromatic.

The primary sequence can adopt stable configurations of local short-range secondary structures mediated by intraresidue peptide bond or "backbone" interactions. Common examples of secondary structures are α-helices, β-sheets, turns, and unstructured (random coil) segments. The structure of an α-helix is stabilized by H-bonding between the nitrogen of the amide and the carbonyl carbons of peptide bonds spaced four residues apart. β-sheets are composed of two or more different linear sequences of at least 5–10 amino acids, stabilized by H-bonding between the nitrogen of the amide of one amino acid and the carbonyl carbons of the amino acid in the adjacent strand. β-sheets can contain strands which are far away from each other in the linear amino acid sequence, but adjacent in the folded structure of the protein. Parallel β-pleated sheets consist of strands that have the same (parallel) orientation on the linear sequence or primary structure of the protein, while in the more common antiparallel β-sheet, the amino acids at the strand–strand interface are coming from opposite directions. Other secondary structures such as bends and turns may exist in specific structures or conformations.

The final functional three-dimensional, or tertiary, structure of a protein is stabilized by a large number of noncovalent interactions that are mediated almost exclusively by side-chain interactions between amino acids that can be from very different regions of the primary structure, but are brought into close proximity as the protein folds. Tertiary structures result from further bonding between side chains within the protein and with any water that may be present around the protein. The "hydrophobic effect," which is a thermodynamic description of the driving forces stabilizing the tertiary or native structure of a protein, is due in large part to the side-chain-specific interaction of H_2O with a protein (Sharp et al. 1991; Murphy and Freire 1992). The hydrophobic effect tends to facilitate the partitioning of polar amino acids on the outside of the folded protein, with nonpolar amino acids clustering in the inside of the protein in its final stabilized structure. Examples of tertiary contacts include hydrophobic (van der Waals) interactions, dipole- and induced-dipole interactions, disulfide bonds, electrostatic and ionic interactions, and hydrogen bonds.

Multidomain proteins are common, where different conserved functional domains are present within a primary sequence and fold into separate domains often connected only by the single strand of amino acids of the primary structure. These domains can show cooperative behavior, with one domain influencing the folding of other domains, or the folding of individual domains may be thermodynamically independent of the rest of the protein. For example, different families of kinases exist with a somewhat conserved catalytic domain but differ in terms of the auxiliary domains that regulate kinase folding, function, and specificity. Likewise, many proteins contain pro-domains that function only to facilitate proper folding, then these are excised via proteolysis to produce the final folded active structure.

Quaternary structure refers to the association of folded domains from different amino acid chains. It is in some ways an extension of tertiary interactions, arising from ordered interactions between two or more polypeptides subunits. The bonds in each monomer are the same as those found in the tertiary structure. The polypeptide subunits are attached either by peptide strand or noncovalent interactions. Common quaternary structures include dimers (transcription factors), trimers (some cytokines), and higher-ordered oligomers. Allostery is one additional dynamic property that is common among multimeric proteins, where different functional properties are observed for unique quaternary structures that are only available to an assembly of monomers. One well-characterized example of allostery is hemoglobin that gives rise to cooperative O_2 binding as well as its differential affinity for O_2 in the lungs and tissues based on the different conditions (e.g., pH) of the surrounding tissues.

Structural analysis of protein therapeutics has been an important focus for the biopharmaceutical industry, as described in the first chapter of this book. The goal of successful protein therapeutic development is the creation of a therapeutic protein with native-like structure, good bioactivity, and stability that elicits a minimal immune response. Moreover, protein conformational stability can correlate phenomenologically with in vivo stability. Protein aggregation and chemical modification are the two most common routes for protein degradation and loss, with aggregation depending more on protein structure and stability (Putnam et al. 2010). Aggregates are heterogeneous mixtures of native folded species, nonnative unfolded, or partially unfolded species, self-associated into larger molecules or polymers. Multi-monomeric aggregates may continue growing in regular shapes such as fibrils, or as amorphous particles, and can range in size from oligomers (nanometers) to subvisible (microns) or visible (hundreds of microns) particulates (Stefani and Dobson 2003).

Manufacturing processes are continually optimized during therapeutic development to improve product yield and product quality, while changes in formulation and delivery methods can also occur in response to information on the target patient population. All of these changes could alter the physicochemical properties of the protein in use, including chemical modification, self-association, and ultimately the tertiary structure and protein folding. Thus, there is a growing need for the development of biophysical tools that are sensitive to subtle changes in protein higher-order structure (HOS) during biopharmaceutical development. This chapter provides a brief review of technologies that are commonly used and available to characterize various higher-order structural properties, with an intended application toward protein biotherapeutics. Applications of these techniques throughout the product lifecycle are provided in many subsequent chapters in this volume.

3.2 Spectroscopic Techniques: Structural Characterization

3.2.1 Absorption Spectroscopy

Spectroscopy and spectroscopic techniques predate our knowledge of the existence of proteins and were in fact instrumental in the earliest characterization of protein chemistry and structure. Some of the earliest examples of absorption spectroscopy in proteins involve porphyrin-containing enzymes (Chance 1953; Chance and Pappenheimer 1954; Stern 1937). Electromagnetic radiation at a certain wavelength λ (or frequency $\nu = c/\lambda$) interacts with a sample, followed by measurement of some properties of the radiation that emerges. Absorbance is the fraction of incident radiation that is absorbed or dissipated. The absorption spectrum of the material represents the fraction of photons that are absorbed by a material over the range of frequencies monitored.

It is challenging to succinctly explain the interaction of light with matter, setting aside a quantum mechanical description. Light is a rapidly oscillating electromagnetic field; likewise, molecules are comprised of bonded atoms with their own distributions of spin states with distinct electromagnetic properties. When the frequency (and energy) of an incident photon matches that of an allowed electronic transition of a molecule, that electron absorbs the photonic energy jumping from its ground state (highest occupied molecular orbital or HOMO) into a transitory excited state (lowest unoccupied molecular orbital or LUMO). This metastable electronic state relaxes back to the ground state through various mechanisms, including radiative ones (fluorescence, phosphorescence), non-radiative pathways such as vibrational modes (Raman spectroscopy, FTIR), or solvent quenching.

Both the amplitude and frequencies of emitted light provide sources of information on the molecular environment and dynamics within a system. These are often summarized in a Jablonski diagram (Fig. 3.1), which schematically depicts which transitions are observed for absorption and relaxation pathways, among the numerous possible transitions. The vertical axis depicts the energy (and frequency) of transitions, showing that the highest energy transition (S_0 S_2) will occur at the lowest frequency, with all other transitions being of lower energy and corresponding higher frequency. What is relevant to an understanding of proteins and their HOS is how the local environment can influence which transitions are or are not observed and how a change in the probability of observing different transitions relates to an understanding of fluctuations within a protein sample. In proteins the amide backbone and aromatic amino acids and disulfide bonds are the intrinsic chromophores involved in absorption spectroscopy. The backbone absorbance spectrum consists of two major electron transitions, the π to π^* transition at 195 nm and the weaker n to π^* transition occurring at about 220 nm, the intensity of which depends on the secondary structure of the protein.

Among applications to macromolecules, there are unique chemistries that afford distinct opportunities for selective interrogation. Most proteins contain aromatic amino acids, whose pi-orbitals have lower energy (higher frequency) absorbance

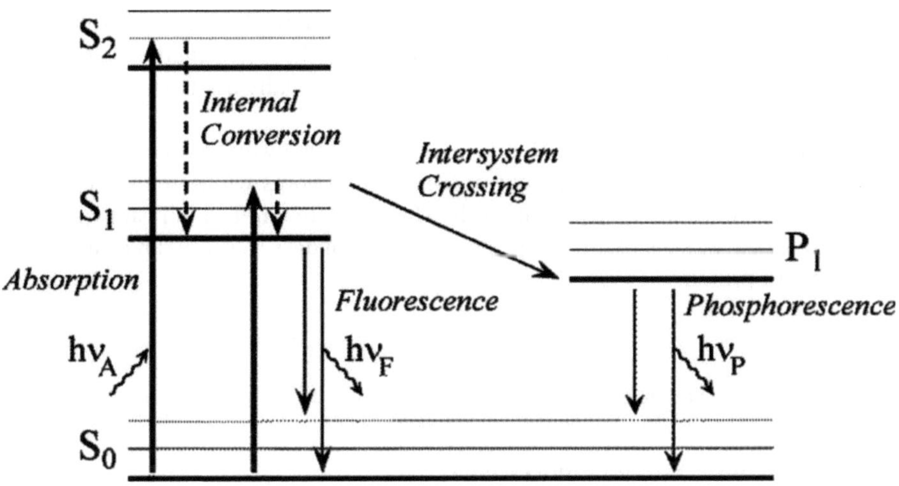

Fig. 3.1 A Jabłoński diagram depicting fluorophore absorption, radiative, and non-radiative relaxation pathways. Singlet ground, first, and second electronic states are represented by S_0, S_1, and S_2. Within each of these energy levels are vibrational energy levels. Transitions between states are shown as vertical lines, representing the instantaneous nature of light absorption, that is, the Franck–Condon principle. Energy losses through non-radiative conversions give rise to the Stokes' shift, a difference in frequency between absorptive and radiative transitions

bands compared to a typical carbon–carbon bond. The indole side chain of tryptophan has the most intense absorbance with a maximum around 280 nm and a weaker transition around 292 nm. Tyrosine has a weaker absorbance, with the strongest transition occurring at 276 and weaker transitions as shoulders at 267 and 280 nm. Phenylalanine has the weakest absorbance which occurs at 250–270 nm. Modifications of any of these residues including oxidation and ionization can affect the absorbance spectra of these chromophores. The disulfide bonds have a very weak absorbance signal from 250 to 300 nm which can also contribute to the intensity of the protein absorbance in this part of the spectrum.

Absorption frequencies are relatively insensitive to molecular environment of these aromatic residues, which provides a useful means of quantifying protein concentrations (Beer's Law or the Beer–Lambert equation) or in any number of applications for detecting proteins (chromatography):

$$A = \lambda o \gamma_{10}\left(\frac{I}{I_0}\right) = \varepsilon \chi \lambda, \tag{3.1}$$

where A is the measured absorbance, I_0 is the original intensity of the light and I is the intensity of the light transmitted through the sample, l is the length of the sample path length the light was shown through, c is the concentration of the chromophores, and ε is the absorbance of a one molar solution of the chromophores, often referred to as the extinction coefficient of a chromophore. Taking the derivative of the spectrum can allow the detection of changes in the wavelength of maximum absorbance due to changes in solvent, and this is occasionally used to follow changes in conformation or aggregation state.

As the subsequent sections will reveal, relaxation pathways can be greatly influenced by the molecular environment of the chromophores within the folded structure of the protein and of the protein itself. These different relaxation pathways have been exploited as sensitive tools to probe the secondary, tertiary, and quaternary structures of proteins.

3.2.2 Fluorescence

3.2.2.1 Intrinsic Protein Fluorescence

Fluorescence is arguably one of the most widely used experimental techniques employed in the study of macromolecules and macromolecular interactions, both in vitro and in vivo. Many thorough reviews representing the array of applications of this technique are available in the literature (Cantor and Schimmel 1980; Lakowicz 2006; Royer 2006). Over the last few decades, continued improvements in experimental techniques and in computational analyses have revolutionized the use of spectroscopic techniques in understanding mechanisms of protein folding and perturbations of protein structure (Bartlett and Radford 2009; Royer 1995; Royer and Scarlata 2008).

The phenomenon of fluorescence was first observed by Sir George G. Stokes in 1852 (Stokes 1852), noting that the fluorescence emission typically occurs at lower energy or longer wavelengths than the incident light. For Stokes, the source of UV excitation was the sun and a blue glass filter (from a stained glass window) which selectively transmitted light $<$400 nm. This incident light was absorbed by a glass of quinine; fluorescence was filtered through a glass of white wine, which selectively transmitted light $>$450 nm and subsequently detected by eye. Energy losses between excitation and emission are universal for fluorescent molecules in solution, giving rise to the Stokes' shift (Fig. 3.1).

In proteins, tryptophan is the strongest intrinsic fluorophore, followed by tyrosine and then phenylalanine. Proteins tend to contain relatively fewer tryptophans than tyrosines and phenylalanines, and due to its strong excited-state dipole, tryptophan fluorescence tends to be more sensitive to general environmental factors. Both Trp and Tyr absorb light around 280 nm, and due to the overlap of their excitation and emission spectra, if the two amino acids are located close enough in the folded protein, this commonly results in energy transfer from Tyr to Trp. Thus, the Trp emission is the one most often observed in proteins.

Due it its aromatic character, tryptophans are often buried or partially buried in the hydrophobic core of protein interiors. Any disruption or destabilization of protein structure can lead to a change in solvent exposure and thus the fluorescent properties of tryptophan residues (Bartlett and Radford 2009; Royer 1995; Royer and Scarlata 2008). This is most clearly observed in systems with a single tryptophan (Epstein et al. 1971; Flanagan et al. 1992; Hynes and Fox 1991; Otto et al. 1994), but also in studies of proteins containing multiple tryptophan residues (LeTilly and Royer 1993; Mann et al. 1993; Royer 1993). While the emission spectrum maximum of a buried tryptophan generally exhibits a red shift (lower energy, higher frequency) upon protein unfolding, the peptide backbone and a number of the amino acid side chains can alter tryptophan fluorescence spectra, generally through excited-state quenching or electron transfer (Adams et al. 2002).

3.2.2.2 Fluorescence and Conformational Dynamics

All proteins are inherently flexible in solution, existing as a distribution of a population of low- and high-energy conformational states that interconvert on different time scales. There are a variety of advanced applications that utilize fluorescence as a probe of conformational dynamics which are described in detail elsewhere (Cantor and Schimmel 1980; Lakowicz 2006; Royer 1995, 2006; Bartlett and Radford 2009; Royer and Scarlata 2008; Kamerzell et al. 2011). Time-resolved fluorescence has been used to measure protein dynamics that occur in time scales ranging from sub-nanoseconds to seconds, via multiple techniques. Some are steady-state measurements, while others are kinetic. In terms of relevance to characterization of higher-ordered protein structure, and fluctuations in structure, their use varies with respect to the simplicity of the measurement and the type of information obtained. Applications include studies of local and global motions, flexibility, heterogeneity, and intermolecular interactions and generally employ a variety of techniques (Kamerzell et al. 2011; Ramsey et al. 2009).

Intrinsic fluorescence is commonly used in time-resolved investigations of protein folding and unfolding. Unfolding is induced by exposure to temperature, chemical denaturants, or other solution conditions. Returning the protein to the original conditions will initiate refolding. The sophistication of these experiments continues to improve, with advances in detection speeds and data analysis. Generally single wavelength intensities are monitored as a function of the folding or unfolding time course via stopped flow or temperature-jump techniques.

Anisotropy measurements provide information on the size and shape of proteins and have been applied extensively in studies of protein–protein interactions. Anisotropy, A, is based on the principle of photoselective excitation of fluorophores by plane-polarized light:

$$A = \frac{(I_{\parallel} - I_{\perp})}{(I_{\parallel} + 2I_{\perp})}. \tag{3.2}$$

Fluorophores that are physically oriented such that their electronic dipole moment is aligned in the plane of incident light are selectively excited. After rotational or vibrational diffusion occurs, some fraction of molecules have their dipoles oriented out of the plane of incident light, giving a decrease in the signal detected parallel to incident light and an increase in the signal detected perpendicular. Steady-state anisotropy employs plane-polarized light, with a pair of polarizing filters before and after the sample; the result is a value of fluorescence anisotropy or polarization that is specific to a sample composition and temperature.

Time-resolved measurements of fluorescence or anisotropy use rapid laser-based excitation and rapid detection to follow the decay kinetics. Also called lifetime measurements, time-resolved experiments are useful in estimating molecular rotational correlation times and protein hydrodynamic radii and how these change in ways that are sensitive to formulations (Ramsey et al. 2009). The relationship between the anisotropy, A, and the lifetime and rotational correlation time is given by the Perrin equation (Perrin 1926):

$$\frac{A_0}{A} - 1 = \frac{\tau}{\tau_c}, \tag{3.3}$$

where A_0 is the limiting anisotropy, τ is the fluorescence lifetime, and τ_c is the rotational correlation time of the macromolecule. For globular (approximately spherical) proteins, the rotational correlation time is a function of the apparent molecular weight, M, of a protein by

$$\tau_c = \frac{\eta V}{RT} = \frac{\eta M}{RT}(\bar{v} + h), \tag{3.4}$$

where η is the viscosity, $\bar{v}$ is the specific volume of the protein, h is the relative hydration, T is temperature (K), and R is the universal gas constant. For a typical protein, $\bar{v} \sim 0.73$ mL/g and hydration levels are $h \sim 0.23$ g of H_2O per gram of protein. This expression predicts that the correlation time of a hydrated protein is on average ~30 % larger than would be predicted for a fully anhydrous protein (spherical approximation).

The apparent volume of a protein can be measured experimentally, solely from measurements of protein anisotropy as a function of varying viscosity; changes in viscosity are typically achieved by variable temperature combined with viscous cosolvents, like glycerol. The Perrin equation is substituted with terms from Eq. 3.4, giving

$$\frac{1}{A} = \frac{1}{A_0} + \left(\frac{\tau RT}{\tau_c \eta V}\right). \tag{3.5}$$

The use of Eq. 3.4 to measure the apparent volume of a protein is one of the earliest applications of fluorescence in a biochemical system (Weber 1952a, b). As discussed in subsequent sections, the apparent volume of a protein can vary significantly in response to changes in formulations, aggregation onset, and protein unfolding.

Another laser-based technique is fluorescence correlation spectroscopy, or FCS, which measures time-dependent fluctuations in fluorescent intensity arising from Brownian motions. This technique is now commonplace in studies of in vivo protein compartmentalization and changes in expression and can also be used to measure protein dynamics in simple in vitro systems (Foldes-Papp et al. 2002). Applications to in vitro systems involve observations of single molecules undergoing folding/unfolding transitions, generally occurring on the microsecond to second time scale.

Design of a correlation spectroscopy experiment to observe fluctuations in protein structure requires that a conformational change gives rise to a change in fluorescent dye intensity. The majority of FCS experiments involve molecules freely diffusing in and out of the small volume being monitored, with fluctuations providing insights into diffusion, either passive or active, within cellular environments.

3.2.2.3 Fluorescence Quenching

Quenching refers generically to any molecular process that decreases the intensity of fluorescence. These include molecular rearrangements, energy transfer, excited-state reactions, and quenching through collisions. Fluorescence quenching has been widely studied as a source of information on biological systems. During collisional quenching, the quenching agent must diffuse to the fluorophore during the lifetime of the excited state. Upon contact, there is energy transfer via interactions of the excited-state dipole and the dipole (or induced dipole) of the quencher, facilitating return of the fluorophore to the ground state without photon emission.

The requirement of intermolecular contact between the quencher and fluorophore has obvious applications to biochemical systems. Quenching measurements can be indicative of the relative accessibility of fluorophores to quenchers. For example, in systems with a few fluorophore groups, either buried in the protein interior of soluble proteins or in lipid environments of transmembrane proteins, quenching rates can be used to determine the permeability of proteins or membranes to the extrinsic quencher. Likewise, quenching rates and lifetimes of fluorophore and quencher groups can be useful in measurements of molecular diffusion. Details on collisional quenching theory and applications are prevalent in the literature (Cantor and Schimmel 1980; Lakowicz 2006).

3.2.2.4 Fluorescence Resonance Energy Transfer

Fluorescence resonance energy transfer (FRET) is a technique that utilizes interactions between fluorophores with overlapping energy states to determine distances between them (Lakowicz 2006). In this technique the two molecules of interest are labeled at specific sites with donor or acceptor fluorophores which have overlapping emission/excitation spectra. As the distance between the molecules decreases, the efficiency of energy transfer from the donor to the acceptor molecule increases. Based on the characteristics of the labels, the actual distance between them can be calculated (dos Remedios and Moens 1995). This technique can also be used to follow changes in the distance between parts of the same polypeptide with changes in solution condition, interactions with other proteins, etc.

Most applications of FRET utilize extrinsic fluorophores, ones that involve cofactors as in the process of photosynthesis or more commonly are covalently attached to a protein via chemical means. These have found widespread applications in labeling of cellular proteins used in studies of cellular trafficking or in vivo protein–protein interactions (Pollok and Heim 1999; Miyawaki 2011; Jares-Erijman and Jovin 2003) and in studies of single molecules (Ferreon et al. 2011; Kubelka et al. 2004). Within proteins there is intrinsic fluorescence energy transfer between the tyrosine, the donor amino acid, and tryptophan, the acceptor molecule. These effects are perhaps most observable in simplified, highly purified protein solutions such as protein therapeutics or in studies of protein folding kinetics, where limits of low signal intensity among Tyr and Trp FRET pairings are not so problematic. In general, however, endogenous fluorophores are not heavily used for FRET-based studies of proteins.

3.2.2.5 Extrinsic Fluorescence

Almost 60 years ago, Weber demonstrated that the quantum yield of 1-anilino-phthalene-8-sulfonic acid (1,8-ANS) increased significantly from a highly quenched state in water (0.004) when bound to bovine serum albumin (0.75) (Daniel and Weber 1966; Weber and Daniel 1966; Weber and Laurence 1954). The mechanism of this effect involves a relief from collisional quenching from H_2O molecules due to direct binding to a hydrophobic drug binding site on albumin. More commonly, the nonspecific binding of ANS to hydrophobic surfaces of proteins has been used as a general indicator of partial or complete protein unfolding.

In contrast to intrinsic protein fluorescence, which arises from the naturally fluorescent amino acids present in a given protein, extrinsic fluorescent dyes offer additional possibilities in the characterization of proteins. These can be covalently attached to proteins (via amines as with lysine or the amino group of the *N*-terminus or via free thiol groups on cysteine) or may interact with a protein via noncovalent-binding interactions. While ANS and other extrinsic fluorescent dye probes are a useful general probe of a change in the conformation of a protein (Hawe et al. 2008, 2011), they provide only a qualitative, nonlocalized means of investigating structural perturbations. As described in Chap. 2, these properties have been exploited for high throughput assays, especially as screening assays, where qualitative ranking of different molecules or conditions is the desired outcome. When more quantitative assessment is needed, extrinsic fluorescence is generally complimented by other techniques.

Extrinsic dyes work via excited-state reactions. These are molecular processes which change the electronic or chemical structure of the excited state prior to relaxation via fluorescence. The best known example of an excited-state reaction is that of phenol deprotonation, which occurs much more readily from its excited state due

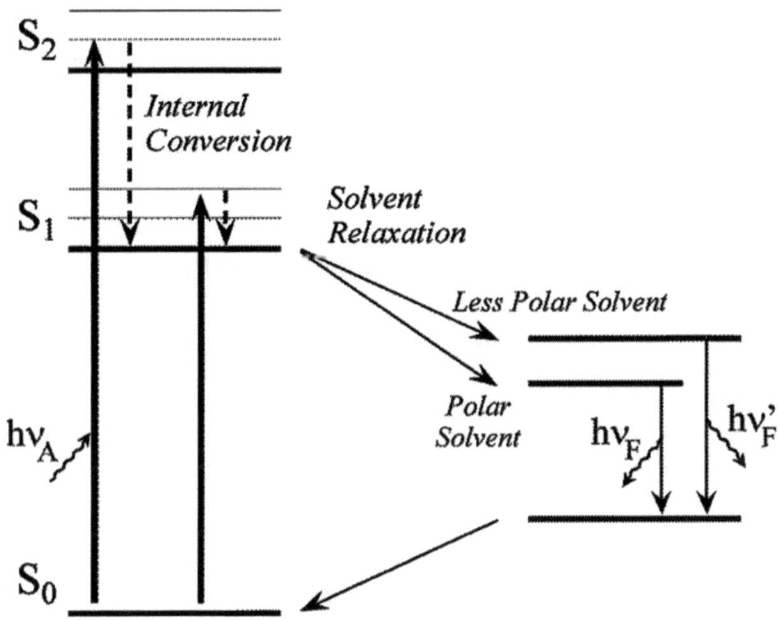

Fig. 3.2 A Jablonski diagram depicting fluorescence with solvent relaxation

to a shift of electrons from the phenolic hydroxyl onto the phenol ring, which acidifies the hydroxyl group. Collisional quenching and solvent relaxation are other well-characterized examples of excited-state reactions, and the ones most relevant to extrinsic dyes.

Solvent relaxation is one aspect of the local environment of a polar fluorophore that can significantly influence its emission spectrum. This is the origin of the Stokes' shift. Spectral shifts result from the interaction of the excited-state fluorophore dipole with solvent dipoles, leading to another energy transfer pathway, depicted in Fig. 3.2. This pathway for energy transfer is due in large part to the kinetics of different phenomena and can result in large Stokes' shifts of ANS and similar dyes. Absorption lifetimes are affectively instantaneous (10^{-15} s) with respect to atomic motions, the Franck–Condon principle. The excited-state lifetime of fluorophores are typical 10^{-9} to 10^{-8} s. Solvent motions are generally on the order of 10^{-10} s, which allows sufficient time for solvent dipoles to align with the fluorophore excited dipole and for non-radiative transfer to occur. Fluorescence still occurs, but at a lower, relaxed energy level for the excited state. With the loss of the fluorophore dipole upon fluorescence emission, the aligned solvent molecules are at a higher energy state, which relaxes to the original ground state through non-radiative processes. The end result is a fluorescence that is red shifted (lower energy) from the original absorption frequency.

Steady-state fluorescence spectroscopy is the most common experimental technique used to follow the interactions of extrinsic dyes with proteins. Protein binding affects the solvent–dye interactions, resulting in changes in emission wavelength or intensity. Other applications discussed above can be used in conjunction with the addition of extrinsic dye for the detection of protein structural changes as well, including time-resolved fluorescence (lifetime measurements), anisotropy (rotational motions), and fluorescence correlation spectroscopy (translation/diffusion).

3.2.3 Fourier Transform Infrared Spectroscopy

The use of Fourier transform infrared (FTIR) spectroscopy to measure the secondary structure of protein molecules has dramatically expanded in the last few decades, encouraged by the need to measure protein secondary structure in samples with various physical forms without sample manipulation. This method can be used to analyze molecules ranging from a few amino acids to large protein complexes, in different physical forms including solutions, lyophilized or crystallized solids, gas, and chemically modified or embedded in polymers (Perez et al. 2002; Carpenter et al. 1998; Haris and Chapman 1995; Chalmers 2002; Griebenow and Klibanov 1995). Additionally, modern FTIR spectrophotometers require only small volumes of solution for analysis, 10–100 μL, though at a relatively high concentration of 10 mg/mL and higher (Perez et al. 2002; Zuber et al. 1992). Compared to other spectroscopic techniques, FTIR can produce high-quality spectra relatively easily without the complications of background fluorescence, light scattering or problems related to the size of the protein. However, the presence of water absorption, buffer components, or other biological molecules produces problems of spectral overlap that can only be successfully subtracted and resolved by mathematical approaches (Susi et al. 1967; Venyaminov and Kalnin 1990a, b; Dong et al. 1990; Dousseau and Pezolet 1990). Thus, FTIR spectroscopy can be used to study various biological systems including proteins in pharmaceutical and non-pharmaceutical formulations (Carpenter et al. 1998; Haris and Chapman 1995; Griebenow and Klibanov 1995; Krimm and Bandekar 1986).

FTIR spectroscopy typically refers to the absorption of infrared light in the mid-IR region (400–4,000 cm^{-1}). The Michelson interferometer (Fig. 3.3) is the central part of an FTIR spectrometer and is unique among other spectroscopic instrumentation. Infrared radiation from the source is collected and collimated before it strikes the beam splitter. The beam splitter ideally transmits one-half of the radiation and reflects the other half. Both transmitted and reflected beams strike mirrors, which reflect the two beams back to the beam splitter. Thus, one-half of the infrared radiation has first been reflected from the beam splitter to the moving mirror (Mirror 2) and then back to the beam splitter prior to being shone through the sample. The other half of the infrared radiation going to the sample first goes through the beam splitter and then is reflected from the fixed mirror (Mirror 1) back to the beam splitter. When these two optical paths are reunited, interference occurs at the beam splitter because of the difference in optical paths caused by the scanning of the moving mirror.

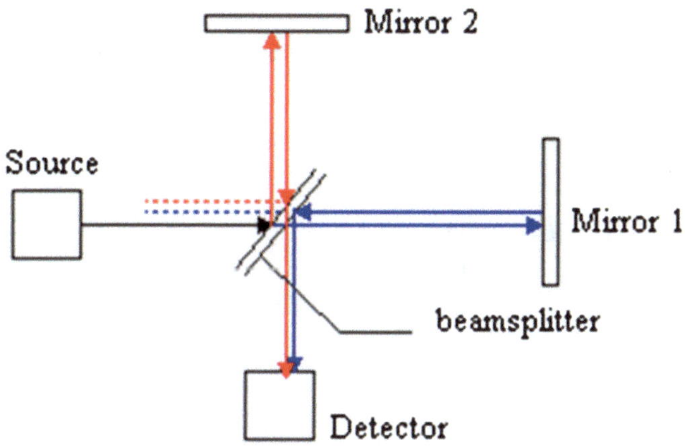

Fig. 3.3 Schematic of the principle of Michelson interferometer in FTIR spectrometer

The interference signal measured by the detector as a function of the optical path length difference is called the interferogram (Chalmers 2002; Griffiths and de Haseth 1986; Smith 1996). In order to extract and present information as an absorbance or transmittance spectrum, the interferogram has to be Fourier transformed, generating a single beam spectrum of IR absorption as a function of wavelength. Instrumental and atmospheric contributions are superimposed in the primary IR spectrum. To eliminate these contributions, a background spectrum obtained without a sample must be recorded and subtracted. Finally, the sample spectrum must be normalized, by dividing the sample spectrum signal, I, and by the background signal, I_0. The result is a %-transmittance spectrum (as $T=I/I_0$). For liquids, transmittance is related to absorbance A as

$$A = \log_{10}\left(\frac{I_0}{I}\right) = \log_{10}\left(\frac{1}{T}\right). \tag{3.6}$$

Physically, FTIR is a type of vibrational spectroscopy measuring the fluctuating covalent bonds within a protein; these give rise to specific wavelengths for different modes of vibration, the intensity relating to the number of bonded atoms in a common chemical environment. There are many modes of vibration a chemical bond can exhibit in a given chemical structure which result in IR absorbance, even for simple molecules such as CO_2 (Fig. 3.4). Given the vast number of normal modes, the vibrational FTIR spectrum is complex with many vibrational bands overlapping. However, it is often possible to select a spectral region that provides specific structural information. Table 3.1 shows spectral regions of common chemical structure vibrations. For analysis the spectra are usually deconvoluted. Better visual resolution of the different bands can be achieved by taking the second derivative, and the data is often presented as the deconvoluted or second derivative of the raw spectrum.

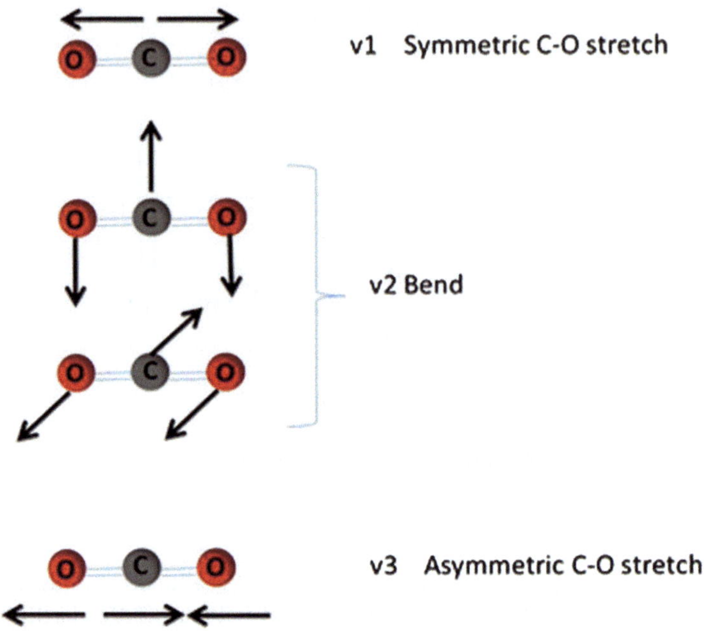

Fig. 3.4 Vibrational modes of CO_2 molecule

Table 3.1 Common vibrations with chemical structure assignments and spectral region

X–H vibrations	Bond	Wave numbers (cm^{-1})
Hydroxyl	O–H	3,610–3,640
Amines	N–H	3,300–3,500
Aromatic rings	C–H	3,000–3,100
Alkenes	C–H	3,020–3,080
Alkanes	C–H	2,850–2,960
Triple bonds		2,500–1,900
Double bonds		1,900–1,500
Heavy atoms		1,500–

Protein secondary structural information is measured by FTIR via monitoring the vibrations exhibited within the peptide backbone. Each vibration mode absorbs IR electromagnetic radiation at specific frequencies depending on the peptide bond orientation and the protein secondary structure associated with a given residue. The vibrational modes of C=O stretching and N–H bending around the peptide bond, called amide I and amide II vibrational modes, are shown in Fig. 3.5. Up to nine different characteristic IR bands, the amide A, B, I, II, III, IV, V, VI, and VII have been reported for protein molecules measured by FTIR (Miyazawa et al. 1956; Bandekar 1992). Amide III and IV are very complex bands resulting from a mixture

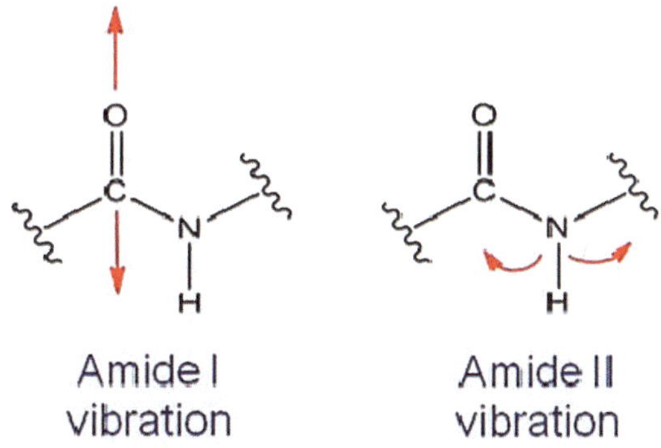

Fig. 3.5 Schematic representation of examples of peptide bond vibrations from amide I and amide II

of several coordinate displacements (Fu et al. 1999; Schweitzer-Stenner et al. 2002). The out-of-plane motions are found in amide V, IV, and VII (Krimm and Bandekar 1986; Bandekar 1992). The amide A band around 3,500 cm^{-1} and amide B around 3,100 cm^{-1} originate from a Fermi resonance between the first overtone of amide II and the N–H stretching vibration (Haris et al. 1986, 1990). The use of these amide vibrations to extract protein secondary structure has been limited due to the complexity of these bands, the possible interference arising from side chains such as the ionized side chains of aspartic acid and arginine and the overlap of different vibrations giving rise to these bands (Dousseau and Pezolet 1990). Therefore, the majority of the work in developing FTIR has been focused on analyzing amide I vibrations to monitor protein secondary structure (D'Antonio et al. 2012).

Absorption associated with the amide I band leads to stretching vibrations of the C=O bond of the amide. This band is sensitive to protein conformation because of its involvement in hydrogen bonding in different elements of secondary structure. Studies with proteins of known structure have been used to systematically correlate features of the amide I band to secondary structure content (Byler and Susi 1986; Surewicz and Mantsch 1988). The second derivative amide I IR spectrum of lysozyme is shown in Fig. 3.6. Assignments of each peak under the amide I band to specific secondary structure elements have been a matter of interest for many researches and have been assigned using model peptides, X-ray crystallography, and mathematical modeling. Table 3.2 shows typical amide I second derivative band assignments to various secondary structures.

In order to quantify the percentage content of each secondary structure element in a protein, a curve fitting procedure can be applied to estimate quantitatively the area of each component representing a type of secondary structure (Dousseau and Pezolet 1990; Byler and Susi 1986; Susi and Byler 1986; Jackson and Mantsch 1995). The percentages of the different secondary structure elements obtained by curve fitting of amide I second derivative IR spectra for various proteins have been in good agreement with the secondary structure information obtained from their X-ray crystallographic structures. An example of the curve fitting of lysozyme amide I FTIR second derivative spectrum is also shown in Fig. 3.6.

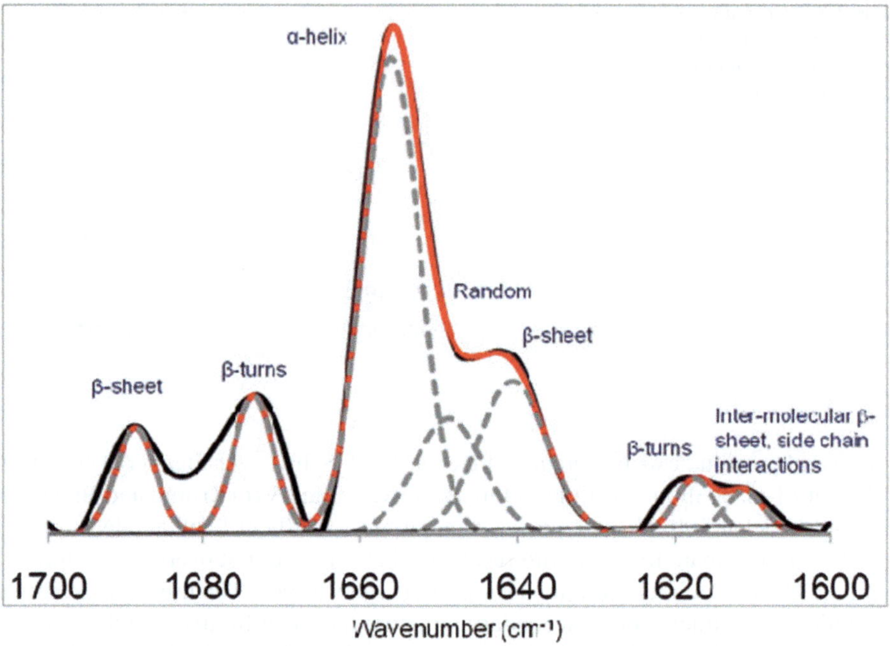

Fig. 3.6 Amide I FTIR spectrum second derivative of lysozyme in 50 mM phosphate buffer at pH 7 and 25°C. Red line is showing the derivative curve fit. *Dotted line* is showing the curve fit of individual peaks

Table 3.2 Common band assignments of protein secondary structure in amide I region

Secondary structure assignment	Spectral region (cm^{-1})
Alpha-helix	1,645–1,662
Beta sheet	1,613–1,637, 1,695–1,682
Unordered	1,645–1,637
Turns	1,682–1,662

FTIR spectroscopy is widely used in the development of protein therapeutics. It has been used to measure the secondary structure of the targeted protein as part of HOS characterization. Quantifying the secondary structure components such as α-helix, β-sheet, bends, and turns has been used to better understand the HOS product quality attributes of the protein under study and used to support process development, formulation development, stability studies, and comparability studies; some of these applications are discussed in the last half of this book. Qualification of this technique for use in comparability and filing documents is described in Chap. 5.

3.2.4 *Circular Dichroism*

Nearly all organic molecules and macromolecules are "optically active," arising from a lack of chemical symmetry. For proteins the symmetry is provided by the three-dimensional structure of the folded protein. The CD spectrum arises from an electronically symmetrical chromophore in an asymmetric environment. The chromophores that absorb light in the UV range are the peptide backbone and the aromatic amino acids and disulfide bonds. The result is electromagnetic interactions between neighboring chromophores that can be detected spectroscopically. There are a number of ways in which optically active samples can alter properties of transmitted light, including linearly polarized light, circularly polarized light, circular birefringence, and circular dichroism (CD) spectroscopy. With CD spectroscopy, one measures the differential absorption of either left-hand or right-hand components of circularly polarized light, the output being a measure of ellipticity, θ, of a sample as a function of different wavelengths. Because this is a difference spectrum, the signal can be either positive or negative, with the blank at zero.

The contribution of different secondary structural elements to the total CD signal is additive, but also wavelength-dependent (Table 3.3). Thus, CD can be used to estimate the relative fraction of different secondary structural elements of a protein, α-helix, β-sheet, and random coil, directly from the CD spectrum without any other experimental input, as described extensively by Greenfield (Greenfield and Fasman 1969; Greenfield 1996, 2006a, b, c, 2007) as well as many others. However, the structural environment around the aromatic residues can result in signals in the far-UV CD spectrum that can complicate this analysis.

In the near-UV CD region the combination of aromatic and disulfide environments present in the protein structure results in a complex spectrum which is like a

Table 3.3 Features of protein CD spectra

Far-UV CD	
Range	(190–240 nm)
Source of signal	Peptide bond in asymmetric environment; protein secondary structure
α-Helix	Negative at 208 nm
	Negative at 222 nm
	Positive at 192 nm
β-Sheet	Negative at 218 nm
	Positive at 196 nm
Random coil	Positive at 212 nm
	Negative at 195 nm
Near-UV CD	(260–320 nm): aromatics in asymmetric environment of folded protein; 3° structure
Range	(260–320 nm)
Source of signal	Aromatics in asymmetric environment; protein tertiary structure

finger print for that particular protein. Each of the aromatic amino acids has a characteristic wavelength profile that corresponds to their absorbance spectra. Trp shows a peak close to 290 nm with fine structure between 285 and 305 nm; Tyr has a peak between 270 and 285 nm, with a shoulder at longer wavelengths often obscured by the Trp band; Phe shows bands with fine structure between 250 and 265 nm. Disulfide bonds also absorb in the near-UV region (weak broad absorption bands from 250 to 280 nm), the changes in the dihedral angle of the disulfide bond will result in a change in the signal in this region of the spectrum. The actual shape and magnitude of the near-UV CD spectrum of a protein will depend on the protein primary sequence, the number of each type of aromatic amino acids present, their mobility, and the nature of their environment (H-bonding, polar groups, and polarizability). While specific residue assignments cannot be made, changes in the protein fold result in changes in the environment of the aromatic amino acids and disulfide bonds and thus in the near-UV CD spectrum. As with fluorescence, the complexity of the aromatic CD spectrum increases with increasing numbers of aromatic groups.

CD spectroscopy has been used extensively to characterize protein folding and unfolding, both as a function of denaturants and temperature. It is a truly general technique that is not limited to constraints of molecular size or chemical composition and is a "label-free" approach that does not require secondary conjugation or chemical modification to obtain data on protein structural composition. Moreover, modern improvements in CD instrumentation and data fitting have made it straightforward to collect full spectra as a function of temperature, providing a means to quantitate the presence of subtle fluctuations in protein structure that occurs at temperatures below the primary protein unfolding transition.

3.2.5 Vibrational Optical Activity

Vibrational optical activity (VOA) is the field of spectroscopy associated with combining infrared absorption or Raman scattering with the optical activity, manifested as either optical rotation or CD. These include infrared vibrational circular dichroism (VCD) and vibrational Raman optical activity (ROA). VCD is an extension of electronic circular dichroism (ECD) described above, using an infrared spectrometer (FTIR) to measure the difference in the IR intensity for left minus right circularly polarized radiation, while ROA measures the difference in intensity of right minus left circularly polarized Raman scattered radiation in several different instrument configurations (Nafie 2011). VOA measurements are complementary to typical spectroscopic methods such as CD and FTIR spectroscopy, but with additional sensitivity in monitoring subtle changes in tertiary structure (Nafie 2011; Nafie and Dukor 2007; Lakhani et al. 2009; Cao et al. 2008; Keiderling et al. 2006; Zhu et al. 2006; Barron 2006).

In a typical optical activity instrument, light emerging from a Michelson interferometer is polarization modulated using a photoelastic modulator (PEM), with light oscillating between left and right circularly polarized states at the PEM frequency

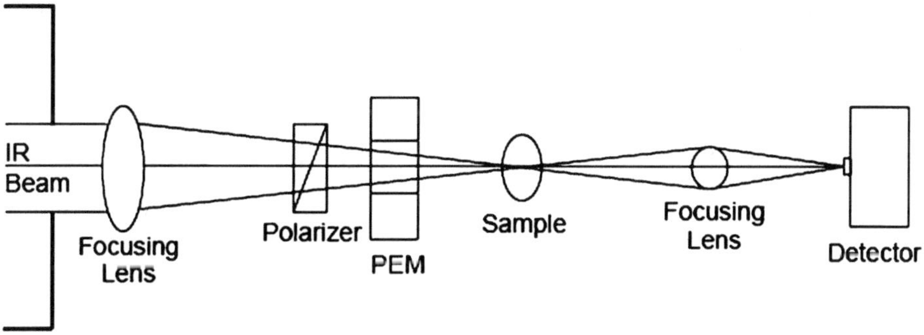

Fig. 3.7 Diagram of a typical FTIR-VCD optical path

(Fig. 3.7). The infrared light is modulated by the Michelson interferometer and then enters the VCD portion of the setup. By passing through a linear polarizer and a PEM, the Fourier-modulated IR beam is further modulated at the PEM frequency. To obtain a VCD spectrum, the doubly modulated signal is first demodulated at the PEM frequency by a lock-in amplifier, then Fourier transformed by a computerized system (Keiderling et al. 1999).

A chiral molecule interacts differently with left and right circularly polarized light, in contrast to classical IR spectroscopy in which vibrational excitation occurs with nonpolarized IR radiation and thus does not vary due to differences in chirality. In certain applications of VCD, it is possible to infer relative absolute configuration between pairs of structurally related molecules. Such correlations are strengthened due to the large number of distinct vibrational bands in a typical mid-infrared and Raman scattering spectrum. This high level of sensitivity to minor structural change gives optical activity technologies, VCD and Raman optical activity, a high level of analytical power to elucidate both absolute molecular structure and conformation (Nafie and Dukor 2007).

Aromatic residues and polysaccharides are known to have overlap in the far-UV CD region, but show good separation in a VCD spectrum (Shi et al. 2006; Dukor and Keiderling 1991). The sensitivity of VCD to interactions of dipoles and the combination of sign, pattern, and band shapes with frequency allow VCD to provide local structural information and to distinguish between different secondary structures (Shi et al. 2006; Dukor and Keiderling 1991). Moreover, VCD, like the other types of vibrational spectroscopy, can be used across different types of samples including high concentration solutions (>50 mg/mL), solids, protein aggregates, and foreign organic particles. VCD is useful for monitoring formation of amyloid-like fibrils, with an order of magnitude increase in the VCD intensity upon fibril formation, reporting on the long-range supramolecular chirality of protein fibrils (Shi et al. 2006; Dukor and Keiderling 1991; Meyer et al. 2004; Baumruk and Keiderling 1993; Yoder et al. 1997). Moreover, VCD spectroscopy should be transparent to many commonly used formulation components that can interfere with CD and FTIR spectroscopy measurements. Glycine is commonly used in protein therapeutic drug product formulations, but the carboxyl group of free glycine in the

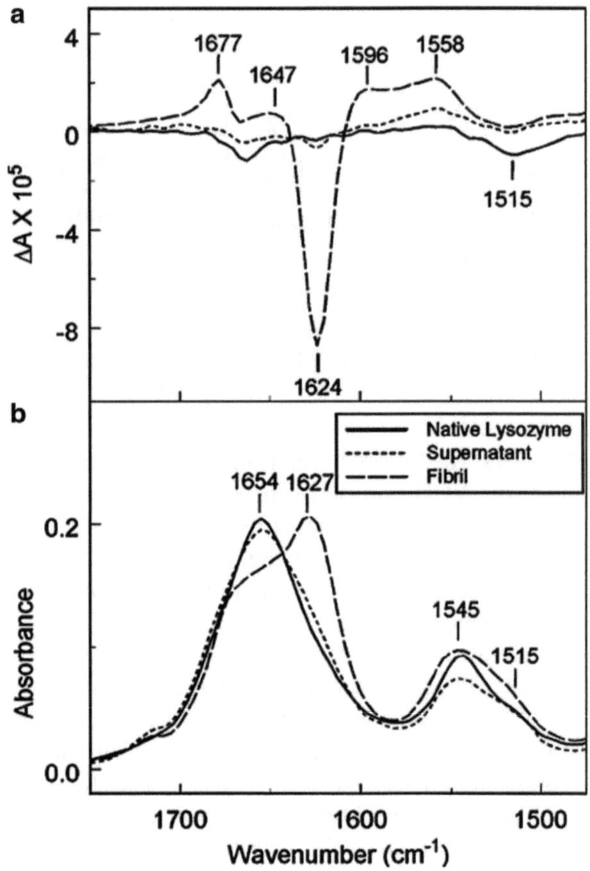

Fig. 3.8 Vibrational circular dichroism (**a**) and FTIR (**b**) spectra of lysozyme; native, centrifuged, and supernatant. Samples were treated at pH 2 and heated at 60°C for 2 days

formulation is particularly problematic for IR spectroscopic characterization of proteins from amide I overlap (Carpenter et al. 1998; Meyer et al. 2004). However, in the case of VCD, glycine does not interfere with the amide I region since its C=O stretch is lower in frequency.

VCD has been used to examine fibril formation in lysozyme and insulin. Under conditions of acidic pH and incubation at 60°C, the fibril formation produced a VCD spectral signature in the amide I centered at 1,627 cm^{-1} and amide II centered at 1,515 cm^{-1} in IR region for both proteins, which was not observed in the corresponding IR spectrum, shown in Fig. 3.8 (Ma et al. 2007; Kurouski et al. 2010). More recent studies have revealed spontaneous interconversion between opposing supramolecular chiral forms of fibril structure and correlations between the signs and magnitudes of the VCD spectra and the morphologies of fibrils as observed by SEM and AFM microscopies (Kurouski et al. 2012).

Raman optical activity (ROA) scattering spectra can also provide very useful information about changes in protein secondary structure. Additional ROA studies have focused on the amide III region where particular sensitivity to partially or

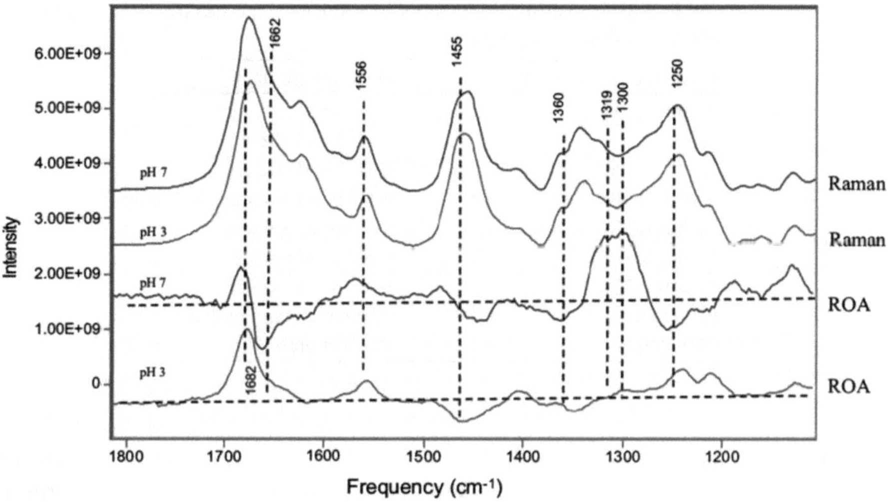

Fig. 3.9 ROA spectra of human IgG2 in solutions at pH 3 and 7

inherently unfolded protein structure can be studied by monitoring protein tertiary structure through signals from the aromatic amino acids. ROA is not impeded by background water scattering, and hence the entire spectral region from 2,000 to 100 cm^{-1} can be used as a diagnostic of overall secondary and tertiary protein structures (Zhu et al. 2006). Several studies have been published illustrating the ability of ROA to monitor aggregate formation in various protein therapeutics in various formulations of monoclonal antibodies. Two classes of human monoclonal antibodies, namely, IgG1 and IgG2, were studied using Raman and ROA at pH 7 and pH 3 (Li and Li 2009). The ROA spectra for both of these two antibodies exhibit large but very similar differences at these two pH values (Fig. 3.9). The data demonstrate the sensitivity of ROA to structural differences in HOS of these two antibodies, with structural differences arising from a net unstructuring or from changes in the stereo-conformational structure of the glycosidic tails (Li and Li 2009). These features are readily measured using ROA, but are not observed by traditional Raman spectroscopy (Li and Li 2009).

Both VCD and ROA are relatively new techniques which are used primarily to provide in-depth characterization of specific proteins, but are not yet widely employed.

3.2.6 *Raman Spectroscopy*

Like infrared (IR) spectroscopy, Raman spectroscopy is a method for probing structures and interactions of molecules by measuring the energies (or frequencies) of molecular vibrations. The main advantage of Raman spectroscopy for applications

Table 3.4 Raman bands features in Raman effect spectra and their vibrational origin

Frequency range (cm^{-1})	Band assignment
2,700–3,100	C–H alkyl free vibration
2,230	C≡N bond stretch
2,190–2,300	C≡C bond stretch
2,100–2,140	C≡C bond stretch
1,650–1,750	C=O bond stretch
1,600–1,675	C=C bond stretch
1,580–1,620	C=C bond stretch
990–1,010	Aromatic ring breathing
650–860	C–Cl stretch

to proteins and their complexes is the fact that liquid water (both H_2O and D_2O) does not confound the Raman effect. Also like FTIR samples in different physical forms can be analyzed including solutions, suspensions, gels, precipitates, fibers, single crystals, and amorphous solids, with no complicated sample preparation necessary. Raman spectroscopy is nondestructive, with high specificity for certain structures and applicability to large supermolecular assemblies (Nafie 1996).

IR and Raman differ fundamentally in the mechanism of interaction between radiation and matter (Benevides et al. 2004). Raman scattering is inelastic light scattering which occurs at wavelengths that are shifted from the incident light by the energies of the molecular vibrations. Typical applications include structure determination, multicomponent qualitative analysis, and quantitative analysis (Nafie 1996; Dong et al. 1998). The vibrational spectrum of a molecule is composed of bands representing active normal vibrations. The spectrum depends on the masses of the atoms in the molecule, the strength of their chemical bonds, and the atomic arrangement. Consequently, groups of atoms connected by certain types of bonds have certain characteristic vibrations in the Raman spectra. Table 3.4 shows a set of common chemical structures with their frequency regions in Raman spectra (Socrates 2004).

A Raman system typically consists of four major components (Fig. 3.10a): an excitation source such as a laser, a sample illumination system, light collection optics, a wavelength selector which can be either a filter or spectrophotometer, and the detector which can be a photodiode array, a charge coupled device, or a photomultiplier tube. A sample is normally illuminated with a laser beam in the ultraviolet (UV) visible (Vis) or near infrared (NIR) range (Dong et al. 1998). Other state-of-the-art Raman spectrometer systems have been described for various biological applications such as UV resonance Raman [UVRR (Brennan et al. 1997; Hashimoto et al. 1993; Asher et al. 1993)], Raman optical activity [ROA (Nafie 1996; Grauw et al. 1997; Barron et al. 1996)], and confocal Raman microscopy (Goldstein et al. 1996). In all cases scattered light is collected with a lens and sent through an interference filter or spectrophotometer to obtain the Raman spectrum of the sample being analyzed. In the case of Raman microscopy, the microscope is

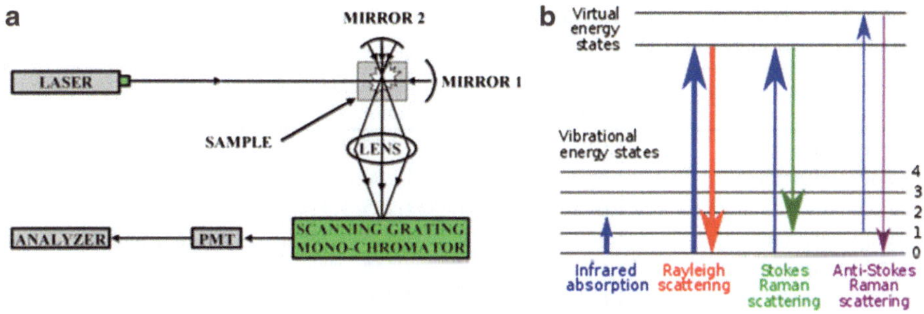

Fig. 3.10 Schematic of (**a**) basic Raman spectroscopy components and (**b**) energy level diagram showing the states involved in Raman signal. The line thickness is roughly proportional to the signal strength from the different transitions

used to visualize the sample and ensure that the spectrum is from a particular particle or region of the sample.

Raman spectra are obtained when the light interacts with the molecule and distorts (polarizes) the cloud of electrons round the nuclei to form a short-lived metastable state from which photons are quickly reradiated (Fig. 3.10b). If only electron cloud distortion is involved in scattering, the photons will be scattered with very small frequency changes. This elastic scattering reaction is called Rayleigh scattering and is the dominant process following electron cloud distortion. However, if light scattering induces relaxation through nuclear motions, then energy will be transferred either from the incident photon to the molecule or from the molecule to the scattered photon. In these cases the process is *inelastic* and the energy of the scattered photon is different from that of the incident photon by one vibrational unit. It is an inherently weak process in that only one in every 10^6–10^8 photons scattered results in Raman scattering. However, it is very sensitive to subtle structural changes and with the development of high-power modern lasers and microscopes this technique is now much more readily available and in use.

Changes in the molecular geometry or conformation that characterize many molecular biological phenomena can produce relatively large changes in Raman band positions, which are usually referred to as frequency shifts. Such shifts in the Raman band frequencies are the basis for applications of the technique to analyze protein secondary structures, monitor tertiary structure changes, determine side-chain conformations, and detect intramolecular interactions. Because the molecular geometry and force field are also sensitive to interactions between molecules, Raman can be effectively applied to probe protein intermolecular interactions, including the formation on specific complexes and large assemblies.

A typical protein spectrum contains virtually all of the fundamental vibrational information on a protein or nucleic acid molecule, except for hydrogen stretch modes which generate bands in the 2,400–3,600 cm^{-1} Raman interval (Barron et al. 2000). Raman intensities associated with protein secondary structure are typically produced by the vibrations of the C=O and C–N chemical groups stretching in the

Table 3.5 Amide I and amide III Raman bands of representative polypeptides and proteins with secondary structure assignments

Molecule	Frequency (cm^{-1}) amide I	Frequency (cm^{-1}) amide III	Secondary structure assignment
α-Poly-l-alanine	1,655	1,265–1,348	α-Helix
α-Poly-l-glutamate	1,652	1,290	α-Helix
α-Poly-l-lysine	1,645	1,295–1,311	α-Helix
β-Poly-l-alanine	1,669	1,226–1,243	β-Strand
β-Poly-l-glutamate	1,672	1,236	β-Strand
β-Poly-l-lysine	1,670	1,240	β-Strand
Poly-l-lysine, pH 4	1,665	1,243–1,248	Irregular
Poly-l-glutamate, pH 11	1,656	1,249	Irregular
cI (Lamda) repressor (1–102)	1,675	1,245	Turns/irregular
Bacteriophage P22 subunit	1,655	1,235	β-Strand/turns

peptide bonds. Stretching vibrations of C–C, C–N, and C=O bonds also produce intense Raman bands, especially if they involve the concerted symmetrical displacements of side-chain skeletons. The Raman bands associated with the bending and stretching modes of individual hydrogens in the protein substituent (e.g., C–H, N–H, and O–H) are generally weak, but their collective spectral intensities, which result from the large numbers of such groups in a protein, can be significant.

Generally, application of Raman spectroscopy to protein secondary structure analysis focuses on changes in the amide I spectral region (1,640–1,680 cm^{-1}), which is primarily a carbonyl stretching mode, as well as the amide III (1,230–1,310 cm^{-1}), which combines both in-plane N–H bending and C–N stretching motions. Table 3.5 lists the Raman frequencies assigned to amide I and amide III bands of representative polypeptides and proteins of differing secondary structures. Other conformation-sensitive bands have been identified in Raman and ultraviolet resonance Raman (UVRR) spectra of peptide model compounds and proteins. For example, amide II (1,500 and 1,600 cm^{-1}), which involves substantial C–N stretching, is useful in studies of protein α-helical content (Wang et al. 1991). Polarized Raman spectra of α-helical proteins also generate a useful secondary structure marker near 1,340–1,345 cm^{-1}, which can be exploited for α-helix orientation analyses (Tsuboi et al. 2000). Additional amide-related modes near 1,390 cm^{-1} have been proposed in UVRR spectra which have been assigned to the UVRR the overtone of amide V and appear to be sensitive to the local conformation of the peptide linkage (Wang et al. 1991).

Certain Raman signatures are unique to protein tertiary structure, notably vibrational frequencies of side-chain conformations. For example, the Raman band between 2,500 and 2,600 cm^{-1} resulting from the cysteine sulfhydryl bond (S–H) stretching vibration is a unique probe of local SH structure and dynamics (Raso et al. 2001). In-plane vibrations of the rings of aromatic side chains such as tryptophan, tyrosine, and phenylalanine are also expected to produce Raman bands of high intensity. In addition, vibrations that involve the displacement of heavy atoms such as sulfur in C–S stretching modes of methionine and cysteine, S–S stretching of cysteine, S–H stretching of cysteine, and Zn–S stretching in zinc metalloproteins are expected to be relatively intense.

The indolyl moiety of tryptophan generates many prominent Raman bands, several of which have been correlated with the local environment and geometry of the tryptophan side chain in proteins (Miura et al. 1991; Siamwiza et al. 1975). For example, normal mode *W*3 generates an intense and sharp Raman band in the 1,540–1,560 cm^{-1} interval. The tryptophan residue also generates a Fermi doublet with components at 1,360 and 1,340 cm^{-1}, with an intensity ratio (*I*1360/*I*1340) that increases with increasing hydrophobicity of the indolyl ring environment and thus serves as an indicator of local hydropathy. The normal mode *W*17 Raman band near 880 cm^{-1} is sensitive to indolyl N–H hydrogen-bond donation which exhibits a relatively high value when the indolyl moiety is located in a highly hydrophobic environment, such as the hydrophobic core of a globular protein. Normal mode *W*18, which is an indole ring-breathing vibration, generates bands near 755 cm^{-1} in Raman spectra of proteins with increased intensity when the hydrophobicity of the indolyl ring environment decreases.

Raman spectroscopy becomes an effective technology for various applications in protein biotechnology due to its capability to simultaneously measure the secondary and tertiary structure of a protein sample and to be applied to various physical forms. In the field of protein therapeutics, Raman spectroscopy has been used to measure protein HOSs in solution, in lyophilized form, or embedded in biocompatible polymer, throughout process and formulation development. Additionally, Raman spectroscopy can be very useful investigating incidents during stability studies and during manufacturing.

3.2.7 *Other Structural Characterization Methods*

X-ray Crystallography. High-resolution atomic structures of proteins have been integral to our understanding of protein function and fundamental principles of protein structure. As much as molecular biology or genomics, X-ray crystallography has been part of the exponential growth and success in biomedical research dating back to our earliest understanding of macromolecular structures. The advent of molecular replacement (non-crystallographic symmetry) has greatly simplified and accelerated crystallographic structure determination and allowed homology modeling to expand structural analysis and comparisons to proteins that have not been (or cannot be) crystallized (Rossmann 2001).

Crystallography has also had an impact in immunology, both in understanding the various antibody–protein interactions and in how to modulate and optimize these interactions from the theoretical to the practical (Sondermann and Oosthuizen 2002; Herr et al. 2003; Davies and Cohen 1996; Chothia et al. 1989; Al-Lazikani et al. 1997). The rapidity with which a structure can be refined once a crystal is available makes crystallography an important tool for understanding the HOS of proteins. The challenges with X-ray crystallography of proteins are the following: (1) many proteins do not crystallize readily, if at all, as is especially the case for the majority of integral-membrane proteins and proteins with flexible regions; (2) the crystallographic state is not always indicative of the functionally relevant state of a protein, in a crowded solution in a cellular environment; and (3) the flexible regions of a protein usually do not show up in the structure derived from X-ray crystallography, even if the protein crystallizes.

Nuclear magnetic resonance (NMR) and Isotope Exchange. NMR spectroscopy is one of the most information-rich experimental techniques, affording atomic-level information on protein structure and dynamics. Like X-ray crystallography, NMR is useful in determining tertiary structures of macromolecules, generally producing ensembles of structures that reflect the solution state of a protein and glycoproteins (Billeter et al. 1992, 2008; Wormald et al. 2002; Markley et al. 2003; Ziarek et al. 2011; Shen et al. 2008). It is often the method of choice for investigating protein motion, as it enables measurement of different dynamic processes from very rapid (ns) fluctuations through slow conformational transitions (μs) and global unfolding (ms-s) (Sapienza and Lee 2010; Tamm et al. 2007; Nashine et al. 2010; Manley and Loria 2012; Chill and Naider 2011; Baldwin and Kay 2009; Boehr et al. 2006; Wand 2001), integral-membrane and membrane-associated proteins (Kielec et al. 2009), and in different enzymatically active or functional states (Boehr et al. 2006; Mittermaier and Kay 2009; Eisenmesser et al. 2005; Loria et al. 2008). Historically, NMR has been more suitable to studies of small (<30 kDa) proteins and protein fragments, but continuing evolution of new pulse sequences and relaxation techniques have led to significant advancements in studies of larger proteins and protein–ligand complexes. For biopharmaceutical development, NMR can provide insights into protein fluctuations, as well as interactions between but also how proteins and excipient and how this affects protein dynamics (Skinner and Laurence 2008, 2010).

Hydrogen–deuterium isotope exchange (HX) is an effective tool for investigating protein fluctuations that relate to unfolding and aggregation. Historically, HX studies employed NMR to monitor the water-catalyzed exchange of peptide backbone amides, providing information on the hydrogen-bonding stability of each and every residue in a protein. Structural fluctuations of local and sub-global unfolding/folding events will expose/protect amides from exchange, with the exchange constant related to the free energy for that residue. These stability values relate to the contribution of each residue towards the global stability of the protein and provide insights into solvent accessibility, regional flexibility, folding pathways, and how these are modified by ligand and excipient binding events (Skinner et al. 2012; Englander et al. 1997; Bai et al. 1995; Englander 2000; Li et al. 2008).

Other methods, including neutron diffraction, mass spectrometry, and FTIR, have been adapted to detect HX in proteins, generally involving trapping of exchangeable amides by pH jumps, followed by detection of peptide fragments generated by low pH proteolysis (Englander 2006; Tsutsui and Wintrode 2007; Hamuro et al. 2005). These techniques are more rapid, since they do not require backbone resonance assignments necessary for NMR-based HX studies, but have diminished resolution with respect to changes in stability on a per peptide, rather than a per residue basis, providing a more heterogeneous view of protein fluctuations.

3.3 Thermodynamic and Hydrodynamic Techniques: Shape and Size

3.3.1 Calorimetric Methods

There are two general calorimetric approaches for investigating the thermodynamics of proteins and protein–ligand (or excipient) interactions: titration calorimetry (Velazquez-Campoy et al. 2004; Ladbury and Chowdhry 1996; Jelesarov and Bosshard 1999; Weber and Salemme 2003) and scanning calorimetry (Jelesarov and Bosshard 1999; Weber and Salemme 2003; Sanchez-Ruiz 2011; Spink 2008; Krishinan and Brandts 1978; Brandts and Lin 1990). All calorimetric techniques are "label-free," that is, they do not require the presence of a particular fluorophore or special labeling, but rather rely on direct measurement of thermodynamic changes that arise during an experiment.

3.3.1.1 Isothermal Titration Calorimetry

With titration calorimetry, a ligand solution is injected into a solution of protein (or vice versa) while the change in heat, either evolved or absorbed, upon binding is observed. These are primarily isothermal titration calorimetry (ITC) experiments, which allow the determination of the binding constant (binding free energy), the binding enthalpy, and the stoichiometry for the interaction. This technique has its origins in understanding ligand-binding energetics and developing an understanding of molecular recognition. It is equally well suited for the study of excipient interactions, to differentiate an effect on protein stability via either a direct binding mechanism or a change in the preferential hydration/dehydration effect. Careful analysis of the heat generated upon mixing of solutions with different compositions must be taken into consideration in order to avoid artifacts during ITC experiments.

3.3.1.2 Differential Scanning Calorimetry

One of the most routine methods for candidate and formulation screening during biopharmaceutical development is differential scanning calorimetry (DSC). In DSC a reference cell (H_2O or buffer) and a sample cell (containing the protein solution being analyzed) are heated at a constant rate of change in temperature, and the difference in the energy provided to the two cells, or heat capacity, in order to achieve this rate is recorded. The output of DSC experiments is an evolution of heat (enthalpy) corresponding to the net thermodynamic loss of tertiary structure and rehydration of unfolded protein and a midpoint temperature, T_m, that represents the point at which the sample is equally populated by folded and unfolded forms. The thermodynamics of a reversible reaction can be obtained from these experiments, while for irreversible reactions, relative differences in conformational stability or stability to aggregation can be obtained from differences in T_m or in the temperature at which unfolding begins. For multidomain proteins such as antibodies, it is often possible to resolve the transition from each domain, providing information about individual parts of the protein that is difficult to obtain by other techniques. The application of DSC is somewhat limited due to the fact that it is a low-throughput method that requires about 1 mg of protein.

DSC has applications in measuring ligand binding as well (Brandts and Lin 1990). Usually binding of a ligand increases the stability of the protein, increasing the protein T_m in a manner that depends both on ligand affinity and on concentration. As described in Chap. 2, a higher-throughput approach that involves heating a solution in the presence of an extrinsic fluorescent probe, which fluoresces only in the presence of unfolded protein, has been developed and is broadly applied when material and time are limited. This technique is variously referred to as differential scanning fluorimetry (Goldberg et al. 2010; He et al. 2009a, b; Li et al. 2011), ThermoFluor™ (Mezzasalma et al. 2007; Matulis et al. 2005), or a thermal shift assay (Kranz and Schalk-Hihi 2011; Cimmperman et al. 2008; Zubriene et al. 2009; Norvaisas et al. 2012; Zhang and Monsma 2010). Each of these has been adapted to plate-based unfolding studies, which reduce sample requirements by several orders of magnitude and allow for parallel unfolding or ligand-binding studies. Thermal shift assays are quite useful in excipient screens, allowing for rapid identification of stabilizing conditions for a given protein. DSC can then be used as a sort of gold standard, providing support for the results obtained with the higher-throughput methods and increasing understanding of the underlying reactions.

3.3.2 Light Scattering

Light scattering studies have been utilized extensively when structural information is needed for protein molecules dissolved in solution (Van Holde et al. 2006). For this purpose, both static light scattering (SLS) and dynamic light scattering (DLS) studies (most commonly using light in the visible range) have been employed; the former is primarily used for molecular weight and dimensional analysis, and the latter is most useful for effective hydrodynamic size measurements. Both these techniques and some key areas of application in protein development are presented below.

3.3.2.1 Static Light Scattering

SLS studies refer to the analysis of the intensity of the light scattered from a molecule in solution. In simple terms, scattering can be defined as deflection of an incident beam of light by a molecule. However, a number of phenomena take place from the instant that light impacts the molecule to the moment it is dispersed. When focused on a molecule, linearly polarized light, an electromagnetic wave moving in a particular direction (e.g., *x*-axis) with its electric vector oscillating perpendicular to this direction (e.g., *z*-axis) induces corresponding oscillation in the electrons of the molecule. Such oscillations of the electron in the molecule result in generation of oscillating dipoles and subsequent emission of some of the incident energy in the form of light in a direction different from that of the incident light. When the size of the scattering molecule is significantly smaller than the wavelength of the incident light beam (about 50-fold), a phenomenon described as Rayleigh scattering occurs, the ratio of the scattered light intensity to that of the incident light can be expressed as either of the following two expressions depending on the angle of observation with respect to either the *z*- or *x*-axis:

$$\frac{i}{I_0} = \frac{16\pi^4\alpha^2 \sin^2\varphi}{r^2\lambda^4}, \tag{3.7}$$

$$\frac{i}{I_0} = \frac{8\pi^4\alpha^2}{r^2\lambda^4}(1+\cos^2\theta), \tag{3.8}$$

where i is the intensity of the scattered light, i_0 is intensity of the incident light, α is the molecular polarizability, π is the angle between the direction of scattered light and the direction of the oscillation of the electric vector (*z*-axis), r is the distance between the point of observation and the position of the scattering particle, λ is the wavelength of the incident light, and θ is the angle between the direction of the scattered light and that of the incident beam (*x*-axis). It is apparent from this expression that (a) there will be no scattering along the direction of the electric vector oscillation of the incident light, that is, when $\pi=0$, (b) the scattering intensity decreases with increasing distance from the molecule and more significantly with increasing wavelength and (c) the scattering intensity will be symmetrical in the forward and backward direction, that is, there is no angular dependence of the scattering light intensity. Equation 3.8, expressing the scattering from a single molecule, can be expanded further to represent the scattering from a collection of such molecules in a dilute solution to

$$\frac{i_\theta}{I_0} = \frac{2\pi^2 n_0^2\left(\dfrac{\mathrm{d}n}{\mathrm{d}c}\right)^2}{r^2\lambda^4 N}cM(1+\cos^2\theta), \tag{3.9}$$

where i_θ is the cumulative scattered intensity from the collection of molecules observed at an angle θ from the direction of the incident light, n_0 is the refractive index of the solvent, dn/dc is the refractive index increment of the solution per unit concentration of the added solute, N is Avogadro's number, c is the molecule or solute concentration, and M is the molecular weight of the solute. Equation 3.9 can be further simplified to the more commonly used expression:

$$R_\theta = \frac{i_\theta}{I_0} \frac{r^2}{(1+\cos^2\theta)} = KcM, \tag{3.10}$$

where R_θ is defined as the Rayleigh ratio. In Eq. 3.10, K is an instrumental constant, since it is fixed for a given instrument and particular solute–solvent system, defined as

$$K = \frac{2\pi^2 n_0^2 \left(\frac{\mathrm{d}n}{\mathrm{d}c}\right)^2}{\lambda^4 N}. \tag{3.11}$$

For a dilute solution consisting of an ensemble of associated states of the solute of interest (for instance aggregated states of different sizes of the protein monomer), Eq. 3.10 is represented as

$$R_\theta = KcM_{\mathrm{w}}, \tag{3.12}$$

with M_{w} being the weight-averaged molecular weight of the solute since each associated state contributes to the average molecular weight based on its total weight fraction instead of its number fraction. Equation 3.7 is commonly employed in protein development studies to determine the molecular weight of protein in solutions as a function of any given solution variable or over the shelf life of the product. It is assumed here that the solution being studied is sufficiently dilute so that the intermolecular interaction between multiple protein molecules does not affect the overall scattering behavior. In other words, the expression applies to ideal solutions. However, this assumption is often invalid, and intermolecular interactions significantly impact the scattering behavior of the molecules and solution.

In protein development, the most common application of static light scattering has been its use for the determination of M_w of the protein in relatively dilute solutions (Wyatt 1993). Since the formation of aggregates and other higher-order species is considered a potential patient safety risk, early detection of propensity to aggregate is critical to successful product development (Goldberg et al. 2011).

In addition to molecular weight measurements, static light scattering is also employed to obtain dimensional information. For this to be possible, the size of the molecule of interest has to be large enough to be treated as a point scatterer. In this case, the scattering behavior for even a single molecule varies depending on the location of the molecule where the incident light strikes. The scattered light thus varies with the angle of measurement. With multiple light waves now capable of striking the molecule simultaneously at different points, interference between the light scattered from these multiple scattering locations needs to be accounted for. The key attribute derived from the scattering data from larger molecules is the radius of gyration (R_G) defined as the root mean square average of the distance of the scattering element from the center of mass of the molecule. For Rayleigh scattering to be valid, λ/R_G should be in excess of about 50. Thus, for instruments equipped with a 600 nm or higher laser, molecules with *a* radius of gyration of less than 12 nm can be treated as point scatterers and will thus not exhibit any angular dependence of the scattering intensity. Obviously in such cases, it will not be possible to determine the radius of gyration of the particle and the primary application of light scattering will be for the determination of molecular weight and the association state (e.g., monomer, dimer). For globular proteins, the average hydrodynamic diameter (discussed in the next section) has been found to be ~3 times the radius of gyration, and thus when using a 600 nm laser, Rayleigh scattering will be valid for molecules with radii as high as 36 nm (Mandel 1993). This is usually the case for most proteins in development for therapeutic application. For example, for immunoglobulins, one of the larger proteins frequently encountered in development with a mass of ~150 kD, the hydrodynamic diameter is ~12 nm. In cases where information regarding the dimensions of the molecular radius of gyration is required, using instruments with incident light at less than 600 nm (ultraviolet light or X-ray) can help. But even this approach has limitations since below 300 nm, protein molecules start absorbing, and below about 160 nm, water itself starts absorbing. X-rays with wavelengths below 10 nm, to which most materials are transparent, have been used at small angles to overcome these limitations for analyzing dimensions of small and globular protein molecules.

Static light scattering has been an excellent tool for characterizing the molecular weight and the higher-order species in protein solutions. However, it has found limited application for effective size measurement in solution given the lack of angular dependence for most of the molecules currently being developed. In instances where hydrodynamic properties need to be measured, biophysical chemists have often relied on the technique of DLS which is the focus of the next section.

3.3.2.2 Dynamic Light Scattering

Unlike static light scattering experiments, which capture the time-averaged scattered light intensity of a solution sample, DLS is concerned with the fluctuations in the intensity of the scattered light over relatively small time periods for a small volume fraction of the total solution. On a molecular level, whereas static measurements track the total scattered intensity from a collection of molecules, dynamic studies track the fluctuations in the intensity of light scattered from a collection of molecules as a function of time. This concept can be expanded to analysis of a small volume element (or fraction) of the total solution wherein a large number of such molecules are in constant motion. This motion is primarily due to Brownian motion, but can also be due to intermolecular interactions. As a result of this constant motion and the constantly changing location and orientation of these molecules in the finite volume element, the light scattered from that volume element exhibits fluctuations in intensity. This fluctuation in scattered intensity is used to calculate the average size of the molecule in solution. Since water (or solvent) molecules closely associate with the molecule of interest and diffuse with it as a single entity, DLS measures the size of the hydrated (or solvated) molecule.

A solution of larger molecules will undergo intensity fluctuations at a slower rate, due to slower movements in and out of the portion of the volume being monitored, compared to a solution of smaller-sized molecules. A DLS instrument tracks these intensity fluctuations over nano- and microseconds and in one (and the most commonly employed) variations of the technique *correlates* the intensity at time, t with that at time $t+\Delta t$, $t+2\Delta t$, and so on through an autocorrelation function represented as

$$g^{(2)}(\Delta t) = \frac{\langle i(t) \cdot i(t+\Delta t)\rangle}{\langle i\rangle^2}, \tag{3.13}$$

where $g^{(2)}$ is the second-order autocorrelation function, $i(t)$ is the scattered intensity at time, t $i(t+\Delta t)$ is the scattered intensity following an interval of Δt, and the brackets (⟨⟩) indicate the averaged product of scattered intensity over time. The symbol $\langle i\rangle$ represents the average values of scattered intensity. At short time intervals (lower multiples of Δt), the correlation between scattered intensity values will be high since the molecules would not have moved significantly from their initial position and there is consistency in scattering states. Thus, higher and more stable values of the autocorrelation function are obtained at shorter time intervals. As the time delays become longer (at higher multiples of Δt), correlation between scattered intensities decreases, since there is decreasingly little correlation between scattering states, and more movement of the molecules in and out of the volume being monitored. When a single component of the solution dominates the scattering from the solution (monodisperse solutions), the autocorrelation function decays exponentially such that

$$g^{(2)}(\Delta t) = 1 + a \cdot \exp(-2q^2 D\Delta t), \tag{3.14}$$

$$\ln\left\{g^{(2)}(\Delta t) - 1\right\} = \ln a - 2q^2 D\Delta t, \tag{3.15}$$

$$q = \frac{4\pi n}{\lambda}\sin\left(\frac{\theta}{2}\right), \tag{3.16}$$

where q is the scattering vector, n represents the solution refractive index, θ is the scattering angle, λ is the wavelength of incident light, a is an instrumental constant, and D is the translational diffusion coefficient (referred to simply as the diffusion coefficient) since it characterizes translational movement of a molecule from one point to another rather than rotational movements about a given axis. Equation 3.15 can be fitted to the system-generated autocorrelation function data over time to calculate the diffusion coefficient of a given component of the solution. It must be noted that, similar to Rayleigh scattering, Eq. 3.15 can be used for the determination of the diffusion coefficient only as long as the dimensions of the scattering molecules are small compared to the wavelength of light. If this condition is not satisfied and the molecular size approaches that of the wavelength of incident light, rotational as well as internal movements of the molecules also contribute to the change in the scattering pattern and intensity with time. In this case, the calculated autocorrelation function does not accurately reflect the translational diffusion coefficient of the molecule.

From the above discussion, it is apparent that DLS directly measures the diffusion coefficient, not the size, of a given solute in solution. The effective or the hydrodynamic size is subsequently derived through the Stokes–Einstein relation assuming a spherical geometry for the molecule:

$$D_{\mathrm{m}} = \frac{kT}{f} = \frac{kT}{3\pi\eta d_{\mathrm{H}}}. \tag{3.17}$$

In Eq. 3.17, f is the frictional coefficient or a measure of drag on the diffusing molecule, k is the Boltzmann constant, T is the absolute temperature, and η is the solution viscosity. Therefore, the diffusion coefficient, sometimes called the self-diffusion coefficient (D_{s}), of the solute, is inversely related to the hydrodynamic diameter (d_{H}) or the effective size of the hydrated solute molecule (Eberstein et al. 1994; Schümmer 1978). It is important to note that the Stokes–Einstein equation is valid for spheres, and thus DLS measurements provide only an apparent hydrodynamic size of the molecule assuming the particle under examination to diffuse in a manner similar to a sphere of the same size. Consequently, globular proteins in DLS measurements exhibit a smaller measured diameter (or larger D_{s}) compared to random coils of the same molecular weight since random coils, due to more complex and interfered motions and interactions with the solvent, diffuse more slowly in the solution. Thus, DLS measurements incorporate the effect of both hydration and shape of the molecule in measuring the size of the molecule. Any change in the shape or hydration of the molecule that results in a change in the diffusion behavior of molecules can be detected by DLS studies. Conformational changes in the protein molecules in solution, occurring as a function of solution environment, can thus be readily detected via this technique.

The use of DLS is usually limited for particles in the submicron size (<1–2 μm). Most common instruments have a practical upper limit of ~1 μm. Fortunately, most protein monomers are significantly smaller, in low nanometer range, and thus can be studied via DLS. Additionally, the formation of higher molecular weight or aggregated species can also be detected since a considerable population of aggregated species can exist at sizes below 1 μm.

The form of the Stokes–Einstein relation expressed in Eq. 3.17 assumes an ideal solution behavior where intermolecular interactions, or in the case of protein solutions protein–protein interactions, do not affect the movement of protein molecules and thus do not impact the measured diffusion coefficient. However, this may not be true in all cases. For ideal solutions, the diffusion coefficient depends on the hydrodynamic size of the protein molecule (in addition to viscosity and temperature). In other words, for systems in which protein–protein interactions are too weak to influence the diffusion of the protein molecule, the diffusion coefficient is independent of the protein concentration, and the measured d_H is an absolute measure of the effective hydrodynamic diameter of the protein. However, for solutions in which protein–protein interactions affect translational movement of the protein molecules, the diffusion coefficient is a function of the protein concentration as well. Thus, DLS measurements should be carried out over a series of protein concentrations instead of a single concentration in order to tease out the effect of molecular interactions on molecular size measurements.

As discussed in several other chapters in this volume, DLS has become a potent tool for use during candidate and formulation screening, where the ability to obtain high throughput, relative qualitative results is perfectly suited for the requirements of these stages of therapeutic development.

3.3.3 Analytical Ultracentrifugation

Analytical ultracentrifugation (AUC) is one of the oldest biophysical techniques used for characterizing the association and conformation of proteins and macromolecules. The basic principles of AUC were pioneered during the 1930s by Svedberg, with few changes or innovations occurring in the technique over the next several decades. With the growth of the biotechnology industry, there has been a concurrent increase in the need for a technique to monitor protein self-association across a large size range. There are two types of AUC analysis currently in use, sedimentation velocity analytical ultracentrifugation (SV-AUC) and sedimentation equilibrium analytical ultracentrifugation (SE-AUC).The development of SV-AUC with software analysis using the c(s) distribution model has resulted in the use of AUC in therapeutic protein development for HOS characterization and as an orthogonal method to size exclusion chromatography for protein aggregation analysis.

3.3.3.1 Sedimentation Velocity

During an SV-AUC experiment, there are three forces acting on a single protein molecule: (1) the gravitational force $m\omega^2 r$ where m is the protein mass, ω^2 is the rotor angular velocity, and r is the distance from the center of rotation; (2) the buoyancy force $-m$v(bar)$\rho\omega^2$ where v (bar) is the protein partial specific volume and ρ is the solvent density; and (3) the frictional force $s(kT/D)\omega^2$ where k is the Boltzmann constant, T is the absolute temperature, and D is the diffusion constant. The sedimentation coefficient s, measured in Svedberg units, is defined as $s = v/\omega^2$ where v is the absolute velocity. The fundamental equation in AUC, the Svedberg equation, is derived from the balance of the three forces acting on a single molecule:

$$\frac{s}{D} = \frac{M(1-\bar{v}\sigma)}{RT}, \quad (3.18)$$

where M is the molar mass, and R is the gas constant. The sedimentation and diffusion of a protein in solution are measured in SV-AUC experiments as a series of sedimentation profiles and can be described by the Lamm equation:

$$\frac{d\chi}{dt} = \left(\frac{1}{r}\right)\frac{d}{dr}\left[rD\left(\frac{d\chi}{dr}\right) - s\omega^2 r^2 \chi\right], \quad (3.19)$$

where χ is the concentration of the protein at a distance r from the center of rotation at time t for a given angular velocity ω. Protein samples are usually not analyzed under standard conditions, that is, aqueous solutions at 20°C; therefore, for the purpose of comparing various samples, the s values obtained under different nonstandard conditions can be converted to the standard conditions using the equation:

$$s_{20,w} = s\left(\frac{\eta_{T,w}}{\eta_{20,w}}\right)\left(\frac{\eta_s}{\eta_w}\right)\left(\frac{1-\bar{v}\rho_{20,w}}{1-\bar{v}\rho_{T,w}}\right), \quad (3.20)$$

where $\eta_{20,w}$ is the viscosity of water at 20°C, $\eta_{T,w}$ is the viscosity of water at the experimental temperature T, η_w is the viscosity of water at the experimental temperature and η_s is the viscosity of the solvent at the same experimental temperature, $\rho_{20,w}$ is the density of water, and $\rho_{T,s}$ is the is the density of the solvent at the experimental temperature. The theory behind protein sedimentation and AUC has been described in detail in multiple books (van Holde et al. 2006; Machtle and Borger 2006; Scott and Schuck 2005; Philo 2005). Advanced AUC techniques and strategies for experimentation and data analysis were described in several critical reviews and tutorial articles (Philo 2006; Liu et al. 2006; Schuck et al. 2002).

In an SV-AUC experiment a protein sample is centrifuged, usually over several hours, while the progress of centrifugation is monitored by a detection system, most commonly UV absorbance. A typical result of SV-AUC analysis is presented in Fig. 3.11. The primary information obtained from an AUC experiment is the sedimentation coefficient. This can be converted (with certain presumptions about shape) into other information better understood by general protein chemists working in formulation development, that is, molecular weight, MW distribution (Fig. 3.11b), or hydrodynamic radius.

There are multiple approaches and also choices that are made during the performance and subsequent data analysis of an AUC experiment that are important for the interpretation of the data. The sample needs to be diluted to be in the linear range of the detector (below optical density of 1.5 for UV absorbance) in buffer. As many protein therapeutics are formulated at high concentration above 100 mg/mL, the dilution can be significant. The first crucial decision is what buffer to use for dilution; two approaches currently exist, to either use buffer without excipients that

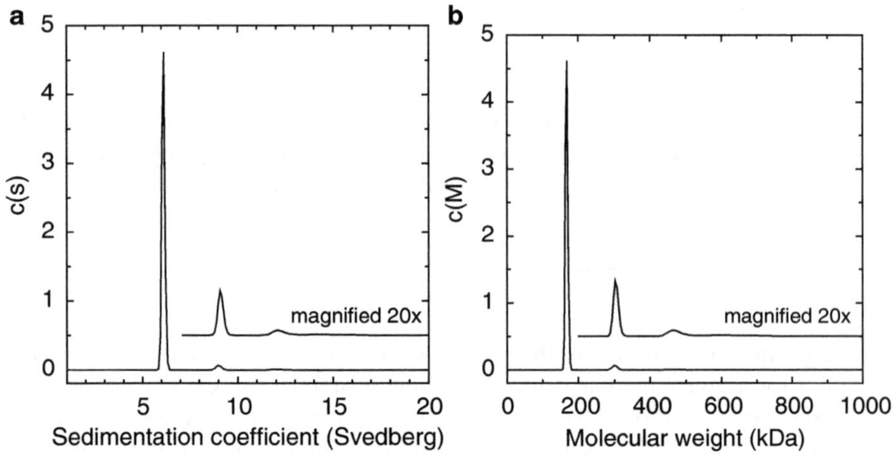

Fig. 3.11 SV-AUC analysis of an IgG analyzed in formulation buffer diluted to 0.5 mg/mL. Panel **a** shows size distribution obtained using continuous c(s) distribution model, and panel **b** presents c(M) distribution model

might interfere with the analysis or to use the original formulation buffer. Many buffers used in protein formulation contain substantial concentrations of excipients, mostly sugars, to enhance long-term stability. Dynamic gradients, caused by the presence of the excipients, can have detrimental effects on the reliability of the data obtained.

The sedimentation coefficient is the crucial parameter for HOS determination and for detection of changes in HOS and/or changes in protein conformation and shape. The value obtained from AUC experiments is in general dependent on protein concentration, as demonstrated in Fig. 3.12, and also solution conditions such as density and viscosity. In order to eliminate the effect of protein concentration on the sedimentation coefficient, it is necessary to extrapolate to zero protein concentration. This is labor intensive because it requires at least several data points and considerable experimental time. A more generally accepted approach is to compare *s* values measured at the same protein concentration, for example, 0.5 mg/mL. Similarly, the effect of density and viscosity on the sedimentation coefficients can be compensated for by extrapolating the sedimentation coefficient s to standard conditions at zero excipient concentration and 20°C ($s_{20,w}$). The sedimentation coefficient of an antibody monomer can be measured by AUC with great precision and therefore changes in sedimentation coefficient values can be used to detect small changes in protein conformation.

There are some situations where measurement of changes in *s* values has valuable application in formulation development: The conformational changes detected by AUC can be also detected by other techniques, such as the spectrophotometric techniques and ITC, described earlier in this chapter. However in some situations, especially when the conformational changes are small, they may not be detected by

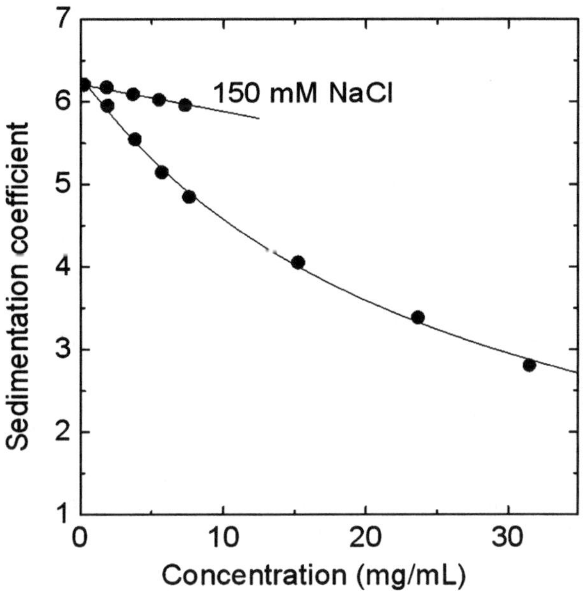

Fig. 3.12 The concentration dependence of the sedimentation coefficient of an IgG antibody. IgG was dialyzed and diluted into very low salt containing buffer at pH 5.2 and in the same buffer with addition of 150 mM of sodium chloride. Data were fit to the equation $s=s_0(1+k_s c)^{-1}$ and values of $k_s=37.8$ cm^3/g for low salt buffer, and $k_s=6.48$ cm^3/g for 150 mM sodium chloride were obtained

spectrophotometric techniques. The most common uses of sedimentation velocity AUC in protein therapeutic development are:

1. To follow subtle changes in conformation due to improper folding under pH or thermal stress, by monitoring subtle changes in the sedimentation coefficient
2. Demonstration of comparability in the conformation of therapeutic proteins from different lots or processes and as an orthogonal method to verify the results of size exclusion chromatography

For both of these applications the molar mass must remain constant; changes in MW, for example, due to protein deglycosylation or clipping, will result in a change of sedimentation coefficient as well.

The most common application of SV-AUC in HOS determination during drug development is demonstrating comparability of conformation and especially self-association for proteins from different lots, or after changes to the manufacturing and purification processes, or the same protein produced at different locations. These comparability studies are a routine part of filing documents. Figure 3.13 demonstrates comparability of an IgG protein formulated in the same formulation buffer but produced at two different locations (same sequence); for the purpose of this comparability, the processes are labeled A and B.

These two processes showed very good comparability in both conformation, analyzed as sedimentation coefficient, and the number and quantity of high molecular species (not presented here). The sedimentation coefficient distributions presented in Fig. 3.13 are summarized in Table 3.6. As can be seen in the Table 3.6, variations of sedimentation coefficients for both processes and also overall are below 0.5 %, which

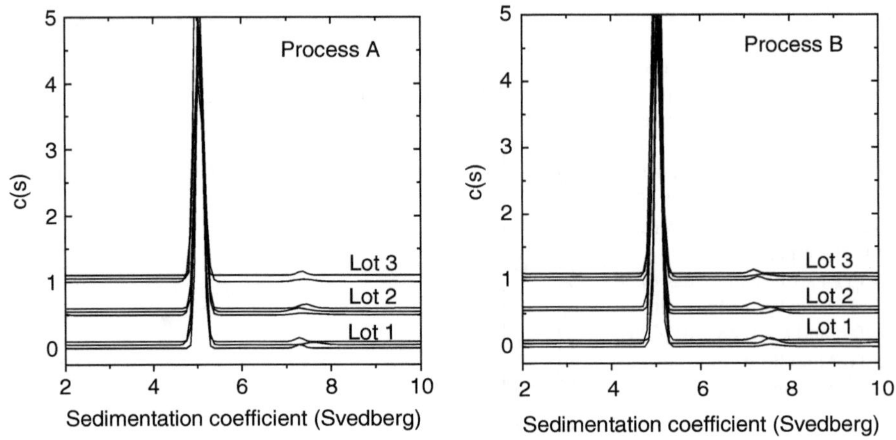

Fig. 3.13 Comparability of size distributions of therapeutic monoclonal antibody samples manufactured by two different processes. Three lots of each process, each lot in triplicate, were analyzed in formulation buffer at identical conditions including protein concentration 0.5 mg/mL

Table 3.6 Comparability of sedimentation coefficients of monoclonal antibody samples manufactured by two different processes. Sedimentation coefficients were obtained as weighted average of sedimentation coefficients of monomers in the range 4S–6S

Process A		Process B	
Lot (#)	Sedimentation coefficient (average ± s.d.)	Lot (#)	Sedimentation coefficient (average ± s.d.)
1	5.000 ± 0.015	1	4.998 ± 0.006
2	5.020 ± 0.024	2	5.021 ± 0.016
3	5.018 ± 0.023	3	5.013 ± 0.009
Process 1 average ± s.d. 5.013 ± 0.021		Process 2 average ± s.d. 5.011 ± 0.014	
Overall average ± s.d. 5.012 ± 0.018			

is remarkable considering that these were produced at two different locations at different points of time and analyzed by AUC in the original formulation buffer containing a considerable amount of excipients.

3.3.3.2 Sedimentation Equilibrium

The SV-AUC method described in the previous section is currently extensively used in protein development for aggregate characterization and for characterization and for studies on the comparability of conformation. SE-AUC is an equilibrium method and therefore does not provide any information on the molecular shape, unfolding,

Fig. 3.14 Sedimentation equilibrium experiment of an IgG antibody. The sample was analyzed at three different concentrations 0.1, 0.2, and 0.4 mg/mL at speeds 6,000, 9,000, and 12,000 rpm. The data were analyzed using single ideal species model, and molecular mass within ±3 kDa of the theoretical value was obtained

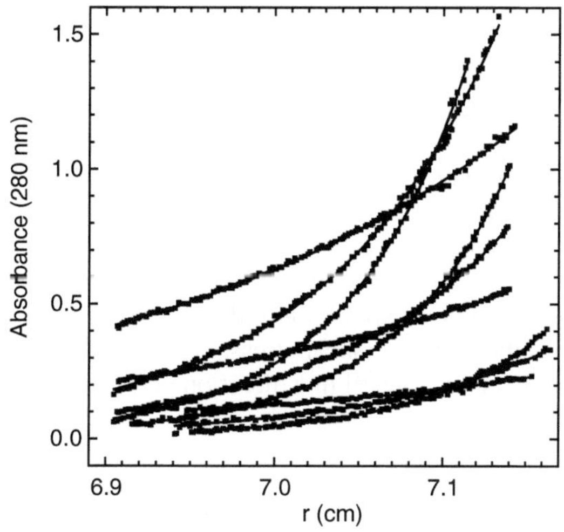

conformation, or changes in protein structure. SE-AUC is a valuable complementary method used to determine the molecular weight of proteins and their oligomeric state.

When equilibrium is reached in an SE-AUC experiment, the equilibrium of a single ideal species is

$$c(r) = c(r_0)e^{(M(1-\bar{v}\rho)\omega^2/2RT)(r^2-r_0^2)}, \tag{3.21}$$

where $c(r)$ is the protein concentration at the radial position r, r_0 is the reference radial position, M is molecular weight, ν (bar) is the protein partial specific volume, ρ is the solvent density, ω^2 is the rotor angular velocity, R is the gas constant, and T is the absolute temperature. If more species are present, dimer or oligomers, the right side of the equation becomes the sum of all species. For proteins that reversibly self-associate, the actual equilibrium association constant can also be obtained from SE-AUC experiments. The strength of this method is that it can detect all types of self-association including those that are very weak and not readily detected by other methods. For situations where non-ideality is present, the molar mass in the equation can be replaced by the apparent molecular mass, dependent on concentration, to compensate for non-ideality.

An SE-AUC experiment is done at much lower rotor speeds compared to sedimentation velocity experiments and a series of protein samples at various concentrations is spun at various rotor speeds (Fig. 3.14). It can take days or weeks, in some situations, for a protein solution to reach sedimentation and diffusion equilibrium during an SE-AUC experiment. In comparison, SV-AUC experiments take only a few hours to complete.

3.3.4 Rheology

Rheology is the study of flow behavior of matter either in the liquid or the solid state. However, for most practical purposes, it applies to liquids and semisolids and other matter which undergo some degree of deformation or flow (viscous) and loss of some or all of the applied energy as opposed to a material in which all of the applied energy can be recovered (elastic, e.g., a perfect spring) (Schümmer 1978). Viscosity (η) of a fluid is its resistance to flow, that is, its "internal friction" and is the simplest expression of the rheological properties of a material. It can be defined as the pressure [shear stress, τ] required to deform or to make a fluid flow at a unit rate [shear rate, γ] in a direction perpendicular to the plane of applied pressure:

$$\eta = \frac{\tau}{\gamma}. \tag{3.22}$$

Most of the fluids that we encounter in daily life exhibit viscosity which is a function of the shear rate, that is, the viscosity will change as one varies the rate at which the fluid is deformed. Such liquids are referred to as non-Newtonian. On the other hand, for Newtonian fluids like water viscosity is not a function of the shear rate but rather remains constant over a broad and practical range of shear rates (Macosko 1994; Shaw 1992).

Rheological behavior of protein dispersions is intimately linked to the characteristics of the protein itself as well as environmental factors. The characteristic properties of the protein that are important in this aspect are its conformation, shape, size, solubility, swellability, and charge which are influenced by environmental conditions like temperature, concentration, pH, and ionic strength (Hermansson 1972, 1979).

The viscosity of protein solutions is largely dependent on the (1) hydrodynamic behavior of the protein molecules in solution and (2) protein–protein interactions between adjacent molecules and their spatial distribution. Consequently, the study of the behavior of protein solutions at infinite dilution of the protein when protein–protein interactions are negligible can provide meaningful insight into the hydrodynamic properties of the protein molecule which in turn can be related to the shape and size or conformational changes in the protein. The problem of relating viscosity of the colloidal dispersions with the fundamental nature of the dispersed particles has been the subject of much experimental investigations and theoretical considerations (Harding 1980, 1981; Lundqvist 1999). Such an effect is defined in terms of viscosity functions such as relative (η_{rel}), specific (η_{sp}), and reduced or reduced specific viscosity (η_{red}):

$$\eta_{rel} = \frac{\eta}{\eta_0}, \tag{3.23}$$

$$\eta_{\text{sp}} = \eta_{\text{rel}} - 1 = \frac{\eta - \eta_0}{\eta_0}, \tag{3.24}$$

$$\eta_{\text{red}} = \eta_{\text{sp}} / c = (\eta_{\text{rel}} - 1) / c, \tag{3.25}$$

where η is the viscosity of the solution of concentration *c* and η_0 is the viscosity of the solvent. Intrinsic viscosity [η] is an intrinsic function of the dissolved/dispersed macromolecule and is defined as the limiting value, at zero concentration, of relative viscosity increment.

Two factors that contribute to this characteristic property of the dispersed particle are particle shape and size/volume as summarized by the following relationship:

$$[\eta] = v \bullet V_s, \tag{3.26}$$

where, ν is the molecular shape factor known as the viscosity increment or universal shape function (Kauzmann 1959; Yang 1961) and V_s is known as the swollen specific volume and is a measure of the solvent associated with the macromolecule. It is defined as the volume of the macromolecule in solution per unit of anhydrous mass of the macromolecule. Intrinsic viscosity is thus indirectly dependent on the axial ratio of the molecule that governs its shape. It increases with increased axial ratio and is therefore minimum for a sphere (Tanford and Buzzell 1956; Harding 1997). As a result of this dependence, intrinsic viscosity has been used as a combined measure of the hydrodynamic volume and molecular shape.

Changes in intrinsic viscosity or in reduced viscosity at low concentration have also been used as a measure of the extent of denaturation, which also reflects the relationship with changing axial ratio (Suryaprakash and Prakash 2000; Ahmad and Salahuddin 1976). Interestingly, for globular proteins unlike random coils, intrinsic viscosity has been found to be independent of the molecular weight and depends solely on the shape and size of the protein. In an excellent review, Harding (Harding 1997) has listed the values of intrinsic viscosity of numerous proteins and peptides. A relatively simple method has been proposed in the literature for calculating the axial ratio of proteins based on viscosity measurements and data fitting (Monkos and Turczynski 1991). Rheological measurements have also been used to determine the unfolding onset for protein in solution in the presence of an increasing concentration of a denaturant like urea (Tu and Breedveld 2005).

A general application of intrinsic viscosity measurements has been to assess the denaturation or conformational change of a protein molecule in solution. For a polypeptide poly-(his-ala-glu) studied in the pH range of 2.97–9.70, the intrinsic viscosity varied from 8 mL/g at pH 2.97 to 33 mL/g at pH 4.99 to 55 mL/g at pH 9.70 indicating an increasing unfolding or conformational change with increase in solution pH. This is consistent with a general increase in the net negative charge on the molecule due to the ionization of the glutamic acid residues and the deprotonation of histidine residues with increasing solution pH. From the example it is apparent that solution pH and ionic strength have a marked effect on the hydrodynamic properties of the protein and therefore on the flow behavior of the system on the whole. This effect is mediated through the effect on charge associated with the protein molecule, its folding and unfolding kinetics, and the solubility of the protein that changes with the pH. In addition to the absolute shape and size effects, electrostatic charge can significantly influence the intrinsic viscosity of a protein molecule in solution, especially if the molecule carries multiple charges in solution (polyelectrolyte) like proteins. In theory three such contributions have been summarized for a compact globular protein molecule: a "primary effect" due to the diffuse double layer surrounding the protein molecule, a "secondary effect" due to repulsion between the double layers of different molecules, and a "tertiary effect" arising from the intermolecular repulsions that affect the shape of the molecule. These three have been collectively referred to as "electroviscous" effects. Under conditions of high net charge on the protein molecule, the electroviscous effect contributes significantly to solution flow behavior. However, under conditions of low molecular charge and/or high charge screening, as is the case with high ionic strength solutions, electroviscous effects are not very prominent.

During the processing and delivery of protein therapeutics, there are multiple steps that can be limited by viscosity, including filtration, chromatography, and delivery through syringes. Thus, viscosity measurements can be important during formulation and candidate screening and during process development.

3.4 Conclusions

A very important characteristic of proteins is the unique, folded structure that is important for function and stability. For a protein, deviations from the preferred structure or ensemble of structures are just as critical to establishing functionality as is the primary amino acid sequence. Understanding these facts and the ability to characterize protein structures provide key tools to develop various protein therapeutics and other biotechnological applications. We have provided an overview of the most common technologies that monitor proteins higher-ordered structural quality attributes, specifically techniques aimed at quantifying changes in secondary structure, tertiary structures, and quaternary structures. The chapter further provided an overview of technologies to characterize protein thermal characteristics such as calorimetric techniques to monitor protein thermodynamic properties, with more attention paid to those approaches that require significant biophysical expertise. Continuous efforts to develop new technologies or improve existing technologies are needed to better understand protein higher-ordered structural distributions in protein biotherapies. In Chaps. 6–10 the applications of these techniques during all stages of protein therapeutic development will be illustrated.

Acknowledgements The authors would like to thank Dr. Lee Olszewski, Dr. Doug Nesta, Dr. Aston Liu, and Dr. Joseph Rinella at GlaxoSmithKline Biopharm Development for supporting this publication. We thank Dr. Rina K. Dukor from BioTools Inc. and Prof. Laurence A. Nafie from Syracuse University for their insights and helpful discussion.

References

Adams PD, Chen Y, Ma K, Zagorski MG, Sonnichsen FD, McLaughlin ML, Barkley MD (2002) Intramolecular quenching of tryptophan fluorescence by the peptide bond in cyclic hexapeptides. J Am Chem Soc 124:9278–9286

Ahmad F, Salahuddin A (1976) Reversible unfolding of the major fraction of ovalbumin by guanidine hydrochloride. Biochemistry 15(23):5168–5175

Al-Lazikani B, Lesk AM, Chothia C (1997) Standard conformations for the canonical structures of immunoglobulins. J Mol Biol 273:927–948

Asher SA, Bormett RW, Chen XG, Lemmon DH, Cho N (1993) UV resonance Raman spectroscopy using a new cw laser source: convenience and experimental simplicity. Appl Spectrosc 47:628–633

Bai Y, Sosnick TR, Mayne L, Englander SW (1995) Protein folding intermediates: native-state hydrogen exchange. Science 269:192–197

Baldwin AJ, Kay LE (2009) NMR spectroscopy brings invisible protein states into focus. Nat Chem Biol 5:808–814

Bandekar J (1992) Amide modes and protein conformation. Biochim Biophys Acta 1120:123–143

Barron LD (2006) Structure and behaviour of biomolecules from Raman optical activity. Curr Opin Struct Biol 16:638–643

Barron LD, Hecht L, Bell AF, Wilson G (1996) Recent developments in Raman optical activity of biopolymers. Appl Spectrosc 50:619–629

Barron LD, Hecht L, Blanch EW, Bell AF (2000) Solution structure and dynamics of biomolecules from Raman optical activity. Prog Biophys Mol Biol 73:1–49

Bartlett AI, Radford SE (2009) An expanding arsenal of experimental methods yields an explosion of insights into protein folding mechanisms. Nat Struct Mol Biol 16:582–588

Baumruk V, Keiderling TA (1993) Vibrational circular dichroism of proteins in H_2O solution. J Am Chem Soc 115:6939–6942

Benevides JM, Overman SA, Thomas GJ Jr (2004) Raman spectroscopy of proteins. Curr Protoc Protein Sci, Chapter 17, Unit 17.18

Billeter M, Vendrell J, Wider G, Aviles FX, Coll M, Guasch A, Huber R, Wuthrich K (1992) Comparison of the NMR solution structure with the X-ray crystal structure of the activation domain from procarboxypeptidase B. J Biomol NMR 2:1–10

Billeter M, Wagner G, Wuthrich K (2008) Solution NMR structure determination of proteins revisited. J Biomol NMR 42:155–158

Boehr DD, Dyson HJ, Wright PE (2006) An NMR perspective on enzyme dynamics. Chem Rev 106:3055–3079

Brandts JF, Lin LN (1990) Study of strong to ultratight protein interactions using differential scanning calorimetry. Biochemistry 29:6927–6940

Brennan JF, Wang Y, Ramachandra RD, Feld MS (1997) Near-infrared Raman spectrometer systems for human studies. Appl Spectrosc 51:201–208

Byler DM, Susi H (1986) Examination of the secondary structure of proteins by deconvolved FTIR spectra. Biopolymers 25:469–487

Cantor RC, Schimmel PR (1980) Biophysical Chemistry: techniques for the study of biological structure and function, vol 2. W. H. Freeman, San Francisco

Cao XL, Dukor RK, Nafie LA (2008) Reduction of linear birefringence in vibrational circular dichroism measurement: use of a rotating half-wave plate. Theor Chem Acc 119:69–79

Carpenter JF, Prestrelski SJ, Dong A (1998) Application of infrared spectroscopy to development of stable lyophilized protein formulations. Eur J Pharm Biopharm 45:231–238

Chalmers J, Griffiths P (eds) (2002) Handbook of vibrational spectroscopy. Wiley, Chichester

Chance B (1953) The carbon monoxide compounds of the cytochrome oxidases. I. Difference spectra. J Biol Chem 202:383–396

Chance B, Pappenheimer AM Jr (1954) Kinetic and spectrophotometric studies of cytochrome b5 in midgut homogenates of cecropia. J Biol Chem 209:931–943

Chill JH, Naider F (2011) A solution NMR view of protein dynamics in the biological membrane. Curr Opin Struct Biol 21:627–633
Chothia C, Lesk AM, Tramontano A, Levitt M, Smith-Gill SJ, Air G, Sheriff S, Padlan EA, Davies D, Tulip WR et al (1989) Conformations of immunoglobulin hypervariable regions. Nature 342:877–883
Cimmperman P, Baranauskiene L, Jachimoviciute S, Jachno J, Torresan J, Michailoviene V, Matuliene J, Sereikaite J, Bumelis V, Matulis D (2008) A quantitative model of thermal stabilization and destabilization of proteins by ligands. Biophys J 95:3222–3231
Daniel E, Weber G (1966) Cooperative effects in binding by bovine serum albumin. I. The binding of 1-anilino-8-naphthalenesulfonate. Fluorimetric titrations. Biochemistry 5:1893–1900
D'Antonio J, Murphy BM, Manning MC, Al-Azzam WA (2012) Comparability of protein therapeutics: quantitative comparison of second-derivative amide I infrared spectra. J Pharm Sci 101:2025–2033
Davies DR, Cohen GH (1996) Interactions of protein antigens with antibodies. Proc Natl Acad Sci USA 93:7–12
Dong A, Huang P, Caughey WS (1990) Protein secondary structures in water from second-derivative amide I infrared spectra. Biochemistry 29:3303–3308
Dong J, Dinakarpandian D, Carey PR (1998) Extending the Raman analysis of biological samples to the 100 micromolar concentration range. Appl Spectrosc 52:1117–1122
dos Remedios CG, Moens PD (1995) Fluorescence resonance energy transfer spectroscopy is a reliable "ruler" for measuring structural changes in proteins. Dispelling the problem of the unknown orientation factor. J Struct Biol 115:175–185
Dousseau F, Pezolet M (1990) Determination of the secondary structure content of proteins in aqueous solutions from their amide I and amide II infrared bands. Comparison between classical and partial least-squares methods. Biochemistry 29:8771–8779
Dukor RK, Keiderling TA (1991) Reassessment of the random coil conformation: vibrational CD study of proline oligopeptides and related polypeptides. Biopolymers 31:1747–1761
Eberstein W, Georgalis Y, Saenger W (1994) Molecular interactions in crystallizing lysozyme solutions studied by photon correlation spectroscopy. J Cryst Growth 143:71–78
Eisenmesser EZ, Millet O, Labeikovsky W, Korzhnev DM, Wolf-Watz M, Bosco DA, Skalicky JJ, Kay LE, Kern D (2005) Intrinsic dynamics of an enzyme underlies catalysis. Nature 438:117–121
Englander SW (2000) Protein folding intermediates and pathways studied by hydrogen exchange. Annu Rev Biophys Biomol Struct 29:213–238
Englander SW (2006) Hydrogen exchange and mass spectrometry: a historical perspective. J Am Soc Mass Spectrom 17:1481–1489
Englander SW, Mayne L, Bai Y, Sosnick TR (1997) Hydrogen exchange: the modern legacy of Linderstrom-Lang. Protein Sci 6:1101–1109
Epstein HF, Schechter AN, Chen RF, Anfinsen CB (1971) Folding of staphylococcal nuclease: kinetic studies of two processes in acid renaturation. J Mol Biol 60:499–508
Ferreon AC, Moran CR, Gambin Y, Deniz AA (2011) Single-molecule fluorescence studies of intrinsically disordered proteins. Methods Enzymol 472:179–204
Flanagan JM, Kataoka M, Shortle D, Engelman DM (1992) Truncated staphylococcal nuclease is compact but disordered. Proc Natl Acad Sci USA 89:748–752
Foldes-Papp Z, Demel U, Domej W, Tilz GP (2002) A new dimension for the development of fluorescence-based assays in solution: from physical principles of FCS detection to biological applications. Exp Biol Med (Maywood) 227:291–300
Fu K, Griebenow K, Hsieh L, Klibanov AM, Langer R (1999) FTIR characterization of the secondary structure of proteins encapsulated within PLGA microspheres. J Control Release 58:357–366
Goldberg DS, Bishop SM, Shah AU, Sathish HA (2010) Formulation development of therapeutic monoclonal antibodies using high-throughput fluorescence and static light scattering techniques: role of conformational and colloidal stability. J Pharm Sci

Goldberg DS, Bishop SM, Shah AU, Sathish HA (2011) Formulation development of therapeutic monoclonal antibodies using high-throughput fluorescence and static light scattering techniques: role of conformational and colloidal stability. J Pharm Sci 100(4):1306–1315
Goldstein SR, Kidder LH, Herne TM, Levin IW, Lewis EN (1996) The design and implementation of a high-fidelity Raman imaging microscope. J Microsc 184:35–45
Grauw CJ, Otto C, Greve J (1997) Linescan Raman microspectrometry for biological applications. Appl Spectrosc 51:1607–1612
Greenfield NJ (1996) Methods to estimate the conformation of proteins and polypeptides from circular dichroism data. Anal Biochem 235:1–10
Greenfield NJ (2006a) Using circular dichroism spectra to estimate protein secondary structure. Nat Protoc 1(6):2876–2890
Greenfield NJ (2006b) Determination of the folding of proteins as a function of denaturants, osmolytes or ligands using circular dichroism. Nat Protoc 1:2733–2741
Greenfield NJ (2006c) Analysis of the kinetics of folding of proteins and peptides using circular dichroism. Nat Protoc 1:2891–2899
Greenfield NJ (2007) Using circular dichroism collected as a function of temperature to determine the thermodynamics of protein unfolding and binding interactions. Nat Protoc 1:2527–2535
Greenfield N, Fasman GD (1969) Computed circular dichroism spectra for the evaluation of protein conformation. Biochemistry 8:4108–4116
Griebenow K, Klibanov AM (1995) Lyophilization-induced reversible changes in the secondary structure of proteins. Proc Natl Acad Sci USA 92:10969–10976
Griffiths PR, de Haseth JA (1986) Fourier transform infrared spectroscopy. Wiley Interscience, New York
Hamuro Y, Weber PC, Griffin PR (2005) High-throughput analysis of protein structure by hydrogen/deuterium exchange mass spectrometry. Methods Biochem Anal 45:131–157
Harding SE (1980) The combination of the viscosity increment with the harmonic mean rotational relaxation time for determining the conformation of biological macromolecules in solution. Biochem J 189(2):359–361
Harding SE (1981) A compound hydrodynamic shape function derived from viscosity and molecular covolume measurements. Int J Biol Macromol 3(5):340–341
Harding SE (1997) The intrinsic viscosity of biological macromolecules. Progress in measurement, interpretation and application to structure in dilute solution. Prog Biophys Mol Biol 68(2–3):207–262
Haris PI, Chapman D (1995) The conformational analysis of peptides using Fourier transform IR spectroscopy. Biopolymers 37:251–263
Haris PI, Lee DC, Chapman D (1986) A Fourier transform infrared investigation of the structural differences between ribonuclease A and ribonuclease S. Biochim Biophys Acta 874:255–265
Haris PI, Chapman D, Harrison RA, Smith KF, Perkins SJ (1990) Conformational transition between native and reactive center cleaved forms of alpha 1-antitrypsin by Fourier transform infrared spectroscopy and small-angle neutron scattering. Biochemistry 29:1377–1380
Hashimoto S, Ikeda T, Takeuuchi H, Harada I (1993) Utilization of a prism monochromator as a sharp-cut bandpass filter in ultraviolet Raman spectroscopy. Appl Spectrosc 47:1283–1285
Hawe A, Sutter M, Jiskoot W (2008) Extrinsic fluorescent dyes as tools for protein characterization. Pharm Res 25:1487–1499
Hawe A, Rispens T, Herron JN, Jiskoot W (2011) Probing bis-ANS binding sites of different affinity on aggregated IgG by steady-state fluorescence, time-resolved fluorescence and isothermal titration calorimetry. J Pharm Sci 100:1294–1305
He F, Hogan S, Latypov RF, Narhi LO, Razinkov VI (2009a) High throughput thermostability screening of monoclonal antibody formulations. J Pharm Sci 99:1707–1720
He F, Phan DH, Hogan S, Bailey R, Becker GW, Narhi LO, Razinkov VI (2009b) Detection of IgG aggregation by a high throughput method based on extrinsic fluorescence. J Pharm Sci 99:2598–2608
Hermansson AM (1972) Functional properties of proteins for foods. Lebensin-Wiss Technol 5:24

Hermansson AM (1979) Aspects of protein structure, rheology and texturization. Food Texture Rheol [Proc Symp] 265–282

Herr AB, Ballister ER, Bjorkman PJ (2003) Insights into IgA-mediated immune responses from the crystal structures of human FcalphaRI and its complex with IgA1-Fc. Nature 423:614–620

van Holde KE, Johnson WC, Ho (2006) Methods for the Separation and Characterization of Macromolecules. In: van Holde KE, Johnson WC, Ho PS. Principles of Physical Biochemistry. 2nd ed. Upper Saddle River, NJ: Pearson Prentice Hall:213–275

Hynes TR, Fox RO (1991) The crystal structure of staphylococcal nuclease refined at 1.7 A resolution. Proteins 10:92–105

Jackson M, Mantsch HH (1995) The use and misuse of FTIR spectroscopy in the determination of protein structure. Crit Rev Biochem Mol Biol 30:95–120

Jares-Erijman EA, Jovin TM (2003) FRET imaging. Nat Biotechnol 21:1387–1395

Jelesarov I, Bosshard HR (1999) Isothermal titration calorimetry and differential scanning calorimetry as complementary tools to investigate the energetics of biomolecular recognition. J Mol Recognit 12:3–18

Kamerzell TJ, Esfandiary R, Joshi SB, Middaugh CR, Volkin DB (2011) Protein-excipient interactions: mechanisms and biophysical characterization applied to protein formulation development. Adv Drug Deliv Rev 63:1118–1159

Kauzmann W (1959) Some factors in the interpretation of protein denaturation. Adv Protein Chem 14:1–63

Keiderling TA, Silva RA, Yoder G, Dukor RK (1999) Vibrational circular dichroism spectroscopy of selected oligopeptide conformations. Bioorg Med Chem 7:133–141

Keiderling TA, Kubelka J, Hilario J (eds) (2006) Vibrational circular dichroism of biopolymers: summary of methods and applications. CRC, Boca Raton

Kielec JM, Valentine KG, Wand AJ (2009) A method for solution NMR structural studies of large integral membrane proteins: reverse micelle encapsulation. Biochim Biophys Acta 1798:150–160

Kranz JK, Schalk-Hihi C (2011) Protein thermal shifts to identify low molecular weight fragments. Methods Enzymol 493:277–298

Krimm S, Bandekar J (1986) Vibrational spectroscopy and conformation of peptides, polypeptides, and proteins. Adv Protein Chem 38:181–364

Krishinan KS, Brandts JF (1978) Scanning calorimetry. Methods Enzymol 49:3–14

Kubelka J, Hofrichter J, Eaton WA (2004) The protein folding "speed limit". Curr Opin Struct Biol 14:76–88

Kurouski D, Lombardi RA, Dukor RK, Lednev IK, Nafie LA (2010) Direct observation and pH control of reversed supramolecular chirality in insulin fibrils by vibrational circular dichroism. Chem Commun (Camb) 46:7154–7156

Kurouski D, Dukor RK, Lu X, Nafie LA, Lednev IK (2012) Spontaneous inter-conversion of insulin fibril chirality. Chem Commun (Camb) 48:2837–2839

Ladbury JE, Chowdhry BZ (1996) Sensing the heat: the application of isothermal titration calorimetry to thermodynamic studies of biomolecular interactions. Chem Biol 3:791–801

Lakhani A, Malon P, Keiderling TA (2009) Comparison of vibrational circular dichroism instruments: development of a new dispersive VCD. Appl Spectrosc 63:775–785

Lakowicz JR (2006) Principles of fluorescence spectroscopy, 3rd edn. Springer, New York

LeTilly V, Royer CA (1993) Fluorescence anisotropy assays implicate protein–protein interactions in regulating trp repressor DNA binding. Biochemistry 32:7753–7758

Li CH, Li T (2009) Application of vibrational spectroscopy to the structural characterization of monoclonal antibody and its aggregate. Curr Pharm Biotechnol 10:391–399

Li Y, Williams TD, Topp EM (2008) Effects of excipients on protein conformation in lyophilized solids by hydrogen/deuterium exchange mass spectrometry. Pharm Res 25:259–267

Li Y, Mach H, Blue JT (2011) High throughput formulation screening for global aggregation behaviors of three monoclonal antibodies. J Pharm Sci 100:2120–2135

Liu J, Andya JD, Shire SJ (2006) A critical review of analytical ultracentrifugation and field flow fractionation methods for measuring protein aggregation. AAPS J 8:E580–E589
Loria JP, Berlow RB, Watt ED (2008) Characterization of enzyme motions by solution NMR relaxation dispersion. Acc Chem Res 41:214–221
Lundqvist R (1999) Molecular weight studies on hydroxypropyl methyl cellulose. Part 2. Intrinsic viscosity. Int J Polym Anal Charact 5(1):61–84
Ma S, Cao X, Mak M, Sadik A, Walkner C, Freedman TB, Lednev IK, Dukor RK, Nafie LA (2007) Vibrational circular dichroism shows unusual sensitivity to protein fibril formation and development in solution. J Am Chem Soc 129:12364–12365
Machtle W, Borger L (2006) Sedimentation velocity. In: Machtle W, Borger L (eds) Analytical ultracentrifugation of polymers and nanoparticles. Heidelberg Springer:47–96
Macosko CW (1994) In: Macosko CW (ed) Rheology: principles, measurements and applications. Wiley-VCH, New York, p 550
Mandel M (1993) Applications of dynamic light scattering to polyelectrolytes in solution. In: Brown W (ed) Dynamic light scattering: the method and some applications. Oxford University Press, New York, pp 319–371
Manley G, Loria JP (2012) NMR insights into protein allostery. Arch Biochem Biophys 519:223–231
Mann CJ, Royer CA, Matthews CR (1993) Tryptophan replacements in the trp aporepressor from *Escherichia coli*: probing the equilibrium and kinetic folding models. Protein Sci 2:1853–1861
Markley JL, Ulrich EL, Westler WM, Volkman BF (2003) Macromolecular structure determination by NMR spectroscopy. Methods Biochem Anal 44:89–113
Matulis D, Kranz JK, Salemme FR, Todd MJ (2005) Thermodynamic stability of carbonic anhydrase: measurements of binding affinity and stoichiometry using ThermoFluor. Biochemistry 44:5258–5266
Meyer JD, Bai SJ, Rani M, Suryanarayanan R, Nayar R, Carpenter JF, Manning MC (2004) Infrared spectroscopic studies of protein formulations containing glycine. J Pharm Sci 93:1359–1366
Mezzasalma TM, Kranz JK, Chan W, Struble GT, Schalk-Hihi C, Deckman IC, Springer BA, Todd MJ (2007) Enhancing recombinant protein quality and yield by protein stability profiling. J Biomol Screen 12:418–428
Mittermaier AK, Kay LE (2009) Observing biological dynamics at atomic resolution using NMR. Trends Biochem Sci 34:601–611
Miura T, Takeuchi H, Harada I (1991) Raman spectroscopic characterization of tryptophan side chains in lysozyme bound to inhibitors: role of the hydrophobic box in the enzymatic function. Biochemistry 30:6074–6080
Miyawaki A (2011) Proteins on the move: insights gained from fluorescent protein technologies. Nat Rev Mol Cell Biol 12:656–668
Miyazawa T, Shimanouchi T, Mizushima S (1956) Characteristic infrared bands of monosubstituted amides. J Chem Phys 24:408
Monkos K, Turczynski B (1991) Determination of the axial ratio of globular proteins in aqueous solution using viscometric measurements. Int J Biol Macromol 13(6):341–344
Murphy KP, Freire E (1992) Thermodynamics of structural stability and cooperative folding behavior in proteins. Adv Protein Chem 43:313–361
Nafie LA (1996) Vibrational optical activity. Appl Spectrosc 50:14A–26A
Nafie LA (2011) Vibrational optical activity: principles and applications. Wiley, Chichester
Nafie LA, Dukor RK (eds) (2007) Applications of vibrational optical activity in the pharmaceutical industry. Wiley, Chichester
Nashine VC, Hammes-Schiffer S, Benkovic SJ (2010) Coupled motions in enzyme catalysis. Curr Opin Chem Biol 14:644–651
Norvaisas P, Petrauskas V, Matulis D (2012) Thermodynamics of cationic and anionic surfactant interaction. J Phys Chem B 116:2138–2144
Otto MR, Lillo MP, Beechem JM (1994) Resolution of multiphasic reactions by the combination of fluorescence total-intensity and anisotropy stopped-flow kinetic experiments. Biophys J 67:2511–2521

Perez C, Castellanos IJ, Costantino HR, Al-Azzam W, Griebenow K (2002) Recent trends in stabilizing protein structure upon encapsulation and release from bioerodible polymers. J Pharm Pharmacol 54:301–313

Perrin F (1926) J Phys Radium 1:390–401

Philo JS (2006) Is any measurement method optimal for all aggregate sizes and types? AAPS J 8:E564–E571

Philo JS (2005) Analytical ultracentrifugation. In: Borchardt RT, Middaugh CR, series eds. Biotechnology: pharmaceutical aspects, Jiskoot W, Crommelin D, vol eds. Volume III: Methods for structural analysis of protein pharmaceuticals. AAPS Press, Arlington, pp 379–412

Pollok BA, Heim R (1999) Using GFP in FRET-based applications. Trends Cell Biol 9:57–60

Putnam WS, Prabhu S, Zheng Y, Subramanyam M, Wang YM (2010) Pharmacokinetic, pharmacodynamic and immunogenicity comparability assessment strategies for monoclonal antibodies. Trends Biotechnol 28:509–516

Ramsey JD, Gill ML, Kamerzell TJ, Price ES, Joshi SB, Bishop SM, Oliver CN, Middaugh CR (2009) Using empirical phase diagrams to understand the role of intramolecular dynamics in immunoglobulin G stability. J Pharm Sci 98:2432–2447

Raso SW, Clark PL, Haase-Pettingell C, King J, Thomas GJ Jr (2001) Distinct cysteine sulfhydryl environments detected by analysis of Raman S-hh markers of Cys–>Ser mutant proteins. J Mol Biol 307:899–911

Rossmann MG (2001) Molecular replacement–historical background. Acta Crystallogr D Biol Crystallogr 57:1360–1366

Royer CA (1993) Understanding fluorescence decay in proteins. Biophys J 65:9–10

Royer CA (1995) Fluorescence spectroscopy. Methods Mol Biol 40:65–89

Royer CA (2006) Probing protein folding and conformational transitions with fluorescence. Chem Rev 106:1769–1784

Royer CA, Scarlata SF (2008) Fluorescence approaches to quantifying biomolecular interactions. Methods Enzymol 450:79–106

Sanchez-Ruiz JM (2011) Probing free-energy surfaces with differential scanning calorimetry. Annu Rev Phys Chem 62:231–255

Sapienza PJ, Lee AL (2010) Using NMR to study fast dynamics in proteins: methods and applications. Curr Opin Pharmacol 10:723–730

Schuck P, Perugini MA, Gonzales NR, Howlett GJ, Schubert D (2002) Size-distribution analysis of proteins by analytical ultracentrifugation: strategies and application to model systems. Biophys J 82:1096–1111

Schümmer P. Mechanics of Non-Newtonian Fluids. Von W. R. Schowalter. Pergamon Press, Oxford–Frankfurt 1978. 1. Aufl., IX, 300 S., zahlr. Abb. u. Tab., geb., $ 35.00. Chemie Ingenieur Technik. 1979;51(7):766–766

Schweitzer-Stenner R, Eker F, Huang Q, Griebenow K, Mosz P, Kozlowski P (2002) Structure analysis of dipeptides in water by exploring and utilizing the structural sensitivity of amide III by polarized visible Raman, FTIR spectroscopy and DFT based normal coordinate analysis. J Phys Chem B 106:4294–4304

Scott DJ, Schuck P (2005) A brief introduction to the analytical ultracentrifugation of proteins for beginners. In: Scott DJ, Harding SE, Rowe AJ (eds) Analytical ultracentrifugation techniques and methods. RSC Publishing, Cambridge, pp 1–25

Sharp KA, Nicholls A, Friedman R, Honig B (1991) Extracting hydrophobic free energies from experimental data: relationship to protein folding and theoretical models. Biochemistry 30:9686–9697

Shaw DJ (1992) Rheology. Colloidal and surface chemistry, 4th edn. Butterworth-Heinemann, Boston

Shen Y, Lange O, Delaglio F, Rossi P, Aramini JM, Liu G, Eletsky A, Wu Y, Singarapu KK, Lemak A, Ignatchenko A, Arrowsmith CH, Szyperski T, Montelione GT, Baker D, Bax A (2008) Consistent blind protein structure generation from NMR chemical shift data. Proc Natl Acad Sci USA 105:4685–4690

Shi Z, Chen K, Liu Z, Kallenbach NR (2006) Conformation of the backbone in unfolded proteins. Chem Rev 106:1877–1897

Siamwiza MN, Lord RC, Chen MC, Takamatsu T, Harada I, Matsuura H, Shimanouchi T (1975) Interpretation of the doublet at 850 and 830 cm^{-1} in the Raman spectra of tyrosyl residues in proteins and certain model compounds. Biochemistry 14:4870–4876

Skinner AL, Laurence JS (2008) High-field solution NMR spectroscopy as a tool for assessing protein interactions with small molecule ligands. J Pharm Sci 97:4670–4695

Skinner AL, Laurence JS (2010) Probing residue–specific interactions in the stabilization of proteins using high-resolution NMR: a study of disulfide bond compensation. J Pharm Sci 99:2643–2654

Skinner JJ, Lim WK, Bedard S, Black BE, Englander SW (2012) Protein dynamics viewed by hydrogen exchange. Protein Sci 21:996–1005

Smith BC (1996) Fundamentals of fourier transform infrared spectroscopy. CRC, Boca Raton

Socrates G (2004) Infrared and Raman characteristic group frequencies: tables and charts, 3rd edn. Wiley, Chichester

Sondermann P, Oosthuizen V (2002) X-ray crystallographic studies of IgG-Fc gamma receptor interactions. Biochem Soc Trans 30:481–486

Spink CH (2008) Differential scanning calorimetry. Methods Cell Biol 84:115–141

Stefani M, Dobson CM (2003) Protein aggregation and aggregate toxicity: new insights into protein folding, misfolding diseases and biological evolution. J Mol Med (Berl) 81:678–699

Stern KG (1937) Spectroscopy of catalase. J Gen Physiol 20:631–648

Stokes GG (1852) On the change of refrangibility of light. Philos Trans R Soc Lond 142:463–562

Surewicz WK, Mantsch HH (1988) New insight into protein secondary structure from resolution-enhanced infrared spectra. Biochim Biophys Acta 952:115–130

Suryaprakash P, Prakash V (2000) Unfolding of multimeric proteins in presence of denaturants. A case study of helianthinin from *Helianthus annuus* L. Nahrung 44(3):178–183

Susi H, Byler DM (1986) Resolution-enhanced Fourier transform infrared spectroscopy of enzymes. Methods Enzymol 130:290–311

Susi H, Timasheff SN, Stevens L (1967) Infrared spectra and protein conformations in aqueous solutions. I. The amide I band in H_2O and D_2O solutions. J Biol Chem 242:5460–5466

Tamm LK, Lai AL, Li Y (2007) Combined NMR and EPR spectroscopy to determine structures of viral fusion domains in membranes. Biochim Biophys Acta 1768:3052–3060

Tanford C, Buzzell JG (1956) The viscosity of aqueous solutions of bovine serum albumin between pH 4.3 and 10.5. J Phys Chem 60:225–231

Tsuboi M, Suzuki M, Overman SA, Thomas GJ Jr (2000) Intensity of the polarized Raman band at 1340–1345 cm^{-1} as an indicator of protein alpha-helix orientation: application to Pf1 filamentous virus. Biochemistry 39:2677–2684

Tsutsui Y, Wintrode PL (2007) Hydrogen/deuterium exchange-mass spectrometry: a powerful tool for probing protein structure, dynamics and interactions. Curr Med Chem 14:2344–2358

Tu RS, Breedveld V (2005) Microrheological detection of protein unfolding. Phys Rev E Stat Nonlin Soft Matter Phys 72(4 Pt 1):041914

Van Holde KE, Johnson WC, Ho PS (2006) Principles of physical biochemistry. Pearson/Prentice Hall, Upper Saddle River

Velazquez-Campoy A, Ohtaka H, Nezami A, Muzammil S, Freire E (2004) Isothermal titration calorimetry. Curr Protoc Cell Biol, Chapter 17, Unit 17.18

Venyaminov S, Kalnin NN (1990a) Quantitative IR spectrophotometry of peptide compounds in water (H_2O) solutions. I. Spectral parameters of amino acid residue absorption bands. Biopolymers 30:1243–1257

Venyaminov S, Kalnin NN (1990b) Quantitative IR spectrophotometry of peptide compounds in water (H_2O) solutions. II. Amide absorption bands of polypeptides and fibrous proteins in alpha-, beta-, and random coil conformations. Biopolymers 30:1259–1271

Wand AJ (2001) Dynamic activation of protein function: a view emerging from NMR spectroscopy. Nat Struct Biol 8:926–931

Wang Y, Purrello R, Jordan T, Spiro TG (1991) UVRR spectroscopy of the peptide bond. J Am Chem Soc 113:6359–6368
Weber G (1952a) Polarization of the fluorescence of macromolecules. II. Fluorescent conjugates of ovalbumin and bovine serum albumin. Biochem J 51:155–167
Weber G (1952b) Polarization of the fluorescence of macromolecules. I. Theory and experimental method. Biochem J 51:145–155
Weber G, Daniel E (1966) Cooperative effects in binding by bovine serum albumin. II. The binding of 1-anilino-8-naphthalenesulfonate. Polarization of the ligand fluorescence and quenching of the protein fluorescence. Biochemistry 5:1900–1907
Weber G, Laurence DJ (1954) Fluorescent indicators of adsorption in aqueous solution and on the solid phase. Biochem J 56:xxxi
Weber PC, Salemme FR (2003) Applications of calorimetric methods to drug discovery and the study of protein interactions. Curr Opin Struct Biol 13:115–121
Wormald MR, Petrescu AJ, Pao YL, Glithero A, Elliott T, Dwek RA (2002) Conformational studies of oligosaccharides and glycopeptides: complementarity of NMR, X-ray crystallography, and molecular modelling. Chem Rev 102:371–386
Wyatt PJ (1993) Light scattering and the absolute characterization of macromolecules. Anal Chim Acta 272(1):1–40
Yang JT (1961) The viscosity of macromolecules in relation to molecular conformations. Adv Protein Chem 16:323–401
Yoder G, Polese A, Silva RAGD, Formaggio F, Crisma M, Broxterman QB, Kamphuis J, Toniolo C, Keiderling TA (1997) Conformational characterization of terminally blocked L-(alpha Me) Val homopeptides using vibrational and electronic circular dichroism. 3(10)-helical stabilization by peptide–peptide interaction. J Am Chem Soc 119:10278–10285
Zhang R, Monsma F (2010) Fluorescence-based thermal shift assays. Curr Opin Drug Discov Devel 13:389–402
Zhu F, Tranter GE, Isaacs NW, Hecht L, Barron LD (2006) Delineation of protein structure classes from multivariate analysis of protein Raman optical activity data. J Mol Biol 363:19–26
Ziarek JJ, Peterson FC, Lytle BL, Volkman BF (2011) Binding site identification and structure determination of protein-ligand complexes by NMR a semiautomated approach. Methods Enzymol 493:241–275
Zuber G, Prestrelski SJ, Benedek K (1992) Application of Fourier transform infrared spectroscopy to studies of aqueous protein solutions. Anal Biochem 207:150–156
Zubriene A, Matuliene J, Baranauskiene L, Jachno J, Torresan J, Michailoviene V, Cimmperman P, Matulis D (2009) Measurement of nanomolar dissociation constants by titration calorimetry and thermal shift assay – radicicol binding to Hsp90 and ethoxzolamide binding to CAII. Int J Mol Sci 10:2662–2680

Chapter 4
Biophysical Techniques for Protein Size Distribution Analysis

Ziping Wei and Alla Polozova

4.1 Introduction

Size distribution analysis of biopharmaceutical protein products is commonly performed to ensure product quality. Protein size heterogeneity usually originates from a combination of the following factors: (1) aggregation, (2) reversible self-association, (3) fragmentation, (4) conformation/shape heterogeneity, and (5) glycosylation. These contributing factors are briefly described below.

Aggregation: Protein aggregation is an association of proteins to form multimeric species that can be soluble aggregates or insoluble particles ranging from several nano meters to millimeters in size (Cromwell et al. 2006a; Mahler et al. 2009). Protein aggregates can be of multiple different types, including covalently or non-covalently linked, reversible or irreversible, and with native or denatured structures.

Self-association: Some protein molecules are prone to transient concentration-dependent reversible self-association, which can lead to significant size heterogeneity (Cromwell et al. 2006b, 2006a).

Fragmentation: Protein fragmentation causes the formation of smaller species in a range of sizes depending on the cleavage site.

Conformation/shape heterogeneity: Proteins often have a broad range of conformational heterogeneity leading to multiple different molecular shapes and a wider distribution of hydrodynamic sizes (Vendruscolo 2007; Manta et al. 2011).

Z. Wei
Analytical Development, Novavax, Rockville, MD 20850, USA
e-mail: wei_ziping@yahoo.com

A. Polozova (✉)
Biopharmaceutical Development, Medimmune, 1 Medimmune Way,
Gaithersburg, MD 20878, USA
e-mail: PolozovaA@MedImmune.com

L.O. Narhi (ed.), *Biophysics for Therapeutic Protein Development*, Biophysics for the Life Sciences 4, DOI 10.1007/978-1-4614-4316-2_4,

Glycosylation: Glycosylation of glycoproteins influences their hydrodynamic sizes, and glycosylation heterogeneity can contribute to size heterogeneity (Wen et al. 1996).

Given the diversity of potential size variants, multiple analytical methods are needed to assess and characterize protein products (Mahler et al. 2009; Philo 2006). Biophysical methods commonly used for size distribution analysis can be classified based on their principles: size exclusion, field-flow fractionation, sedimentation, light scattering, microscopy, light obscuration, electrical zone sensing, and mass spectrometry. This chapter describes these common size distribution analysis methods, including their advantages and limitations (Table 4.1).

4.2 Analytical Methods for Size Distribution Analysis

4.2.1 Size-Exclusion Chromatography

High-performance size-exclusion chromatography (HPSEC) is one of the most frequently used analytical methods for detection and quantitation of soluble protein aggregates and fragments. High-throughput capability and a good resolution between monomers and soluble aggregates are important advantages of HPSEC.

HPSEC separates molecules based on partitioning between porous stationary phase and aqueous mobile phase. The protein molecules are eluted according to their hydrodynamic size rather than actual molecular weight. Monomeric conformational variants of monomer may have different hydrodynamic sizes and, therefore, elute as separate peaks. For example, partially denatured monomers with an elongated shape have been shown to elute before the monomer peak (Philo 2006). Protein glycosylation with large oligosaccharides often affects hydrodynamic size and can have a noticeable effect on the peak position in the HPSEC profile (Wen et al. 1996).

Reversible self-association can be detected by HPSEC as either broadening of monomer peaks or through the presence of shoulder peaks. Although HPSEC can be useful for analysis of protein self-association (Moore et al. 1999), the results should always be interpreted with caution. In some cases nonspecific interactions with the column matrix can cause distorted peak shape (Wang et al. 1996).

Other limitations of HPSEC include the potential loss of some aggregate species due to selective absorption on the stationary phase (Liu et al. 2006; Gabrielson et al. 2007a) or retention by the column of very large aggregate species exceeding the upper HPSEC limit of about 0.1 μm. In addition, higher ionic strength, choice of salts, or different pH of HPSEC mobile phase may change protein conformation, induce new aggregates, or cause dissociation of weakly associated aggregates and thus bias the results (Washabaugh and Collins 1986; Philo 2006).

Table 4.1 Analytical methods for protein size distribution analysis

Method category	Method	Size	Pros and cons	Methods references
Size-exclusion chromatography	HPSEC	<0.1 μm	Pros: fast, high throughput Cons: limited size range, limited choice of buffers, possible matrix interactions	Liu et al. (2006), Gabrielson et al. (2007a), Philo (2006)
Field-flow fractionation	AF4	<1 μm (normal mode) and 1–50 μm (steric mode)	Pros: high throughput, broad dynamic range, works with wide range of buffers Cons: requires highly trained analysts. Membrane interactions are possible. Separation mode crossover at ~1 μm	Gabrielson et al. (2007a), Litzen et al. (1993), Liu et al. (2006), Cao et al. (2009a, b)
Sedimentation	AUC	<0.1 μm	Pros: works with wide range of buffers and conditions, no matrix interactions Cons: low throughput, requires highly trained analysts	Liu et al. (2006), Gabrielson et al. (2007b), Rivas et al. (1999)
Light scattering	Static light scattering	<1 μm	Pros: absolute measurement Cons: requires highly trained analysts, does not resolve different sizes	Follmer et al. (2004), Oliva et al. (2004)
	Dynamic light scattering	<1 μm	Pros: fast, easy, can be high throughput Cons: sensitive to dust, does not resolve different sizes in heterogeneous samples	Jossang et al. (1988), Follmer et al. (2004), Philo (2006, 2009)
	Turbidity	NA	Pros: compendial test, fast, easy Cons: no quantitative information on size, counts	Tatford et al. (2004)
	Nanoparticle tracking analysis	0.02–1 μm	Pros: unbiased size measurements in heterogeneous samples Cons: limited dynamic range (~10^7–10^9 particles/mL)	Filipe et al. (2010)
Light obscuration	Light obscuration	1–150 μm	Pros: compendial test. Counts one particle at a time, does not rely on distribution models Cons: sizing of protein particles is not reliable. Does not work well with high protein concentrations, turbid and viscous samples	Fujishita et al. (1995), Narhi et al. (2009)

(continued)

Table 4.1 (continued)

Method category	Method	Size	Pros and cons	Methods references
Electrical sensing zone	Electrical sensing zone	0.4–1,200 μm	Pros: wide dynamic range, counts single particle at a time Cons: limited concentration range, requires electrolyte solution and multiple measurements with different apertures to cover the entire range	Narhi et al. (2009)
Microscopy	EM	<100 μm	Pros: high resolution Cons: sample preparation artifacts possible	Roux (1999)
	AFM	<1 μm	Pros: high resolution Cons: sample preparation and imaging artifacts possible	Thomson (2005), Fechner et al. (2009)
	Light microscopy	>0.5 μm	Pros: fast, easy Cons: low precision, difficult to detect translucent particles	Barber et al. (1989)
Flow microscopy	Flow microscopy	>0.5 μm	Pros: fast, provides information on counts, size, and morphology Cons: difficult to detect translucent particles	Narhi et al. (2009), Sharma et al. (2007), Huang et al. (2009), Wuchner et al. (2010)
Mass spectrometry	Electrospray or MALDI MS	>0.1 μm	Pros: detailed structural information, high sensitivity and resolution Cons: requires highly trained analysts. Potential dissociation of non-covalently linked aggregates Limitation of m/z range. Qualitative or semiquantitative method	Krutchinsky et al. (2000), Kukrer et al. (2010), Liu et al. (2008), North et al. (2009), Wang et al. (2011)

New technology	Gas-phase electrophoretic mobility	0.003–0.25 μm	Pros: fast, direct measurement, high resolution Cons: limited concentration range, electrospray buffer different from formulation buffer	Pease et al. (2008)
	Resonant mass measurement	>0.1 μm	Pros: measures single particle at a time, direct mass measurement, can distinguish different types based on density Cons: low sensitivity to particles with density close to that of surrounding solution	Burg et al. (2007)

4.2.2 Field-Flow Fractionation

Asymmetric flow field-flow fractionation (AF4) is an important orthogonal separation-based matrix-free technique with a wide dynamic range, capable of analysis of fragments, monomers, and aggregates up to 50 μm in size (Gabrielson et al. 2007a; Litzen et al. 1993; Liu et al. 2006; Cao et al. 2009a). AF4 separation is carried out in a flat trapezoidal channel equipped with a semipermeable membrane. Analytes are separated based on their hydrodynamic size by a cross flow through the semipermeable membrane, perpendicular to the elution flow. The pore size of the semipermeable membrane is selected to retain the analytes, but it allows a free flow of the mobile phase. In the typical AF4 separation, the elution flow is the fastest in the middle of the channel and slower near the top and the bottom. Cross flow pushes larger species closer to the semipermeable membrane at the bottom of the channel; therefore, larger species elute later, after smaller species. Recent publications demonstrated that AF4 can be successfully applied for analysis of large protein aggregates, and collected data are consistent with other aggregate analysis methods such as HPSEC and flow microscopy (Cao et al. 2009b).

One of the important advantages of AF4 is a wider selection of mobile phases compared to HPSEC, because the former is compatible with a broader range of pH and salt concentrations. Other advantages include a wide dynamic range for separation of different species and the absence of a stationary phase.

Limitations of AF4 include potential nonspecific interactions of certain analytes with the semipermeable membrane, lack of information on particle counts, and a crossover of the separation mode in the 1 μm range described below. Nonspecific interactions of analytes with the semipermeable membranes may lead to distorted elution profiles. The possibility of such interactions should be carefully assessed for each sample type. If needed, mobile phase should be optimized to minimize the nonspecific interactions. Separation mode crossover may pose a significant challenge for AF4 analysis of highly heterogeneous samples. For species larger than 1 μm, the steric elution mode becomes dominant. It results in a faster elution of larger analytes, which is the opposite of the normal elution order in the typical AF4 profile (Yohannes et al. 2011). Therefore, the analysis of samples containing both sub- and above-micron species can be compromised, which can limit AF4 practical applicability to submicron range.

4.2.3 Analytical Ultracentrifugation

Analytical ultracentrifugation (AUC) is widely used for measurements of aggregate and fragment levels, determination of aggregate stoichiometry, and analysis of reversible self-association (Liu et al. 2006; Gabrielson et al. 2007b; Rivas et al. 1999). AUC can also provide information on distribution of protein conformations, if they are related to different molecular shapes. Two major AUC methods are sedimentation velocity and sedimentation equilibrium. In a typical sedimentation velocity experiment, AUC cell channels loaded with a sample and a reference buffer are scanned by either absorbance or interference optics as analytes sediment to the bottom of the cells at relatively high speed, about 45,000 rpm. Collected scans contain information on the shape of the sedimentation boundary as a function of time. Sedimentation boundaries depend on both the size and shape of the sedimenting molecules. Molecular weight distribution can be derived from the analysis of multiple sedimentation boundaries if certain assumptions about the shape of the molecule can be made.

Sedimentation equilibrium experiments involve spinning at lower speeds until sedimentation-diffusion equilibrium is established. Usually, several different speeds and sample concentrations are evaluated in one sedimentation equilibrium experiment. Molecular weight distribution information is derived from the analysis of scans by fitting to theoretical distribution models (e.g., monomers, monomers and non-dissociable aggregates, reversible dimers, and trimers).

Similarly to AF4, important advantages of AUC are the minimization of matrix interactions, which may result in loss of some aggregate species, and compatibility with many buffers in a broad pH range. In addition, AUC allows sample analysis at predetermined concentrations, and it is the method of choice for studying self-association behavior. Limitations of AUC include the following: a relatively high (about 1 %) limit of quantitation for small aggregates and fragments, a relatively narrow (up to approximately 0.1 μm) dynamic range, low throughput, and the requirement for highly trained analysts (Narhi et al. 2009).

4.2.4 Light Scattering

Multiple static and dynamic light scattering techniques are widely used for protein size heterogeneity analysis, including sizing and quantitation of aggregates (Follmer et al. 2004; Philo 2006, 2009), distribution of glycosylation (Wen et al. 1996), and reversible self-association (Wu et al. 1997; Attri and Minton 2005). In methods based on static or classical light scattering, the molecular weight of molecules in solution is derived from the light scattering theory equations. These equations link scattered light intensity and concentration to the molecular weight. Multiangle light scattering (MALS) applications are most commonly used for proteins in solution (Oliva et al. 2004). In a typical MALS measurement, light scattering intensity measured at three or more angles is determined and used to calculate the molecular weight. For molecules larger than 10 nm, the radius of gyration can be also calculated based on the dependance of the light scattering intensity on the scattering angle.

Dynamic light scattering (DLS) relies on the analysis of an autocorrelation function derived from measurements of scattered light intensity fluctuations due to Brownian motion of molecules in solution (Jossang et al. 1988). The translational diffusion coefficient $D\tau$ of protein molecules in the samples is derived from the autocorrelation function. The hydrodynamic radius R_h is calculated using the Stokes–Einstein equation linking $D\tau$, temperature, and viscosity of the solution.

Both static and DLS methods can be used to determine the presence of protein aggregates in solution and to measure aggregates up to 1 μm in size, including high-throughput screening applications. A composition gradient technique based on static light scattering measurements of increasing protein concentrations can be applied for analysis of self-associating systems (Attri and Minton 2005).

The light scattering methods work particularly well for sizing of monodisperse samples. However, they have significant limitations in the analysis and quantitation of size distributions in polydisperse samples. Results can be skewed to larger sizes, because scattering intensity depends on the sixth power of the size, and thus, larger species scatter significantly more light. If separation of different species (e.g., by HPSEC or AF4) is carried out prior to analysis, it significantly improves the sizing capabilities of the light scattering methods. MALS or DLS coupled with HPSEC or AF4 as on-line detectors are powerful methods for the characterization of aggregates. In particular, MALS and DLS detectors in tandem can provide information on the molecular weight, size, and shape of aggregates (Rambaldi et al. 2011). Proteins with high levels of glycosylation can be analyzed with a triple-detector HPSEC method with consecutively attached MALS, UV, and refractive index detectors. This method allows on-line determination of proportions and distributions of oligosaccharide across different fractions (Wen et al. 1996).

Solution turbidity measurement is a low-resolution light scattering-based technique that can provide a rough assessment of size heterogeneity levels. Solution turbidity can be determined quantitatively by intensity loss of the transmitted beam or by the intensity of the light scattered at a given angle. The presence of protein aggregates or particles often results in an increase in turbidity. Therefore, the degree of turbidity in a protein solution may be used to qualitatively assess the extent of aggregation (Tatford et al. 2004).

Nanoparticle tracking analysis (NTA) is a new light scattering-based technique that is rapidly gaining use for the analysis of submicron particles (Filipe et al. 2010). Particle size measurements are based on tracking the distance traveled by individual particles over a given amount of time and do not rely on the scattering intensity. Due to this, the NTA technique can provide relatively unbiased measurements of particle sizes even in samples with heterogeneous populations. A rather narrow dynamic range of particle counts suitable for the measurements (~10^7–10^9 particle/mL) is a limitation to its applicability.

4.2.5 *Light Obscuration*

Light obscuration techniques (Fujishita et al. 1995; Narhi et al. 2009) can be applied to the enumeration and size determination of large aggregates and particles (micrometer size). Measurements are based on blockage of laser light by individual particles passing the detector in the flow cell. The change in light intensity reaching the detector is converted to the particle size based on calibration with known particle standards. Light obscuration is often used as a routine monitoring test for injectable drugs to detect levels of subvisible particles (≥10 μm). Counting of a single particle at a given time is an important advantage of this method, because it offers a direct measurement of particle counts that does not require distribution models. However, high-concentration protein solutions are not always compatible with this method. For example, the presence of large amounts of background protein can reduce contrast between the particles and the solution and interfere with measurements. In addition, sizing of protein particles by light obscuration methods is often not reliable because they have different optical properties and block different amounts of light compared to commonly used polystyrene bead calibration standards. Introduction of calibration standards similar in optical properties to large protein aggregates would improve the sizing accuracy; however, such standards are still under development and not yet commercially available.

4.2.6 Electrical Sensing Zone Analysis

Electrical sensing zone (ELS) analysis can be applied to large aggregates ranging in size from 0.4 μm up to 1,200 μm (Narhi et al. 2009). Particle volume is a primary parameter measured by this method. The sample is suspended in a weak electrolyte solution. When large aggregates and particles pass through a small orifice with applied voltage, they displace a volume of the electrolyte solution equivalent to their volume. This results in a decrease in electric current, registered as a pulse. The size information is derived from the magnitude of the pulse, which is directly proportional to the particle volume. In addition to the size information, the ELS technique also provides information on the particle counts for a given sample volume. The requirements for an electrolyte solution and the use of multiple apertures to analyze the entire aggregate and particle distribution may limit the applicability of this technique.

4.2.7 Microscopy

Electron microscopy (EM) and atomic force microscopy (AFM) offer a high-resolution capability for the characterization of conformational heterogeneity and aggregates at the molecular level (Roux 1999; Thomson 2005). In electron microscopy methods, the image is formed by a beam of electrons either passing through (transmission EM) or reflecting off the specimen (scanning EM), while sets of magnets serve as condenser and objective lenses. Because an electron beam is used for imaging, EM offers very good spatial resolution. The theoretical resolution limit is less than 10 Å; however, in practical applications with biological samples, the resolution limit is ~1 nm, which still allows a detailed analysis of structure (Roux 1999). The most promising high-resolution tool for detailed analysis of conformational heterogeneity and aggregate structure is cryo-transmission EM tomography (Sandin et al. 2004; Banyay et al. 2004). A small droplet of aqueous protein solution is flash-frozen to form a vitreous ice and directly analyzed in the TEM without further sample manipulation or preparation. The three-dimensional images reconstructed by computer software from a set of tilted-plane images provide detailed information about aggregate order and structure. However, this technique is still evolving, and, at this time, it is not routinely applied to the analysis of heterogeneity in protein solutions.

In AFM, the image is obtained by scanning the surface, usually in a raster pattern, with very thin tip attached to a cantilever. The spatial resolution and sensitivity of AFM depend on the mechanical properties of the cantilever and the sharpness of the tip. Usually, 1 nm or better resolution can be achieved (Engel and Muller 2000). AFM can provide valuable information on the structure of individual protein molecules (Thomson 2005), as well as aggregates (Lee et al. 2011). However, results should be interpreted with caution because this method might be prone to sample preparation artifacts arising from adsorption on the support surfaces and sample-tip interactions (Fechner et al. 2009).

Optical microscopy is one of the oldest techniques available for analysis of very large aggregates, and it can be a useful screening tool for an initial assessment. Resolution of optical microscopy is defined by the diffraction limit, which is ~0.5 μm for visual light. Large aggregates can be analyzed directly in solution; however, some translucent particles might be difficult to see due to their low contrast with the surrounding solution. Alternatively, large aggregates can be captured on a filter and detected under oblique light illumination (Barber et al. 1989), although some thin and fragile particles might be lost during the sample preparation process.

4.2.8 Flow Microscopy

Over the past several years, flow microscopy has become an increasingly popular tool for the analysis of large micron-size aggregates in protein solutions (Narhi et al. 2009; Sharma et al. 2007; Huang et al. 2009). This technique offers the combined advantages of a particle counter and an optical microscope integrated into one instrument (Sharma et al. 2007; Wuchner et al. 2010). The sample solution flows through a narrow flow cell, typically 100–300 μm thick. Images magnified by a 10–20× microscope objective are captured by a high-speed camera operating at 8–10 frames per second and automatically analyzed by the image processing software for presence of particles. Images of all detected particles in the sample are collected in a database, and information on counts, particle size, transparency, and morphology can be generated by the software. Large aggregates (particles) in the 2–300 μm range can be detected and analyzed. The ability to provide information on morphology, shape, and optical properties of large aggregates are important advantages of flow microscopy. However, results depend on the contrast between the protein particles and the solution, which may limit the sensitivity of this method to translucent particles, especially for smaller particle sizes.

4.2.9 *Mass Spectrometry*

Mass spectrometry (MS) separates analytes based on their mass and charge (m/z ratio). Electrospray ionization (ESI) and matrix-assisted laser desorption ionization (MALDI) techniques are commonly used for the analysis of proteins. MS offers high sensitivity and good mass resolution and has been a powerful tool used in a large number of studies for determining protein molecular weights and investigating protein complexes (Bich and Zenobi 2009). Although MS has been recently reported to be a useful online detection tool for HPSEC analysis of protein aggregates (Kukrer et al. 2010), there have been very limited applications of this technique to study protein aggregation for several reasons (Wang et al. 2011). Firstly, the potential dissociation of non-covalently linked aggregates during ionization or transfer into the gas phase could limit the applicability of MS-based techniques for size distribution analysis. Secondly, most MS mass analyzers have upper-range limitations in analysis of high molecular weight species. A time-of-flight (TOF) mass analyzer provides a relatively wide m/z range and is often used for protein aggregate analysis (Krutchinsky et al. 2000). Specialized detectors can be used to make the m/z range of MALDI MS suitable for analysis of larger proteins and complexes (Bich and Zenobi 2009; Bahr et al. 1996; Wenzel et al. 2005). Lastly, MS is usually limited to a qualitative analysis of small oligomeric protein aggregates. However, a recent study showed that MS was capable of probing the early stage of protein aggregation in a semiquantitative fashion (Wang et al. 2011).

MS is a powerful tool to determine the structures, sequences, and sites of N-linked and O-linked glycosylation (North et al. 2009). MS is widely used to determine the molecular weights of protein fragments and is also capable of identification of cleavage sites of fragments based on the accurate molecular mass and known posttranslational modifications of fragments (Liu et al. 2008). Therefore, MS provides more detailed structural information than other techniques for glycoprotein size analysis and protein fragment analysis.

4.2.10 *New Technologies*

4.2.10.1 Gas-Phase Electrophoretic Mobility

Electrospray differential mobility analysis, also known as gas-phase electrophoretic mobility molecular analysis, is a new method for measuring low-order soluble protein aggregates based on the separation of charge-to-aerodynamic size ratio

(Pease et al. 2008). Size distribution of protein species in the range of 3–250 nm can be obtained, including distinct peaks for IgG monomers to pentamers. It provides a fast and quantitative readout of the aggregate distribution. The limitations of the method are that it works within a limited protein concentration range and the electrospray step cannot be performed in buffer containing nonvolatile salts. Additionally, the electrospray buffer may affect the aggregation state of proteins.

4.2.10.2 Resonant Mass Measurement

Resonant buoyant mass measurement of individual aggregates and particles is a novel technique that is capable of covering a broad range of size distributions for aggregates and particles larger than 0.1 μm (Burg et al. 2007). Particles flowing through a thin microfluidic channel are detected via a change in the resonance frequency. This technique measures one particle at a time and it can accommodate viscous solutions (up to 50 cP) without dilution. However, this technique relies on the density difference between the particles and the surrounding solution. Because the density of some protein aggregates and particles can be close to that of the aqueous solutions containing them, the sensitivity of this method can be impaired in this situation.

4.2.11 Discussion

Therapeutic protein products can have a very broad heterogeneous size distribution, making size distribution analysis a challenging task. Because no single analytical method is capable of covering the entire size range of all types of size variants, it is critical to apply complementary tools to monitor and characterize the entire size distribution. Depending on the specific requirements for a product, different information, such as molecular weight, distribution of different species, size, counts, or morphology may be needed. Multiple biophysical methods presented in this chapter are capable of providing this information. Each method may provide a different type of information depending on a measurement principle, and correlation between orthogonal methods should be thoroughly assessed and understood.

As each analytical method has its own strengths and weaknesses, the choice of analytical method(s) should be tailored to the sample type and to the intended purpose. To ensure robust and meaningful size distribution measurements and to avoid false conclusions, care must be taken in method selection and result interpretation. Regardless of the method(s) chosen, the use of appropriate controls is critical because sample preparation and handling may have a significant impact on the results. Overall, a good knowledge of factors which may influence size heterogeneity during sample handling and a thorough understanding of physical principles of the chosen analytical techniques are keys for reliable measurements of size distribution in protein solutions.

References

Attri AK, Minton AP (2005) Composition gradient static light scattering: a new technique for rapid detection and quantitative characterization of reversible macromolecular hetero-associations in solution. Anal Biochem 346:132–138

Bahr U, Rohling U, Lautz C, Strupat K, Schurenberg M, Hillenkamp F (1996) A charge detector for time-of-flight mass analysis of high mass ions produced by matrix-assisted laser desorption/ionization (MALDI). J Mass Spectrom Ion Process 153:9–21

Banyay M, Gilstring F, Hauzenberger E, Ofverstedt LG, Eriksson AB, Krupp JJ, Larsson O (2004) Three-dimensional imaging of in situ specimens with low-dose electron tomography to analyze protein conformation. Assay Drug Dev Technol 2:561–567

Barber TA, Lannis MD, Williams JG (1989) Method evaluation: automated microscopy as a compendial test for particulates in parenteral solutions. J Parenter Sci Technol 43:27–47

Bich C, Zenobi R (2009) Mass spectrometry of large complexes. Curr Opin Struct Biol 19:632–639

Burg TP, Godin M, Knudsen SM, Shen W, Carlson G, Foster JS, Babcock K, Manalis SR (2007) Weighing of biomolecules, single cells and single nanoparticles in fluid. Nature 446:1066–1069

Cao S, Pollastrini J, Jiang Y (2009a) Separation and characterization of protein aggregates and particles by field flow fractionation. Curr Pharm Biotechnol 10:382–390

Cao S, Jiao N, Jiang Y, Mire-Sluis A, Narhi LO (2009b) Sub-visible particle quantitation in protein therapeutics. Pharmeur Bio Sci Notes 2009:73–79

Cromwell ME, Hilario E, Jacobson F (2006a) Protein aggregation and bioprocessing. AAPS J 8:E572–E579

Cromwell MEM, Felten C, Flores H, Liu J, Shire SJ (2006b) Self-association of therapeutic proteins: implications for product development. In: Murphy RM, Tsai AM (eds) Misbehaving proteins: protein (Mis)folding, aggregation, and stability. Springer, New York

Engel A, Muller DJ (2000) Observing single biomolecules at work with the atomic force microscope. Nat Struct Biol 7:715–718

Fechner P, Boudier T, Mangenot S, Jaroslawski S, Sturgis JN, Scheuring S (2009) Structural information, resolution, and noise in high-resolution atomic force microscopy topographs. Biophys J 96:3822–3831

Filipe V, Hawe A, Jiskoot W (2010) Critical evaluation of Nanoparticle Tracking Analysis (NTA) by NanoSight for the measurement of nanoparticles and protein aggregates. Pharm Res 27:796–810

Follmer C, Pereira FV, Da Silveira NP, Carlini CR (2004) Jack bean urease (EC 3.5.1.5) aggregation monitored by dynamic and static light scattering. Biophys Chem 111:79–87

Fujishita O, Sendo T, Hisazumi A, Otsubo K, Aoyama T, Oishi R (1995) The evaluation of sizing accuracy of particle counters for parenteral drugs. PDA J Pharm Sci Technol 49:267–271

Gabrielson JP, Brader ML, Pekar AH, Mathis KB, Winter G, Carpenter JF, Randolph TW (2007a) Quantitation of aggregate levels in a recombinant humanized monoclonal antibody formulation by size-exclusion chromatography, asymmetrical flow field flow fractionation, and sedimentation velocity. J Pharm Sci 96:268–279

Gabrielson JP, Randolph TW, Kendrick BS, Stoner MR (2007b) Sedimentation velocity analytical ultracentrifugation and SEDFIT/c(s): limits of quantitation for a monoclonal antibody system. Anal Biochem 361:24–30

Huang CT, Sharma D, Oma P, Krishnamurthy R (2009) Quantitation of protein particles in parenteral solutions using micro-flow imaging. J Pharm Sci 98:3058–3071

Jossang T, Feder J, Rosenqvist E (1988) Photon correlation spectroscopy of human IgG. J Protein Chem 7:165–171

Krutchinsky AN, Ayed A, Donald LJ, Ens W, Duckworth HW, Standing KG (2000) Studies of noncovalent complexes in an electrospray ionization/time-of-flight mass spectrometer. In: Chapman J (ed) Mass spectrometry of proteins and peptides. Humana Press, Totowa

Kukrer B, Filipe V, Duijn EV, Kasper PT, Vreeken RJ, Heck AJR, Jiskoot W (2010) Mass spectrometric analysis of intact human monoclonal antibody aggregates fractionated by size-exclusion chromatography. Pharm Res 27:2197–2204
Lee H, Kirchmeier M, Mach H (2011) Monoclonal antibody aggregation intermediates visualized by atomic force microscopy. J Pharm Sci 100:416–423
Litzen A, Walter JK, Krischollek H, Wahlund KG (1993) Separation and quantitation of monoclonal antibody aggregates by asymmetrical flow field-flow fractionation and comparison to gel permeation chromatography. Anal Biochem 212:469–480
Liu J, Andya JD, Shire SJ (2006) A critical review of analytical ultracentrifugation and field flow fractionation methods for measuring protein aggregation. AAPS J 8:E580–E589
Liu H, Gaza-Bulseco G, Lundell E (2008) Assessment of antibody fragmentation by reversed-phase liquid chromatography and mass spectrometry. J Chromatogr B Analyt Technol Biomed Life Sci 876:13–23
Mahler HC, Friess W, Grauschopf U, Kiese S (2009) Protein aggregation: pathways, induction factors and analysis. J Pharm Sci 98:2909–2934
Manta B, Obal G, Ricciardi A, Pritsch O, Denicola A (2011) Tools to evaluate the conformation of protein products. Biotechnol J 6:731–741
Moore JM, Patapoff TW, Cromwell ME (1999) Kinetics and thermodynamics of dimer formation and dissociation for a recombinant humanized monoclonal antibody to vascular endothelial growth factor. Biochemistry 38:13960–13967
Narhi LO, Jiang Y, Cao S, Benedek K, Shnek D (2009) A critical review of analytical methods for subvisible and visible particles. Curr Pharm Biotechnol 10:373–381
North SJ, Hitchen PG, Haslam SM, Dell A (2009) Mass spectrometry in the analysis of N-linked and O-linked glycans. Curr Opin Struct Biol 19:498–506
Oliva A, Llabres M, Farina JB (2004) Applications of multi-angle laser light-scattering detection in the analysis of peptides and proteins. Curr Drug Discov Technol 1:229–242
Pease LF 3rd, Elliott JT, Tsai DH, Zachariah MR, Tarlov MJ (2008) Determination of protein aggregation with differential mobility analysis: application to IgG antibody. Biotechnol Bioeng 101:1214–1222
Philo JS (2006) Is any measurement method optimal for all aggregate sizes and types? AAPS J 8:E564–E571
Philo JS (2009) A critical review of methods for size characterization of non-particulate protein aggregates. Curr Pharm Biotechnol 10:359–372
Rambaldi DC, Reschiglian P, Zattoni A (2011) Flow field-flow fractionation: recent trends in protein analysis. Anal Bioanal Chem 399:1439–1447
Rivas G, Stafford W, Minton AP (1999) Characterization of heterologous protein-protein interactions using analytical ultracentrifugation. Methods 19:194–212
Roux KH (1999) Immunoglobulin structure and function as revealed by electron microscopy. Int Arch Allergy Immunol 120:85–99
Sandin S, Öfverstedt L-G, Wikström A-C, Wrange Ö, Skoglund U (2004) Structure and flexibility of individual immunoglobulin G molecules in solution. Structure 12:409–415
Sharma DK, King D, Moore PD, Oma P, Thomas D (2007) Flow microscopy for particulate analysis in parenteral and pharmaceutical fluids. Eur J Parenter Pharm Sci 12:97–101
Tatford OC, Gomme PT, Bertolini J (2004) Analytical techniques for the evaluation of liquid protein therapeutics. Biotechnol Appl Biochem 40:67–81
Thomson NH (2005) Imaging the substructure of antibodies with tapping-mode AFM in air: the importance of a water layer on mica. J Microsc 217:193–199
Vendruscolo M (2007) Determination of conformationally heterogeneous states of proteins. Curr Opin Struct Biol 17:15–20
Wang YJ, Shahrokh Z, Vemuri S, Eberlein G, Beylin I, Busch M (1996) Characterization, stability, and formulations of basic fibroblast growth factor. Pharm Biotechnol 9:141–180
Wang G, Johnson AJ, Kaltashov IA (2011) Evaluation of electrospray ionization mass spectrometry as a tool for characterization of small soluble protein aggregates. Anal Chem 84:1718–1724

Washabaugh MW, Collins KD (1986) The systematic characterization by aqueous column chromatography of solutes which affect protein stability. J Biol Chem 261:12477–12485

Wen J, Arakawa T, Philo JS (1996) Size-exclusion chromatography with on-line light-scattering, absorbance, and refractive index detectors for studying proteins and their interactions. Anal Biochem 240:155–166

Wenzel RJ, Matter U, Schutheis L, Zenobi R (2005) Analysis of megadalton ions using cryodetection MALDI time-of-flight mass spectrometry. Anal Chem 77:4329–4337

Wu G, Hummel D, Herman A (1997) Analysis of the solution behavior of protein pharmaceuticals by laser light scattering photometry. ACS Symp Ser 675:168–185

Wuchner K, Buchler J, Spycher R, Dalmonte P, Volkin DB (2010) Development of a microflow digital imaging assay to characterize protein particulates during storage of a high concentration IgG1 monoclonal antibody formulation. J Pharm Sci 99:3343–3361

Yohannes G, Jussila M, Hartonen K, Riekkola ML (2011) Asymmetrical flow field-flow fractionation technique for separation and characterization of biopolymers and bioparticles. J Chromatogr A 1218:4104–4116 [illegible] (2006) [illegible] Pharm [illegible]

Wu C-Y, [illegible] classification [illegible]

[illegible]

Yee S [illegible] (1997) [illegible] Caco-2 [illegible]

[illegible] Pharm Sci [illegible]

[illegible] the use of transporters [illegible] Pharm Res [illegible]

Chapter 5
Qualification of Biophysical Methods for the Analysis of Protein Therapeutics

Yijia Jiang, Cynthia Li, and John Gabrielson

5.1 Introduction

As described elsewhere (Jiang and Narhi 2006) and in other chapters of this book, biophysical methods are routinely used by the biopharmaceutical industry to gain information on protein higher-order structure and stability throughout the therapeutic protein development life cycle. These methods include SV-AUC and SE-HPLC-LS for size distribution analysis and molecular weight determination of species separated by SE-HPLC; CD, FTIR, and DSC for secondary and tertiary structure and overall conformation and thermal stability analysis; and light obscuration (LO)- and micro-flow imaging (MFI)-based methods for subvisible particle analysis. The results of these biophysical analyses have been included in regulatory filings as required by ICH guidelines M4Q (R1), Q5E, and Q6B (ICH M4Q 2001; ICH Q5E 2005; ICH Q6B 1999). These results are included in the filing documents to demonstrate that the therapeutic protein shows the predicted and expected properly folded higher-order structure and retains the higher-order structure after manufacturing process, site, and/or formulation changes. ICH guidelines M4Q (R1) (ICH M4Q 2001) on common technical document (CTD) for the registration of pharmaceuticals for human use and Q6B (ICH Q5E 2005) on specifications state that for desired

Y. Jiang (✉)
Process and Product Development, Amgen Inc., One Amgen Center Drive, 30E-1-C, Thousand Oaks, CA 91320, USA
e-mail: yjiang@amgen.com

C. Li
Process and Product Development, Amgen Inc., One Amgen Center Drive, Thousand Oaks, CA 91320, USA
e-mail: cynthial@amgen.com

J. Gabrielson
Analytical Sciences, Amgen Inc., 4000 Nelson Road, Longmont CO 80503, USA
e-mail: jgabriel@amgen.com

L.O. Narhi (ed.), *Biophysics for Therapeutic Protein Development*, Biophysics for the Life Sciences 4, DOI 10.1007/978-1-4614-4316-2_5,

product and product-related substances, details should be provided on primary, secondary, and higher-order structure when relevant. ICH guideline Q5E (ICH Q6B 1999) specifies: "following a manufacturing process change, manufacturers should attempt to determine that higher order structure (secondary, tertiary, and quaternary structure) is maintained in the product."

Unlike the methods used for product release testing, for which the qualification and validation are regularly performed as governed by ICH guideline Q2 (R1) (ICH Q2 (R1) 2005) to show precision, accuracy, linearity, LOD, and LOQ of the methods, little has been done and published to standardize an approach to qualify characterization methods in general, and the biophysical in particular, until recently. The lack of qualification of biophysical methods used is mainly due to the qualitative nature of some of these methods. In recent years, in response to increasing regulatory requirements and scrutiny, more effort has been invested in understanding the precision and sensitivity of the biophysical techniques, which has resulted in several publications and presentations (Jiang et al. 2011; Li et al. 2011; Wen et al. 2011; Gabrielson et al. 2010). In this chapter we review the qualification of the biophysical methods including AUC, CD, FTIR, DSC, SE-HPLC-LS, MFI, and LO-based methods and demonstrate how qualification of these methods enables better knowledge of the methods and objective interpretation of the results. The general considerations described here can be applied to other biophysical methods as appropriate as well.

5.2 Qualification of CD

Circular dichroism (CD) spectroscopy is a technique that has been widely used in structural biology for examining secondary structure, tertiary structure, conformational changes, and folding and binding interactions involving protein molecules. It is also used by the biopharmaceutical industry as a characterization tool to study the effect of manufacturing processes, formulation compositions and conditions, storage conditions, and delivery systems on protein conformation (Johnson 1988, 1999a; Woody 1994, 1996; Manning and Woody 1989; Greenfield and Fasman 1969; Yang et al. 1986). This information is often included in regulatory filings following changes in any of these manufacturing conditions (ICH Q5E 2005; ICH Q6B 1999). In recent years, a new extension of the conventional CD method, synchrotron radiation circular dichroism (SRCD), has been shown to have a better sensitivity for the higher-order structure of a protein by producing data to lower wavelengths with higher information content and improved signal-to-noise levels relative to those obtained using conventional benchtop instruments (Wallace 2005; Lees and Wallace 2002; Miles et al. 2007; Wallace and Janes 2010). SRCD spectroscopy can provide important static and dynamic structural information on proteins in solution, including secondary structures of intact proteins and their domains, protein stability, the differences between wild-type and mutant proteins, the identification of natively disordered regions in proteins, and the dynamic processes of protein folding and membrane insertion and the kinetics of enzyme reactions (Miles and Wallace 2006; Meersman et al. 2010; Drechsler et al. 2009). A new web-based bioinformatics resource, the Protein Circular Dichroism Data Bank (PCDDB), has also been created recently by Dr. Wallace's laboratory which enables archiving, access, and analyses of CD and SRCD spectra and supporting data (Powl et al. 2010; Whitmore and Wallace 2004; Wallace et al. 2006; Lees et al. 2006). Unfortunately many of the formulation buffers used for therapeutic proteins interfere with the SRCD, especially at the lower wavelengths required, and this technique can usually not be applied to comparability evaluations or similar studies.

Even though CD spectroscopy has received considerable attention and been widely used for biological molecules over the last 20 years by both industry and academic scientists, the technique is mostly used in research and development laboratories rather than quality control laboratories. This is mainly due to the complexity of the analysis, the need for an experienced spectroscopist to run the studies, and the qualitative nature of the spectra. The lack of defined best practices in making the CD measurement and the variability in the data obtained from different laboratories also make this an inappropriate technique for routine use in quality control laboratories. There have been a few publications describing standardization of best practices in making CD measurements (Van Stokkum et al. 1990; Kelly et al. 2005).

To understand the extent of comparability of CD data, the National Physical Laboratory (NPL) has conducted an interlaboratory CD spectroscopy study in recent years. Schiffmann et al. published a report titled "CD spectroscopy: an interlaboratory study" (Jones et al. 2004) and Ravi et al. (2010) published two papers about the international comparability in spectroscopic measurements of protein structure by CD (Schiffmann et al. 2004; Ravi et al. 2010). Both studies involved academic and industry laboratories worldwide. The studies have highlighted some of the current difficulties in obtaining comparable CD spectra from different instruments and possible solutions to some of these problems, for instance, the iterative spectral alignment method (ISAM). Both studies show that the uncertainties involved in a CD measurement are rather complex, involving components relating to calibration, path length, sample concentration, and instrumentation (Ravi et al. 2010; Whitmore and Wallace 2008; Miles et al. 2003, 2004, 2005). Therefore, a careful calibration of the CD instrument and exact path lengths of the cuvettes must be carried out to ensure equivalence of CD measurements when using different instruments and cuvettes. The calibration and standardization of CD instruments has been described by Dr. Wallace et al. (Whitmore and Wallace 2008; Miles et al. 2003). The authors suggested the procedure for proper calibration of a CD spectrophotometer and proposed a method for standardization of protein spectra obtained on any instrument. For example, calibrations should be done at more than one wavelength in order to cover the wavelength range measured in a protein spectrum, and the linearity of the response should be demonstrated; the ratios of the measured and expected ellipticity values at each wavelength should be used to develop a model function for corrections.

For application of the method in the biopharmaceutical industry, one of the important questions is the degree of comparability of the CD data between different lots of protein generated from different processes or formulation conditions. In Ravi and Schiffmann's studies, they used multivariate statistical methods, such as principal component analysis (PCA) and soft independent modeling of class analogies (SIMCA), to investigate the comparability of CD spectroscopic data (Jones et al. 2004; Schiffmann et al. 2004; Ravi et al. 2010). These methods can also be used in an industrial environment to assess the comparability of different lots of products. In the last few years, more numerical methods to compare the entire spectra have been developed and used by researchers in the biopharmaceutical industry, for example, using 3× standard deviation of reference spectra (3× SD) to assess spectral similarity (Li and Coworkers, personal observation, unpublished data). The method uses historical data to assess the precision of the instrument. These methods are suitable for analytical product comparability studies, and the data obtained are compared to a reference spectrum. However, none of these methods have been used to systematically assess the sensitivity of the CD method.

Another method commonly used by researchers in the field is the deconvolution of the far-UV CD spectrum of a protein to contributions of individual secondary structural elements (alpha-helix, beta-sheet, beta-turn, and unordered conformation, as well as the contribution in the 195–240 nm region from aromatic amino acids) followed by determination of the percentage of each component in the protein secondary structure. Comparability is then determined by comparing changes in the percentage of each component (Greenfield 1996; Johnson 1999b; King and Johnson 1999; Perczel et al. 1991; Perczel and Fasman 1993; Venyaminov et al. 1991). However, the method is not optimized for assessing overall similarity of protein structure between samples from different processes or formulations. Also, the quantification of the secondary structure composition can vary depending on the methods and parameters used and involves considerable mathematical manipulations including spectra manipulation, curve fitting, area integration, and normalization. There is no analogous deconvolution of tertiary structure components that give rise to spectra in the near-UV CD region due to the lack of defined and pure side-chain conformational CD features in these spectra. Therefore, deconvolution of CD spectra is not routinely carried out for protein comparability assessments in the biopharmaceutical industry.

There are very limited published data on the systematic qualification of CD methods to evaluate the precision and sensitivity of the CD measurement. In a recent publication by Li and colleagues, the authors used Thermo Electron OMNIC software QC compare function (Cover and Hart 1967; TQ Analyst Algorithm 2007–2010) to compare the entire CD spectra in both the near- and far-UV CD regions to systematically qualify a CD method based on its precision and sensitivity (Li et al. 2011).

The *Thermo Electron* OMNIC software QC compare function was used by Li et al. to quantitatively assess the spectral similarity of different CD spectra, in order to determine the precision and sensitivity of near- and far-UV CD spectroscopy (Fig. 5.1). Proteins from different structural families in their stable storage buffer or under stress conditions were analyzed. The results from this precision study showed that (1) for the near-UV CD, the analysis has good precision in terms of spectral

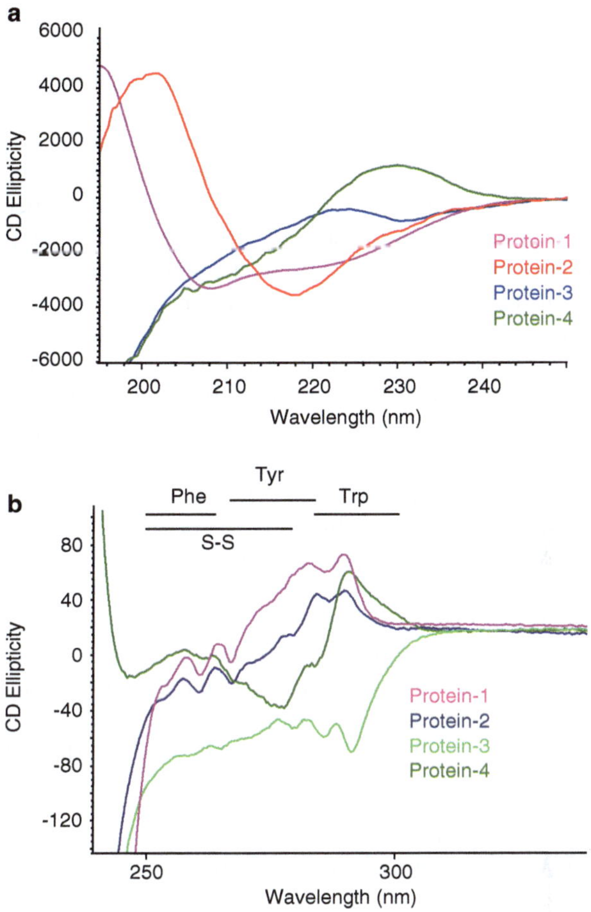

Fig. 5.1 (**a**) Far-UV CD spectra of protein 1–4 (spectral similarity compared to that of protein 1 is 100% protein 1; 5.0% protein 2; 44.9% protein 3; 28.7% protein 4). (**b**) Near-UV CD spectra of protein 1–4 (spectral similarity compared to that of protein 1 is 100% protein 1; 41.0% protein 2; 3.1% protein 3; 12.9% protein 4)

similarity when comparing the spectra obtained with the same instrument and cuvette independent of the analyst, location, and date; and (2) for far-UV CD, the precision varies depending on the structural nature of the protein and the buffer composition. For proteins with helical secondary structure in buffers with little interference in the spectra, far-UV CD analysis shows low variability in spectral similarity. For proteins with predominantly beta-sheet secondary structure and in buffers where the excipients have absorbance or CD signals in this region, far-UV CD analysis shows higher variability in spectral similarity. The sensitivity assessment of CD spectroscopy is challenging because the unfolding of a protein by some downstream processing conditions, for example, low pH or Gdn, is often reversible or partially reversible when the stresses are removed. Thus a standard curve that correlates spectral similarity to the percentage of denatured form of the protein cannot be generated empirically. To address this issue, Li et al. used a computer-simulated blending study of the spectra of the native protein and the unfolded form of the same protein to generate a standard curve that correlates the spectral similarity of the measured spectrum to the native spectrum with the amount of unfolded form of the protein present. To determine the sensitivity of CD spectroscopy, one would have to combine the precision assessment of the technique under the required test conditions (e.g., protein concentration and buffer) with the spectral similarity correlation of the simulated blending study of denatured and native proteins. The sensitivity of the method can be expressed as the percentage of denatured form of the protein when the spectral similarity of the sample spectrum compared to the native/control protein exceeds the precision of the method. Under different conditions and with a different protein, the precision and sensitivity of the CD analysis may also change.

The results from Li et al.'s qualification study indicate that CD spectroscopy can be qualified for characterizing protein secondary and tertiary structures and for comparability studies of biopharmaceuticals. With proper calibration of the instruments and cuvettes, and standardized good operating practices, CD spectroscopy is quite reproducible and able to detect changes induced in the secondary and tertiary structure of a protein by lower pH and other possible denaturing reagents that are present during therapeutic protein manufacturing processes. The sensitivity of the technique to detect the changes in conformation that might be induced by the manufacturing process is dependent on the experimental variability and the nature and extent of the structural changes in the particular protein being analyzed.

The results shown above provide the ground work on how the CD method should be calibrated and qualified and demonstrate that the method is sufficiently precise to monitor protein structure changes beyond the variability of the method.

5.3 Qualification of FTIR

Fourier transform infrared (FTIR) spectroscopy can be used to obtain information about the higher-order structure of a protein both in solution and in the solid state (Carpenter et al. 1998; Fu et al. 1999; Haris and Chapman 1995; Kong and Yu 2007). It is routinely employed as a characterization method in the biopharmaceutical industry to determine the higher-order structure of protein therapeutics (Gross and Zeppezauer 2010). It is often included as part of the biophysical characterization methods to assess the secondary structure of protein therapeutics before and after changes to process, formulation, manufacturing, and storage conditions, and the data are included in regulatory filings (Jiang and Narhi 2006; Jiang et al. 2008). However, the qualification of the FTIR method for protein secondary structure analysis has mostly remained elusive, mainly due to the lack of a consistent method applied to quantitatively determine the similarity of the FTIR spectra and hence the comparability of the protein samples analyzed. Visual comparisons of a sample spectrum to a reference have typically been used as a way to verify proper folding of a protein sample. To objectively qualify the FTIR method, a mathematical algorithm is needed to compare the FTIR spectra quantitatively and demonstrate the precision and sensitivity of the method for protein secondary structure analysis.

Work has been done to quantify the percentages of different secondary structural components for proteins in solution by FTIR (Vonhoffa et al. 2010; Susi and Byler 1987). There are two primary ways of doing this. The first uses curve fitting of the second-derivative or self-deconvoluted spectra to obtain the relative amounts of different types of secondary structure based on the band areas (Susi and Byler 1986; Byler and Susi 1986). The second involves peak fitting of the non-deconvolved and baseline-corrected amide I bands and then obtaining the percentage of secondary structures by correlating with the shape and intensity using an interval partial least squares algorithm (Vedantham et al. 2000; Surewicz and Mantsch 1988). These methods are helpful in determining the major secondary structure components and estimating their percentages in a protein. However, these methods are not optimized for assessing overall similarity of protein structure between samples from different processes or formulations. The quantification of the secondary structure composition can vary depending on the methods and parameters used, and it involves considerable mathematical manipulations including spectra manipulation, curve fitting, area integration, and normalization. Therefore, they are not routinely carried out for protein secondary structure assessment and qualification of FTIR by analysts in the biopharmaceutical industry.

One other mathematical approach explored by Prestrelski et al. (1993) is to compare the FTIR spectra quantitatively by using the correlation coefficient function. Occasionally, the correlation coefficient value of the uncorrected spectra did not agree with a visual assessment of the spectral similarity due to an offset in baselines which led to an artificially low value. Conversely, if the spectra were baseline corrected and peak positions were similar, but there were differences in relative peak heights, the correlation value would be unreasonably high. To avoid this inconsistency, Kendrick has developed a method to quantify the overlap of second-derivative FTIR spectra to determine structural similarity between proteins (Kendrick et al. 1996). In this approach area-normalized second-derivative spectra were used and compared. The authors found that quantifying the area of overlap between area-normalized spectra provides a reliable, objective method to compare overall spectral similarity and avoids the problems associated with calculation of the correlation coefficient. However, due to normalization of the spectra, the area of overlap approach cannot distinguish between true differences in the magnitude of FTIR signals. Recently, D'antonio et al. 2012 showed a modified area of overlap method that improved the differentiation power for quantitative comparison of FTIR spectra. The authors also compared four different algorithms including the correlation coefficient and area of overlap and summarized their strength and weakness for studying comparability of protein therapeutics.

A new approach using *Thermo Electron* OMNIC software QC compare function (Cover and Hart 1967; TQ Analyst Algorithm 2007–2010) has been applied to directly compare FTIR spectra with minimal mathematical manipulation (Fig. 5.2), which led to the identification of important performance characteristics for the qualification of the FTIR method (Jiang et al. 2011). The QC compare function was originally intended and used as a library searching tool for small-molecule identification by FTIR (TQ Analyst Algorithm 2007–2010). It correlates the spectral

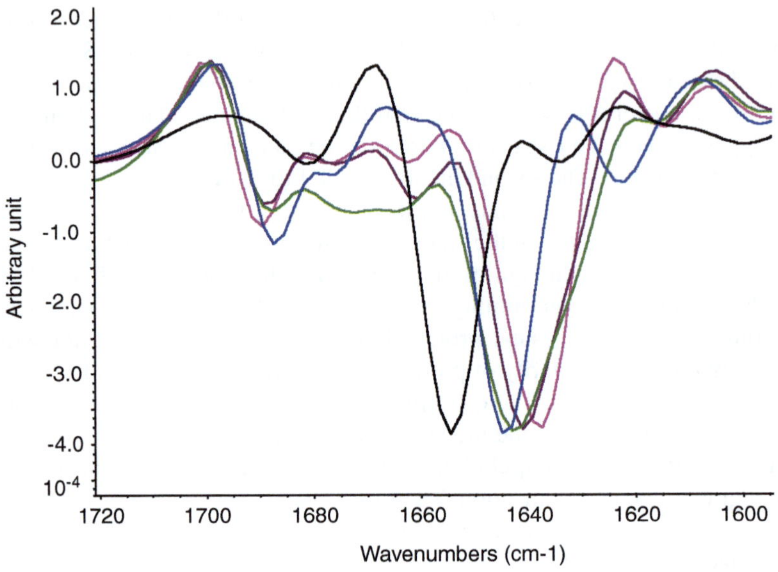

Fig. 5.2 Second-derivative FTIR spectra of protein 1–5 (spectral similarity compared to that of protein 1 is 100% protein 1 (*pink*); 46.3% protein 2 (*purple*); 36.3% protein 3 (*green*); 9.7% protein 4 (*blue*); 12.1% protein 5 (*black*))

information in the specified region of the sample and the reference spectra to determine the similarity between the two. The results of the method are reported as a value between 0% and 100%, which indicates how well the sample spectrum matches the reference spectrum. A spectral similarity value of 100% indicates the two spectra are identical.

The selection of an appropriate approach to numerically compare the spectra should depend on the capability of the method, the ease of access and use of the method, and the purpose of the study.

With the OMNIC QC compare algorithm, Jiang and colleagues evaluated the precision and sensitivity of FTIR for the analysis of protein secondary structure through a multisite/instrument and multi-analyst study in an effort to qualify the method (Jiang et al. 2011). Proteins containing different types of secondary structures such as alpha-helical and beta-sheet were included to ensure that the precision assessment would apply to secondary structure analysis of all proteins regardless of the specific structural type. In addition inter-day repeatability and the effect of protein concentration differences on the precision of the method were also evaluated. The overall FTIR spectral similarities of these analyses on the same proteins were compared. The sensitivity of the method was evaluated by comparing the reference spectrum of a protein to that of the partially or fully unfolded protein after exposure to the denaturing condition and also by blending studies where the spectrum of a reference was mixed with that of a denatured protein. Standard curves were generated where the spectral similarity was plotted as a function of the percentage of the unfolded protein.

Results by Jiang et al. (2011) demonstrate that the FTIR method for the analysis and characterization of protein secondary structure is precise with the standard deviations of the characteristic FTIR band frequencies <1.1 cm^{-1} and the spectral similarity >93% for replicate measurements. The FTIR spectra remain the same whether collected on the same day or during a three-day period and when the protein concentration deviates from the target concentration (≥30 mg/mL) by ≤10% regardless of the maker of the instrument used or the structural family of the protein. The authors also demonstrate that the method is sensitive and appropriate for assessing protein secondary structure and changes in conformation resulting from stresses that can be encountered during common biopharmaceutical manufacturing processes. The sensitivity of the FTIR method is dependent on the extent of the structural changes induced and the magnitude of the resulting changes in the spectra.

The precision of the FTIR method is closely related to the signal-to-noise ratio of the instrument, the sample concentration, the buffer components, and the spectral region of interest. The sensitivity of the method to detect structural change depends on the nature of the sample and relative spectral changes corresponding to the structure changes. Therefore, protein-specific qualification/verification of the method may need to be performed to ensure the understanding of precision and sensitivity of the method.

5.4 Qualification of DSC

DSC has been used widely to study thermal dynamics of protein folding. The melting/thermal transition temperature(s) of proteins can be used to understand the energy barrier for protein unfolding and thermal stability of proteins (Pyrpassopoulos et al. 2006; Johnson et al. 1995). DSC has also been used to study the binding interactions of proteins with other molecules and the effect of mutations and the presence of carbohydrates on protein thermal stability (Celej et al. 2006; Bruylants et al. 2005; Johnston et al. 2011; Protasevich et al. 2010; Wen et al. 2008). The application of DSC has been reviewed extensively in the area of characterization of macromolecules and their interactions during pharmaceutical product development (Chiu and Prenner 2011; Bruylants et al. 2005). However, even though the method has been used extensively to monitor and show changes in protein thermodynamic parameters such as enthalpy and melting temperature (T_m), there are very limited published data on the systematic qualification of the method to show that the DSC measurements are precise, accurate, and sensitive for the analysis of protein conformation and thermal stability. A protocol on measuring protein thermostability by DSC was published by Makhatadze (Coligan et al. 2001). It provides a detailed procedure for conducting DSC analysis including sample preparation and interpretation of the results, as well as calibration of the instrument and maintenance of DSC cells. Calibration of DSC was described by Gmelin and Sarge (Gmelin and Sarge 1995). They provided recommendations for instrument-independent calibration of temperature, heat, heat flow rate of the scanning calorimeter, operation procedures, calibration substance, and data treatment algorithms which can be used for all instruments.

In the early days of DSC analysis, various data analytical algorithms were developed and used to assess the effect of instrument configuration and scan rate on DSC measurement results (Lopez and Freire 1987). During the use of DSC for purity analysis of drug product, instrumental differences, sample size, heating rate, and details of the calculation of such analysis were examined (Yoshii 1997). It was found that changing the sample size or heating rate resulted in a difference in the effect of purity value between two instruments. Collectively controlling the heating rate during DSC measurement appears to be critical in obtaining reproducible results.

The precision of DSC measurements across different instruments and labs using the same sample (polymeric material) and parameters has been assessed. The study was organized by Empa (Swiss Federal Institute for materials testing and research) (Schmid 2012; Affolter et al. 2001). It collected measured data on glass transition (T_g) and melting point (T_m) and evaluated those using statistical methods to assess the precision of DSC. The results show that variability of DSC measurements is mainly caused by differences in the analyst, instrument, and calibration of the instrument. For one-point temperature measurements such as T_g and T_m, good agreements were observed with a standard deviation of repeatability at 0.3–1.0°C and that of reproducibility at 1.0–2.1°C. The standard deviation increases significantly with user-defined data evaluation.

A recent paper by Wen et al. (2011) explored the qualification of DSC for its applications in thermal stability analysis of proteins. The authors assessed the precision and sensitivity of the DSC method through a multisite, instrument, and analyst study using several proteins from different structural families including monoclonal antibodies and cytokines and parameters including T_m and profile similarity. The same experimental parameters such as heating/scan rate and similar protein concentrations were used for measurement on all instruments with the exception of one. The results show that the T_m values obtained for the same protein by the same or different instruments and/or analysts are quite reproducible, varying generally <1.1°C with the same instrument and <1.5°C across different instruments. The profile similarity values obtained using the OMNIC QC compare function (discussed above in the CD and FTIR sections) for the same protein from the same instrument are also high (> ~95%). The variability of inter-day DSC measurements is very similar to that of the intra-day measurements (only slightly higher by ~0.1°C). The sensitivity of the DSC method for assessing protein thermal stability and conformational changes was evaluated by several experiments including analyzing samples perturbed by pH 3 or 6 M guanidine HCl (Gdn). The results show that DSC is able to detect changes in protein conformation caused by low pH and denaturant to levels as low as 10% denatured protein present in the sample. The authors then applied DSC to the analysis of pH stability and buffer screening of a protein and candidate screening. They demonstrated that DSC is an appropriate method for assessing protein thermal stability and conformational changes that may result from manufacturing processes, formulation, and storage conditions and can also be used to compare relative stability of candidate molecules.

Validation of DSC for pharmaceutical analysis under cGMP was described by Weissburg et al. (2002). The validation included development of a validation plan (i.e., scope of validation, criterion for completion of validation), design documents, user requirements, calibration and maintenance procedures, IQ/OQ, standard operation procedures, and test scripts. The authors provided details about validation of the computer-controlled system to be 21CFR Part 11 compliant and of the thermal accuracy and precision using a NIST-certified indium standard at 0.5 and 50°C per minute scan rate employing the DSC testing script. Validation of DSC was achieved when the test results obtained met the specifications. The validation showed that the DSC instrument and the computer program and system used met 21 CFR Part 11, cGMP, and the user predefined requirements.

The cases shown above demonstrate that DSC can be qualified and validated for the purpose of assessing protein higher-order structure changes for regulatory filings. The qualification and/or validation should focus on demonstrating the precision of the method and that the method is suitable for its applications. When the instrument is properly calibrated and maintained, DSC is accurate, precise, and sufficiently sensitive to detect changes in proteins resulting from process and formulation changes and point mutations of the protein primary structure.

5.5 Qualification of SV-AUC

Sedimentation velocity analytical ultracentrifugation (SV-AUC) is used as a characterization tool in biotherapeutic development primarily to assess product purity; that is, it measures the relative abundance of size variants in a protein solution. Although nonroutine, it is commonly applied in product comparability studies, protein reference standard qualifications, product quality investigations, and other product characterization activities (Berkowitz 2006; Shire 1994).

The SV-AUC analysis is an orthogonal method to size-exclusion (SE) HPLC, and it is in this context that it is perhaps most useful, because it does not suffer from many of the potential limitations of SE-HPLC: physical disruption of the sample, extensive dilution into what is often a different solution environment, and lack of separation of product size variants close to the column exclusion limit (Gabrielson et al. 2007a; Philo 2006; Carpenter et al. 2010). Because it employs a fundamentally different mode of separation, SV-AUC can be used to verify the accuracy of SE-HPLC methods when they are developed and remediated and to confirm the suitability of SE-HPLC for routine use.

In addition to measuring the relative concentration of size variants, SV-AUC can be used to determine their mass and shape. The sedimentation boundary profiles provide information about the sedimentation and diffusion rates of size variants present in the sample (Dam et al. 2004; Laue et al. 1992; Lebowitz et al. 2002; Schuck et al. 2002). The sedimentation coefficient is directly measured from the time-dependent displacement of the boundaries. For a size variant of known mass, consistency of the measured sedimentation coefficient provides confirmation that the overall shape and charge of the size variant has been maintained when multiple samples are measured under the same solution conditions. However, the accuracy and precision of the measured sedimentation coefficient decreases as the concentration of the size variant decreases (Gabrielson et al. 2007b). In very pure samples with low concentrations of fragmented and aggregated size variants (<1% of the total protein mass), only the sedimentation coefficient of the most abundant species (typically monomer) can be reliably determined.

In the biopharmaceutical industry, SV-AUC is primarily used as an orthogonal method to SE-HPLC to quantitatively measure protein purity and secondarily used to measure protein monomer sedimentation coefficients (Gabrielson et al. 2010). Qualification of an SV-AUC method should reflect the fact that it will be used to quantitatively measure these two attributes. Furthermore, two performance characteristics of an SV-AUC method should be considered during qualification: (1) evaluation of the method's precision and (2) confirmation that the method is sufficiently sensitive to provide quantitative results, thereby allowing meaningful conclusions to be drawn.

Assessing the precision of SV-AUC, and determining which type(s) of precision to evaluate, should be governed by how the method will be used. For example, if SV-AUC analysis can typically be completed by one analyst using one instrument for product characterization studies required during biotherapeutic development, then it may not be necessary to study the reproducibility of the method. However, if the number of samples requires multiple runs on different days, then understanding the intermediate precision of the method is critical.

Experimental factors that impact the precision of SV-AUC measurements have been reviewed elsewhere (Gabrielson et al. 2007b, 2010; Schuck et al. 2002; Gabrielson and Arthur 2011; Arthur et al. 2009; Pekar and Sukumar 2007) and include the instrument, rotor, centerpiece and other cell components, overall alignment of channels, temperature equilibration, analysis software and settings, and sample characteristics. Even under the tightest experimental control currently achievable, normal variation of these factors leads to variability in results of approximately 0.4% for size variant quantification (Gabrielson and Arthur 2011; Arthur et al. 2009) and about 0.02 S for sedimentation coefficient measurements (Gabrielson and coworkers, personal observation, unpublished data). Much of the imprecision arises from the specific centerpieces used (Arthur et al. 2009; Pekar and Sukumar 2007) and from the alignment of cells (Pekar and Sukumar 2007), factors that can vary as much within a run as they do across runs or even across laboratories. A robust qualification design for an SV-AUC method should therefore account for normal variations from all relevant experimental factors, that is, those factors that reflect how the method will be applied during development of a biotherapeutic.

Studies to evaluate SV-AUC method performance indicate that performance parameters including linear range, limit of detection (LOD), limit of quantitation (LOQ), and measurement precision are, in general, independent of the specific protein product analyzed (Gabrielson et al. 2007b; Gabrielson and Arthur 2011; Arthur et al. 2009). Variability in measured results is primarily due to experimental factors, such as centerpiece differences, and depends to a lesser extent on sample characteristics. Therefore, after sufficient data have been acquired to account for relevant experimental factors, it is possible to qualify SV-AUC analysis of a new protein product by simply verifying that the precision of sedimentation coefficient measurements and the precision of size variant measurements are within a range expected from previous experience with the method. Additional qualification experiments may be warranted if the expected method performance is not achieved for a particular product or sample type, for example, protein samples with reversible self-association. In those cases, protein-specific intermediate precision should be more rigorously defined.

The ability of SV-AUC to detect changes to the product, both to its size distribution profile and to the sedimentation coefficient of the monomer, is essential for successful application of the method in biotherapeutic development. Therefore, method qualification should include experiments demonstrating the suitability of SV-AUC for detecting such changes. This may be accomplished by manipulating a sample to perturb the protein structure and to produce higher levels of size variants. A suitable amount of degradation can usually be obtained from thermal, UV, or pH stresses. For example, subjecting a monoclonal antibody sample to low pH conditions will often cause the sedimentation coefficient of the monomer to shift and induce formation of high molecular weight size variants (aggregates), which can be used to confirm the sensitivity of the method.

The sensitivity of the method can best be determined when changes to the measured output are correlated with changes deliberately made to the product. Changes to the sedimentation coefficient of a protein and the amount of aggregation measured by SV-AUC for a different protein, both induced by exposure to low pH, are shown in Fig. 5.3.

The data presented in Fig. 5.3 are instructive for several reasons. First, the response of the method to a given stress condition may or may not be linear. In panel A, the sedimentation coefficient responds nearly linearly to pH changes from 5.5 to 3.5 and then declines nonlinearly at more acidic pH, whereas in panel B the maximum aggregation levels are measured in a narrow pH range near 3.5; aggregation levels decline markedly at higher and lower pH. Second, it is often convenient to express the method sensitivity by normalizing the response, either by dividing by the range of the independent variable (i.e., by calculating the slope if the response is linear) or by dividing by an estimate of the variability of the method to express the change in response as a number of standard deviations. For example, in panel A of Fig. 5.3, a pH change from 5.5 to 3.5 results in a 0.2 Svedberg shift in the sedimentation coefficient. The sensitivity of the method to pH could then be expressed as 0.1 Svedberg per pH unit (or five standard deviations per pH unit when normalized by the method variability of 0.02 S). Finally, method sensitivity studies are most useful when they evaluate a range wider than what the product will likely encounter during manufacture, shipping, and long-term storage. This allows interpolation of the response, rather than extrapolation to conditions which have not been studied.

Any number of stress conditions can be used to demonstrate the sensitivity of SV-AUC. These may include thermal degradation, light/UV exposure, pH modification, chemical denaturation, or primary structural modifications (e.g., oxidation).

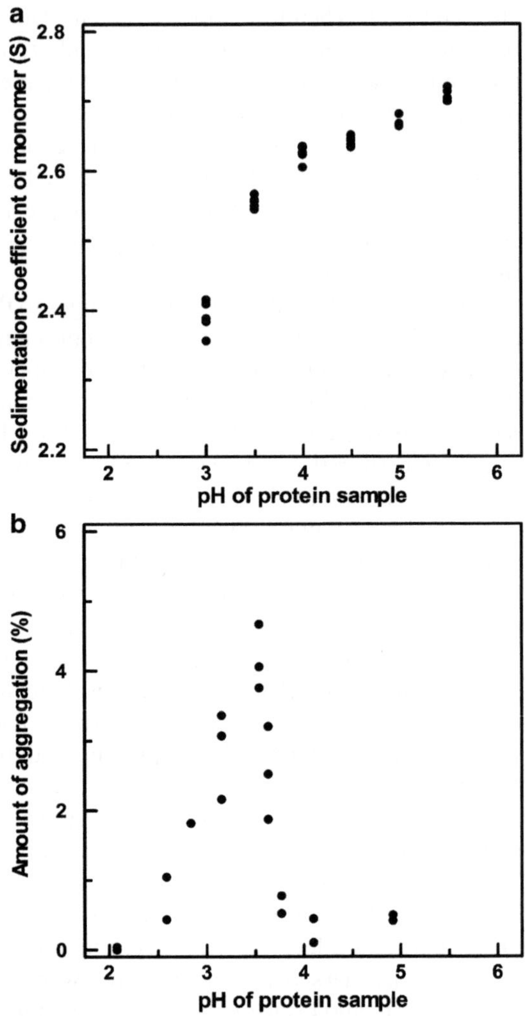

Fig. 5.3 Sedimentation coefficients (**a**) and aggregation levels (**b**) measured by SV-AUC for protein product samples across the pH range indicated

Many considerations should inform the selection of product stress conditions with which to determine the sensitivity of the method, including the likelihood that the product will encounter a certain stress during manufacturing and the intended change to the product. That is, a general stress like elevated temperature may be desired to produce aggregation, whereas a more targeted stress like oxidation may be useful to correlate the sedimentation rate of the protein with changes to its primary structure. And finally, the intended purpose of the method should also be considered when selecting product stress conditions. A different stress condition may be desirable if the method will be used exclusively to characterize a drug substance manufacturing process compared to if the method will be used to evaluate drug product comparability.

5.6 Qualification of SE-HPLC-LS

During biotherapeutic development, it is often valuable to measure the molecular weights of size variants present in product samples. Molecular weight determination with sufficient accuracy facilitates identification of these size variants (such as dimer and larger aggregate species), which is required for comprehensive product characterization.

Addition of light scattering (LS) detection to a size separation method like SE-HPLC makes it possible to determine the molecular weights of size variants after elution from the column (Kendrick et al. 2001; Wen et al. 1996). Under some conditions, the LS detection is also more sensitive for large size variants than conventional detection (such as UV, RI) alone. Both of these LS capabilities arise from the fact that the intensity of the light scattered by an eluting species is proportional not only to its concentration but also to its molecular weight (Wen et al. 1996; Wyatt 1993). However, because the LS detector response is a function of two variables, molecular weight and concentration, neither can be calculated directly from the LS signal alone. The molecular weight of each eluting species can only be determined by using a separate measurement of protein concentration (i.e., a simultaneously collected UV or RI chromatogram) to deconvolute the molecular weight and concentration information for each peak.

LS analysis is commonly applied as a characterization tool in product comparability studies, characterization studies to support product licensure, and product quality investigations. For molecular weight measurements by LS, the accuracy of the measured result is the key method performance parameter that should be evaluated during method qualification. The precision of the method may be determined in a straightforward manner and is useful to include in the design of qualification experiments. Accounting for different columns and mobile phases in these experiments is critical if characterization studies might be performed over multiple days and analysis sequences. Nonetheless, it is the accuracy of the method, that is, how much the measured mass differs from the true mass, which dictates the usefulness of the method for product characterization. Therefore, the method should be qualified to demonstrate that it is sufficiently accurate for the intended purpose of determining the molecular weights of size variants separated by SE-HPLC (or another flow separation method such as field flow fractionation).

Comparison of the measured mass of a protein standard to its known mass is an important way to assess the accuracy of SE-HPLC-LS. This confirms that the system is capable of providing results with sufficient accuracy. Nevertheless, the accuracy of results for unknown species cannot be inferred from the accuracy of a protein standard because the measurement accuracy depends on factors which may be different for each protein and size variant, such as the extinction coefficient and concentration. Therefore, propagation of error analysis provides a convenient approach by which to assess the accuracy of the measured result for any protein size variant.

First, the signal and noise associated with each detector (LS, UV, and/or RI) can be determined directly from a chromatogram. If a multiangle LS detector is used, signal and noise estimates should be determined from the 90° detector. (For isotropic scatters, such as proteins, the 90° detector has the highest signal-to-noise ratio.) Each detector's signal is taken to be the maximum signal (in volts) of the peak of interest. The noise for each detector is estimated from a region without protein signal (i.e., a baseline region) using any one of several noise calculation approaches such as the square root of the mean squared error.

The governing equations that relate the signals of each detector to the physical attributes of the protein are provided in Eqs. 5.1, 5.2, and 5.3 (Wen et al. 1996):

$$S_{\mathrm{LS}} = k_{\mathrm{LS}} M c \left(\frac{\mathrm{d}n}{\mathrm{d}c} \right)^2, \tag{5.1}$$

$$S_{\mathrm{UV}} = k_{\mathrm{UV}} \varepsilon l c, \tag{5.2}$$

$$S_{\mathrm{RI}} = k_{\mathrm{RI}} \left(\frac{\mathrm{d}n}{\mathrm{d}c} \right) c, \tag{5.3}$$

where k_{LS} (V mol L mL^{-2}), k_{UV} (V), and k_{RI} (V g mg^{-1}) are detector proportionality constants and l (cm) is the path length of the UV cell, also a constant. The concentration, c (mg mL^{-1}), of the size variant can be determined from either Eq. 5.2 or Eq. 5.3, depending on the choice of detector, when the extinction coefficient ε (mL mg^{-1} cm^{-1}) and the refractive index increment dn/dc (mL g^{-1}) are provided. The molecular weight of the size variant, M (g mol^{-1}), is then determined from Eq. 5.1.

The error of the molecular weight measurement can be estimated by applying a propagation of error analysis to Eq. 5.1. The generalized form of the error propagation formula (Taylor 1997) is shown in Eq. 5.4 for a response, f, that is a function of independent variables, x, y, and z:

$$e_f = \sqrt{\left(\frac{\partial f}{\partial x} e_x \right)^2 + \left(\frac{\partial f}{\partial y} e_y \right)^2 + \left(\frac{\partial f}{\partial z} e_z \right)^2}, \tag{5.4}$$

where e_f is the expected error in the calculated value of f and e_x, e_y, and e_z are error estimates for variables x, y, and z, respectively.

5.6.1 Case 1: LS-UV Detectors

When LS and UV detectors are connected in series, the size variant concentration is calculated from Eq. 5.2, which is then used to calculate the size variant molecular weight from Eq. 5.1. It is assumed that some degree of uncertainty exists in the protein extinction coefficient and refractive index increment, both of which are provided by the analyst. Propagation of error using Eq. 5.4 leads to the following expression for the expected error in the molecular weight calculation (Eq. 5.5):

$$\frac{e_M}{M} = \sqrt{\left(\frac{e_{\mathrm{SLS}}}{S_{\mathrm{LS}}}\right)^2 + \left(\frac{e_{\mathrm{SUV}}}{S_{\mathrm{UV}}}\right)^2 + \left(\frac{e_\varepsilon}{\varepsilon}\right)^2 + 4\left(\frac{e_{\mathrm{d}n/\mathrm{d}c}}{\mathrm{d}n/\mathrm{d}c}\right)^2}. \tag{5.5}$$

As expected, the molecular weight accuracy depends on the error in the LS and UV detector signals as well as uncertainty in the extinction coefficient and refractive index increment, which may not be explicitly known for peaks in the chromatogram. The error in each detector's signal can be approximated by its noise (i.e., $e_{\mathrm{SLS}} \approx N_{\mathrm{LS}}$) by assuming that inter-detector band broadening is either negligible or accounted for by the data analysis software.

5.6.2 *Case 2: LS-RI Detectors*

It is possible to determine the size variant concentration from Eq. 5.3 when RI detection is used instead of UV detection. In this case, the extinction coefficient is not required for the calculation, but the refractive index increment is required. Uncertainty in this parameter, along with errors in the LS and RI detector signals, is propagated using Eq. 5.4 to determine an expression for the molecular weight error (Eq. 5.6):

$$\frac{e_M}{M} = \sqrt{\left(\frac{e_{\mathrm{SLS}}}{S_{\mathrm{LS}}}\right)^2 + \left(\frac{e_{\mathrm{SRI}}}{S_{\mathrm{RI}}}\right)^2 + \left(\frac{e_{\mathrm{d}n/\mathrm{d}c}}{dn/dc}\right)^2}. \tag{5.6}$$

In this case, the error expression has only three terms: error in each detector's signal and uncertainty in the protein refractive index increment.

5.6.3 *Case 3: LS-UV-RI Detectors*

In the special case where all three detectors are connected in series, the refractive index increment can be measured using the RI detection. This can be shown by reexpressing Eq. 5.1 after solving for *M* and using Eqs. 5.2 and 5.3 to eliminate the d*n*/d*c* term, as shown in Eq. 5.7:

$$M = k\frac{S_{\mathrm{LS}}S_{\mathrm{UV}}}{S_{\mathrm{RI}}^2\varepsilon}, \tag{5.7}$$

where *k* is a collection of constants. Using all three detectors, an a priori estimate of the refractive index increment is not required, but the extinction coefficient is required to determine the concentration. Following similar error propagation analysis, an expression for the error of the molecular weight calculation can be determined:

$$\frac{e_M}{M} = \sqrt{\left(\frac{e_{\mathrm{SLS}}}{S_{\mathrm{LS}}}\right)^2 + \left(\frac{e_{\mathrm{SUV}}}{S_{\mathrm{UV}}}\right)^2 + 4\left(\frac{e_{\mathrm{SRI}}}{S_{\mathrm{RI}}}\right)^2 + \left(\frac{e_\varepsilon}{\varepsilon}\right)^2}. \tag{5.8}$$

Because all three detectors are used, errors in each of their signals appear in the overall error expression. Uncertainty in the protein extinction coefficient will also impact the accuracy of the molecular weight determination.

Using the appropriate error expression based on the choice of detectors used, and accounting for uncertainty in the protein extinction coefficient and/or refractive index increment, allows an assessment of molecular weight accuracy to be performed for any peak in any chromatogram. Thus, an evaluation of accuracy is not made at a single point in time but is a dynamic assessment that varies with protein load, size variant mass, and system noise. The accuracy also improves as analysts apply more accurate estimates of extinction coefficients and refractive index increments.

A graphical representation of the molecular weight error estimates from the three detector arrangements is provided in Fig. 5.4. Fractional error in the molecular weight (e_M/M) is plotted as a function of the size variant concentration for different levels of uncertainty in the extinction coefficient and/or refractive index increment.

At high protein concentrations (where $S \gg N$), the fractional errors in the detector signal terms in Eqs. 5.5, 5.6, and 5.8 are negligible. Therefore, for cases 2 and 3 in which LS and RI detection are used, the fractional error in molecular weight approaches the fractional uncertainty in either d*n*/d*n* (LS-RI, Eq. 5.6, Fig. 5.4b) or ε (LS-UV-RI, Eq. 5.8, Fig. 5.4c) at high protein concentrations. When UV detection is used without RI detection, the limiting error in the molecular weight is higher than when RI detection is used because uncertainty in both d*n*/d*n* and ε contribute to the error estimate (LS-UV, Eq. 5.5, Fig. 5.4a). At low concentrations (where $S \sim N$), the LS signal error term dominates the molecular weight error expression (all panels of Fig. 5.4).

To show that the instrumentation is performing properly and the calculated molar masses are meaningful, the measured masses by LS should be compared to a standard of known mass such as bovine serum albumin. At least one such standard should be analyzed within each sequence to confirm the system's suitability. Then, the molecular weights of size variants in the sample elution profile can be reported with associated error estimates derived from Eqs. 5.5, 5.6, or 5.8.

An understanding of both the precision and accuracy of measured molecular weights using LS detection should be considered when determining whether the method will be appropriate for specific applications in the biopharmaceutical industry. Addition of LS detection to an SE-HPLC method may be useful for product profile characterization, contaminant identification, and product comparability assessments. Nevertheless, the performance required of the method may be different for each of these applications, which is why qualification of the method is critical to enable selection of appropriate uses of LS detection.

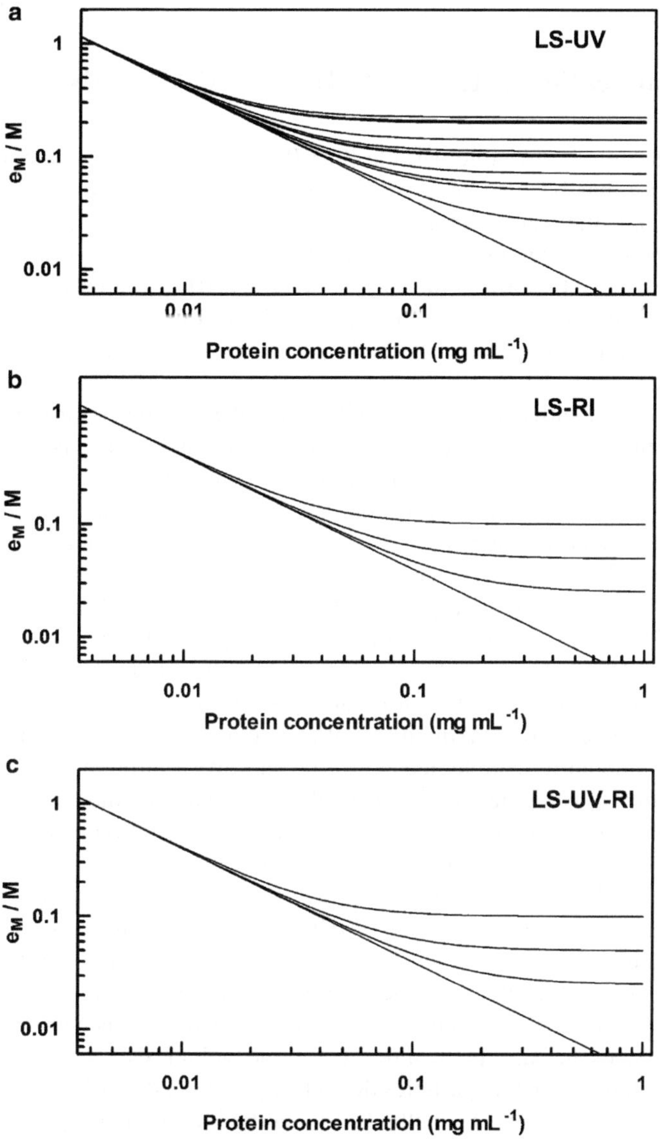

Fig. 5.4 Molecular weight fractional error estimates derived from error propagation analysis (Eqs. 5.5, 5.6, and 5.8) and plotted across a range of protein concentrations. In each panel, different curves represent different levels of uncertainty for ε and/or d*n*/d*c*. The curves were generated with 0, 2.5%, 5%, and 10% error for ε and d*n*/d*c* to encompass the expected range of uncertainty in these parameters. (**a**) LS-UV detectors (Eq. 5.5). Fractional errors in ε/d*n*/d*c* from lowest curve to highest: 0/0 (lowest curve); 0.025/0 (curve 2); 0.05/0 and 0/0.025 (indistinguishable, curve 3); 0.025/0.025 (curve 4); 0.05/0.025 (curve 5); 0/0.05, 0.1/0, and 0.025/0.05 (indistinguishable, curve 6); 0.05/0.05 and 0.1/0.025 (indistinguishable, curve 7); 0.1/0.05 (curve 8); 0/0.1, 0.025/0.1, and 0.05/0.1 (indistinguishable, curve 9); and ε = 0.1, d*n*/d*c* = 0.1 (highest curve). (**b**) LS-RI detectors (Eq. 5.6). Fractional errors in d*n*/d*c* from lowest curve to highest: 0, 0.025, 0.05, and 0.1. (**c**) LS-UV-RI detectors (Eq. 5.8). Fractional errors in ε from lowest curve to highest: 0, 0.025, 0.05, and 0.1. Data shown in all panels were generated using the following values: $N_{UV} = 0.00035$ V; $N_{LS} = 0.0055$ V; $N_{RI} = 0.00025$ V; $k_{UV} = 10$ V; $k_{LS} = 0.0015$ V mol L mL^{-2}; $k_{RI} = 20$ V g mg^{-1}; $M = 34{,}000$ g mol^{-1}; $\varepsilon = 0.48$ mL mg^{-1} cm^{-1}; $l = 1$ cm; and d*n*/d*n* = 0.165 mL g^{-1}

5.7 Qualification of LO- and MFI-Based Methods

The importance of subvisible particle (SbVP) analysis of protein therapeutics has increased as it has become an area of intense regulatory scrutiny due to concerns around the potential safety impact of protein aggregates in the micron size range, especially potential immunogenicity (Carpenter et al. 2009). The available methods have been reviewed (Narhi et al. 2009; Cao et al. 2012; Mahler et al. 2009; Zölls et al. 2011), with the authors showing the utilities of LO-, MFI-based, and conductivity and light scattering-based methods and microscopy for the analysis of subvisible particles.

The amount of subvisible particles allowed in therapeutics is currently governed by the pharmacopeia of the individual countries in which a drug is marketed, such as the USP <788> (United States Pharmacopeial Convention 2009) for the USA and EP 2.9.19 for the EU (Council of Europe 2007). These compendial methods for the quantification of subvisible particles were written for application to small-molecule therapeutics delivered intravenously using the LO technique, with the primary concern being the occlusion of capillaries by foreign particles due to contamination (Cao et al. 2009; Haines-Nutt and Munton 1984; Taylor and Spence 1983; Alexander and Veltman 1988). Applying USP <788> or EP 2.9.19 to protein therapeutics has met some significant issues including false-positive and negative results caused by trapped air bubbles and breaking down of fragile protein particles (Cao et al. 2009, 2010). The amount of sample needed for testing SbVP throughout the therapeutic protein product development life cycle by the compendial method is also prohibitive since the therapeutic protein is generally at high concentration and the availability of the material at early development stages is very limited.

A modified LO method for SbVP quantification in protein therapeutics was developed and qualified by Cao et al. (2010) recently. The authors described a modified LO method that can be used in procedures to analyze and control subvisible particles in protein therapeutics. They introduced appropriate sample handling such as vacuum degassing to minimize the false-positive and negative results. The method includes the ability to analyze sample volumes as small as 1 mL and quantification of subvisible particles down to 2 μm instead of 10 μm as specified by USP <788> or EP 2.9.19. The method was qualified for accuracy, precision (including repeatability and intermediate precision), and linearity using polystyrene particle counting standards (generally $<10\%$ RSD with sufficient amount of particles). The USP is currently in the process of establishing a biologics-specific chapter <787> that will closely resemble this modified LO method.

Protein samples can only be used to demonstrate precision and linearity, due to the lack of protein particle counting standards. Much of the qualification and validation of methods for SbVP determination must therefore be done with particles of known size and quantity made of polymers, The differences in morphology/shape, refractive index, and other optical properties of protein particles and polystyrene particle counting standards have made it difficult to quantify accurately the amount of protein particles without an appropriate standard, especially at the lower size range (i.e., <5 μm). The effect of the difference in refractive index between the polystyrene calibration spheres and test particles on the apparent size of particles determined by LO-based methods has been observed for other types of particles such as lipids as well (Driscoll 2004). However, the demonstrated precision and linearity, as well as the broadened size range of the modified LO method for protein samples, enable the use of the method for monitoring effectiveness of filtration and clearance of SbVPs during downstream processing steps, assessing the effect of various process stresses on SbVP formation, and the contribution of primary container components to the SbVP present (Cao et al. 2009, 2010).

The validation of the particle size method and instrument qualification have been described by Rawle in the book on "Practical Approaches to Method Validation and Essential Instrument Qualification"(Chan et al. 2011) and in the pharmacopeia (United States Pharmacopeial Convention 2009; Council of Europe 2007) from different country/regions. The procedures for instrument calibration, performance verification, and the use of particle counting standards are specified. Different instrument vendors essentially follow the same calibration procedure for LO-based instruments. This ensures the precision and accuracy of the measurements when polystyrene particle counting standards are measured. The performance of the LO method for protein therapeutics needs to be verified as shown by Cao et al. before it is used for testing.

Even though LO-based methods have been demonstrated to be precise and are widely used by the pharmaceutical industry to monitor SbVPs in their injectable products, the information obtained from these methods is limited to the size and concentration of the particles. In order to gain more understanding of the nature of SbVP, MFI-based techniques were developed recently (Mahler et al. 2009; Huang et al. 2009; Sharma et al. 2010a; Wuchner et al. 2010). Flow imaging techniques allow the characterization of particle shape and morphology in addition to size and concentration. It has been shown to be more sensitive than LO methods for measuring subvisible proteinaceous particles in protein formulations.

Due to the potential challenges in detecting particles in opalescent mAb formulations, the accuracy of MFI in determining size and particle counts in opalescent solutions was investigated and compared to LO and membrane microscopy methods by Sharma et al. (2010b). They used proteinaceous monoclonal antibody (mAb) particles, generated either by chemical denaturation or agitation stress, polystyrene, and glass particles as model systems for measurements in opalescent mAb solutions. The authors demonstrate that the sizing and counting accuracies of MFI were unaffected by the opalescence of the medium. Using glass particles as a model system for proteinaceous particles, MFI was able to detect relatively low particle concentrations (approximately 10 particle/mL) in opalescent solutions. MFI showed excellent linearity ($R(2)=0.9969$) for quantifying proteinaceous particles in opalescent solutions over a wide range of particle concentrations (approximately 20–160,000 particle/mL). The authors also show that the intensities of MFI particle image are significantly different with respect to the transparency of proteinaceous particles as a function of their size and mode of generation. The LO method significantly underestimated proteinaceous particles, particularly those in the 2–10 μm size range. The less opaque proteinaceous particles, the more underestimated by the LO method. Similar phenomenon was observed by Driscoll for lipid particles (Driscoll 2004).

Polystyrene bead standards are generally used as a reference for calibrating and validating particle analyzers. However, these standards, unlike protein particles, are easily detected and do not challenge the sensitivity of optical instruments. Instruments calibrated and verified only with beads can still exhibit significant differences in measuring concentrations of protein particles. To ensure the accuracy of the analysis of the size and concentration of suspended protein particles in protein therapeutics, standards should resemble protein particles with respect to their transparency and refractive index. Sharma et al. (2011) described the work on evaluating a potential standard and the use of it to harmonize the performance of MFI instruments. They have shown that the use of a suitable standard can significantly increase measurement consistency when multiple instruments are used to characterize the same protein particle suspension.

Similar to MFI, a micro-flow digital imaging (MDI) method was developed, optimized, and qualified to detect and monitor protein particles in a high-concentration IgG1 monoclonal antibody formulation by Wuchner et al. (2010). The MDI method was qualified as a characterization assay with respect to accuracy and precision of particle size and number using polystyrene standards (5–300 μm) and protein particles (2 to >100 μm). The authors then used it to monitor the stability profile of a 90 mg/mL IgG1 formulation stored at 2–8°C and −70°C for up to 18 months. The MDI assay results showed improved sensitivity in detecting subvisible particles (≥5 μm) compared to the conventional LO-based method and a good overall correlation with the amount of visible protein particles (>70 μm). The MDI results showed an accumulation of protein particles in certain size ranges and an increase in the overall particle size distribution over time. The combined MDI results and protein characterization studies have provided an enhanced understanding of protein particulate formation during long-term storage of a high-concentration IgG1 monoclonal antibody formulation.

In summary, the LO- and MFI-based methods can and should be qualified for SbVP analysis in specific protein solutions. The methods offer sensitive and reproducible measurement of particles in protein therapeutics solutions which enables proper monitoring and control of SbVP levels during protein therapeutics development, manufacturing, transportation, storage, and delivery.

5.8 Conclusion

The qualification of the biophysical methods including CD, FTIR, DSC, SV-AUC, SE-HPLC-LS, and LO- and MFI-based methods has been reviewed. Consistent instrument calibration and maintenance and proper sample handling are required to ensure successful qualification of these methods for protein higher-order structure and size distribution analysis. A numerical method was applied to compare the overall spectra or profile similarity for the qualification of some of the methods such as CD and FTIR. AUC and SE-HPLC-LS are first principle methods; therefore, the qualification of these methods does not require the use of a standard. However, for the proper qualification of the LO- and MFI-based methods, suitable particle standards that closely resemble the optical properties and morphologies of protein particles are still needed. For all of the biophysical methods, qualification is focused on the precision of the methods and demonstration that the methods are suitable for their intended applications. Successful qualification enables the understanding of the method capability and reliable determination and control of product attributes. Therefore, qualification of biophysical methods should be undertaken for protein therapeutics to ensure the best possible knowledge of the product and its manufacturing process.

References

Affolter S, Ritter A, Schmid M (2001) Interlaboratory tests on polymers by differential scanning calorimetry (DSC): determination of glass transition temperature (Tg). Macromol Mater Eng 286(10):605–610

Alexander DM, Veltman AM (1988) Particulate contamination in solutions of antibiotics packed as dry powders in vials. J Pharm Pharmacol 40(5):358–359

Arthur KK et al (2009) Detection of protein aggregates by sedimentation velocity analytical ultracentrifugation (SV-AUC): sources of variability and their relative importance. J Pharm Sci 98(10):3522–3539

Berkowitz SA (2006) Role of analytical ultracentrifugation in assessing the aggregation of protein biopharmaceuticals. AAPS J 8(3):590–605

Bruylants G, Wouters J, Michaux C (2005) Differential scanning calorimetry in life science: thermodynamics, stability, molecular recognition and application in drug design. Curr Med Chem 12(17):2011–2020

Byler DM, Susi H (1986) Examination of the secondary structure of proteins by deconvoluted FTIR spectra. Biopolymers 25:469–487

Cao S, Jiao N, Jiang Y, Mire-Sluis A, Narhi LO (2009) Sub-visible particle quantitation in protein therapeutics. Pharmeuropa Bio & Scientific Notes 73–79

Cao S, Jiang Y, Narhi LO (2010) A light obscuration method specific for quantifying subvisible particles in protein therapeutics. US Pharmacoepial Forum 36(3):824–834

Cao S, Narhi LO, Jiang Y, Rajan RS (2012) Analytical methods to measure sub-visible particulates. In: Mahler HC, Jiskoot W (eds) Analysis of aggregates and particles in protein pharmaceuticals. Wiley, New York

Carpenter JF, Prestrelski SJ, Dong A (1998) Application of infrared spectroscopy to development of stable lyophilized protein formulations. Eur J Pharm Biopharm 45(3):231–238

Carpenter JF et al (2009) Overlooking subvisible particles in therapeutic protein products: gaps that may compromise product quality. J Pharm Sci 98(4):1201–1205

Carpenter JF et al (2010) Potential inaccurate quantitation and sizing of protein aggregates by size exclusion chromatography: essential need to use orthogonal methods to assure the quality of therapeutic protein products. J Pharm Sci 99(5):2200–2208

Celej MS, Dassie SA, González M, Bianconi ML, Fidelio GD (2006) Differential scanning calorimetry as a tool to estimate binding parameters in multiligand binding proteins. Anal Biochem 350(2):277–284

Chan CC, Lam H, Zhang XM (2011) Practical approaches to method validation and essential instrument qualification. Wiley, New York

Chiu MH, Prenner EJ (2011) Differential scanning calorimetry: an invaluable tool for a detailed thermodynamic characterization of macromolecules and their interactions. J Pharm Bioallied Sci 3(1):39–59

Coligan JE et al (2001) Measuring protein thermostability by differential scanning calorimetry. In: Makhatadze GI (ed) Current protocols in protein science/editorial board. Wiley, New York

Council of Europe (2007) Particulate contamination: sub-visible particles, general chapter 2.9.19. Council of Europe, Strasbourg

Cover TM, Hart PE (1967) Nearest neighbor pattern classification. IEEE Trans Inf Theory IT-13(1):21–27

D'antonio J, Murphy BM, Manning MC, Al-Azzam WA (2012) Comparability of protein therapeutics: quantitative comparison of second-derivative amide I infrared spectra. J Pharm Sci. doi:10.1002/jps.23133

Dam J et al (2004) Calculating sedimentation coefficient distributions by direct modeling of sedimentation velocity concentration profiles. In: Methods in enzymology. Academic. pp 185–212

Drechsler A, Miles AJ, Norton RC, Wallace BA, Separovic F (2009) Effect of lipid on the conformation of the n-terminal region of equinatoxin II: a synchrotron radiation CD study. Eur Biophys J 39:121–127

Driscoll DF (2004) Examination of selection of light-scattering and light-obscuration acceptance criteria for lipid injectable emulsions. Pharm Forum 30:2244–2253

Fu K, Griebenow K, Hsieh L, Klibanov AM, Langer R (1999) FTIR characterization of the secondary structure of proteins encapsulated within PLGA microspheres. J Controlled Release 58:357–366

Gabrielson JP, Arthur KK (2011) Measuring low levels of protein aggregation by sedimentation velocity. Methods 54(1):83–91

Gabrielson JP et al (2007a) Quantitation of aggregate levels in a recombinant humanized monoclonal antibody formulation by size-exclusion chromatography, asymmetrical flow field flow fractionation, and sedimentation velocity. J Pharm Sci 96(2):268–279

Gabrielson JP et al (2007b) Sedimentation velocity analytical ultracentrifugation and SEDFIT/c(s): limits of quantitation for a monoclonal antibody system. Anal Biochem 361:24–30

Gabrielson JP, Arthur KK, Stoner MR et al (2010) Precision of protein aggregation measurements by sedimentation velocity analytical ultracentrifugation in biopharmaceutical applications. Anal Biochem 396(2):231–241

Gmelin E, Sarge St.M (1995) Calibration of differential scanning calorimeters. IUPAC, Pure Appl Chem 67:1789–1800

Greenfield NJ (1996) Methods to estimate conformation of proteins and polypeptides from CD data. Anal Biochem 235(1):1–10

Greenfield NJ, Fasman GD (1969) Computed circular dichroism spectra for the evaluation of protein conformation. Biochemistry 8:4108–4116

Gross PC, Zeppezauer M (2010) Infrared spectroscopy for biopharmaceutical protein analysis. J Pharm Biomed Anal 53:29–36

Haines-Nutt RF, Munton TJ (1984) Particle size measurement in intravenous fluids. J Pharm Pharmacol 36(8):534–536

Haris PI, Chapman D (1995) The conformational analysis of peptides using Fourier transform IR spectroscopy. Biopolym (Pept Sci) 37:251–263

Huang CT, Sharma D, Oma P, Krishnamurthy R (2009) Quantitation of protein particles in parenteral solutions using micro-flow imaging. J Pharm Sci 98(9):3058–3071

ICH M4Q (2001) The common technical document for the registration of pharmaceuticals for human use: quality. US Department of Health and Services, FDA, CDER

ICH Q2(R1) (2005) Validation of analytical procedures: test and methodology. In: ICH Harmonised Tripartite guideline in international conference on harmonisation of technical requirements for registration of pharmaceuticals for human use

ICH Q5E (2005) Comparability of biotechnological/biological products subject to changes in their manufacturing process. US Department of Health and Services, FDA, CDER & CBER
ICH Q6B (1999) Test procedures and acceptance criteria for biotechnological and biological products
Jiang Y, Narhi LO (2006) Applying selective biophysical techniques in assessing the comparability of protein therapeutics – case studies. J Am Pharm Rev 9(5):34–43
Jiang Y, Li C, Ramachander R, Wen J, Narhi LO (2008) Meeting new regulatory expectations of characterization methods-Qualification of biophysical analyses. J Am Pharm Rev February, online issue, Russell Publishing. http://pharmoutsourcing.com/ViewArticle.aspx?ContentID=2221. Accessed 7 Feb 2012
Jiang Y, Li C, Nguyen X, Muzammil S, Towers E, Gabrielson JP, Narhi LO (2011) Qualification of FTIR spectroscopic method for protein secondary structural analysis. J Pharm Sci 100(11):4631–4641
Johnson WC Jr (1988) Secondary structure of proteins through circular dichroism spectroscopy. Annu Rev Biophys Biophys Chem 17:145–166
Johnson WC Jr (1999a) Analyzing protein circular dichroism spectra for accurate secondary structures. Proteins 35(3):307–312
Johnson WC (1999b) Analysing protein circular dichroism spectra for accurate secondary structures. Proteins 35(3):307–312
Johnson CR, Morin PE, Arrowsmith CH, Freire E (1995) Thermodynamic analysis of the structural stability of the tetrameric oligomerization domain of p53 tumor suppressor. Biochemistry 34(16):5309–5316
Johnston MJW, Frahm G, Li X, Durocher Y, Hefford MA (2011) O-linked glycosylation leads to decreased thermal stability of interferon alpha 2b as measured by two orthogonal techniques. Pharm Res 28(7):1661–1667
Jones C et al (2004) Val-CiD Best Practice Guide: CD spectroscopy for the quality control of biopharmaceuticals. National Physical Laboratory. DQL-AS 008 ISSN: 1744–0602. British Library Document Supply Centre, Boston Spa, Wetherby, West Yorkshire, LS23 7BQ
Kelly SM, Jess TJ, Price NC (2005) How to study proteins by circular dichroism. Biochim Biophys Acta 1751:119–139
Kendrick BS, Dong A, Allison SD, Manning MC, Carpenter JF (1996) Quantitation of overlap of infrared second derivative spectra to determine structural similarity between proteins. J Pharm Sci 85:155–158
Kendrick BS et al (2001) Online size-exclusion high-performance liquid chromatography light scattering and differential refractometry methods to determine degree of polymer conjugation to proteins and protein–protein or protein–ligand association states. Anal Biochem 299(2):136–146
King SM, Johnson WC (1999) Assigning secondary structure from protein coordinate data. Proteins 35:313–320
Kong J, Yu S (2007) Fourier transform infrared spectroscopic analysis of protein secondary structures. Acta Biochim Biophys Sin 39(8):549–559
Laue TM et al (1992) Computer-aided interpretation of analytical sedimentation data for proteins. In: Harding SE, Rowe AJ, Horton JC (eds) Analytical ultracentrifugation in biochemistry and polymer science. Royal Society of Chemistry, Cambridge, pp 90–125
Lebowitz J, Lewis MS, Schuck P (2002) Modern analytical ultracentrifugation in protein science: a tutorial review. Protein Sci 11(9):2067–2079
Lees JG, Wallace BA (2002) Synchrotron radiation circular dichroism and conventional circular dichroism spectroscopy: a Comparison. Spectroscopy 16:121–125
Lees JG, Miles AJ, Wien F, Wallace BA (2006) A reference database for circular dichroism spectroscopy covering fold and secondary structure space. Bioinformatics 22:1955–1962
Li C, Nguyen X, Narhi LO, Chemmalil L, Towers E, Muzammil S, Gabrielson JP, Jiang Y (2011) Applications of circular dichroism for structural analysis of proteins: qualification of near- and far-UV CD for protein higher order structural analysis. J Pharm Sci 100(11):4642–4654

Lopez MO, Freire E (1987) Dynamic analysis of differential scanning calorimetry data. Biophys Chem 27(1):87–96

Mahler HC, Friess W, Grauschopf U, Kies S (2009) Protein aggregation: pathways, induction factors and analysis. J Pharm Sci 98:2909–2934

Manning MC, Woody RW (1989) Theoretical study of the contribution of aromatic side chains to the circular dichroism of basic bovine pancreatic trypsin inhibitor. Biochemistry 28:8609–8613

Meersman F, Atilgan C, Miles AJ, Bader R, Shang W, Matagne A, Wallace BA, Koch MHJ (2010) Consistent picture of the reversible thermal unfolding of hen egg-white lysozyme from experiment and molecular dynamics. Biophysical J 99:2255–2263

Miles AJ, Wallace BA (2006) Synchrotron radiation circular dichroism spectroscopy of proteins and applications in structural and functional genomics. Chem Soc Reviews 35:39–51

Miles AJ, Wien F, Lees JG, Rodger A, Janes RW, Wallace BA (2003) Calibration and standardisation of synchrotron radiation circular dichroism (SRCD) amplitudes and conventional circular dichroism (CD) spectrophotometers. Spectroscopy 17:653–661

Miles AJ, Wien F, Wallace BA (2004) Redetermination of the extinction coefficient of camphor-b-sulfonic acid, A calibration standard for circular dichroism spectroscopy. Anal Biochem 335:338–339

Miles AJ, Wien F, Lees JG, Wallace BA (2005) Calibration and standardisation of synchrotron radiation and conventional circular dichroism spectrometers. Part 2: factors affecting magnitude and wavelength. Spectroscopy 19:43–51

Miles AJ, Hoffman SV, Tao Y, Janes RW, Wallace BA (2007) Synchrotron radiation circular dichroism (SRCD) spectroscopy: new beamlines and new applications in biology. Spectroscopy 21:245–255

Narhi LO, Jiang Y, Cao S, Benedek K, Shnek D (2009) A critical review of analytical methods for subvisible and visible particles. Curr Pharm Biotechnol 10(4):373–381

Pekar A, Sukumar M (2007) Quantitation of aggregates in therapeutic proteins using sedimentation velocity analytical ultracentrifugation: practical considerations that affect precision and accuracy. Anal Biochem 367:225–237

Perczel A, Fasman GD (1993) Effect of spectral window size on circular dichroism spectra deconvolution of proteins. Biophys Chem 48:19–29

Perczel A, Hollosi M, Tusnady G, Fasman GD (1991) Convex constraint analysis: a natural deconvolution of circular dichroism curves of proteins. Protein Eng 4:669–679

Philo JS (2006) Is any measurement method optimal for all aggregate sizes and types? AAPS J 8(3):564–571

Powl AM, O'Reilly AO, Miles AM, Wallace BA (2010) Synchrotron radiation circular dichroism spectroscopy-defined structure of the C-terminal domain of NaChBac and its role in channel assembly. Proc Natl Acad Sci 107:14064–14069

Prestrelski S, Tedeschi N, Arakawa T, Carpenter J (1993) Biophys J 65(2):661–671

Protasevich I, Yang Z, Wang C, Atwell S, Zhao X, Emtage S, Wetmore D, Hunt JF, Brouillette CG (2010) Thermal unfolding studies show the disease causing F508del mutation in CFTR thermodynamically destabilizes nucleotide-binding domain 1. Protein Sci 19(10):1917–1931

Pyrpassopoulos S, Vlassi M, Tsortos A, Papanikolau Y, Petratos K, Vorgias CE, Nounesis G (2006) Equilibrium heat-induced denaturation of chitinase 40 from *Streptomyces thermoviolaceus*. Proteins 64(2):513–523

Ravi J, Rakowska PD, Garfagnini T et al (2010) International comparability in spectroscopic measurements of protein structure by circular dichroism: CCQM-P59.1, Metrologia 47(6): 631–641

Ravi J, Schiffmann D, Tantra R et al (2010) International comparability in spectroscopic measurements of protein structure by circular dichroism: CCQM P59, Metrologia 47:1A

Schiffmann D, Butterfield DM, Yardley RE, Knight A, Windsor SA, Jones C (2004) Val-CiD Appendix A: CD spectroscopy: an inter-laboratory study. DQL-AS 009 ISSN: 1744–0602

Schmid M (2012) Precision of DSC measurements: results of interlaboratory tests. http://www.eurostar-science.org/MAP1/Schmid.pdf. Accessed 7 Feb 2012

Schuck P et al (2002) Size-distribution analysis of proteins by analytical ultracentrifugation: strategies and application to model systems. Biophys J 82(2):1096–1111

Sharma DK, King D, Oma P, Merchant C (2010a) Micro-flow imaging: flow microscopy applied to sub-visible particulate analysis in protein formulations. AAPS J 12(3):455–464

Sharma DK, Peter O, Pollo MJ, Sukumar M (2010b) Quantification and characterization of subvisible proteinaceous particles in opalescent mab formulations using micro-flow imaging. J Pharm Sci 99(6):2628–2642

Sharma DK, King D, Merchant C (2011) Reference material development for calibration and verification of image-based particle analyzers. Int J Pharm 416(1):293–295

Shire SJ (1994) Analytical ultracentrifugation and its use in biotechnology. In: Schuster TM, Laue TM (eds) Modern analytical ultracentrifugation. Birkhauser, Boston

Surewicz WK, Mantsch HH (1988) New insight into protein conformation from infrared spectroscopy. Biochim Biophys Acta 952:115–130

Susi H, Byler M (1986) Resolution enhanced fourier transform infrared spectroscopy of enzymes. Meth Enzymol 130:290–311

Susi H, Byler M (1987) Fourier transform infrared study of proteins with parallel b-chains. Arch Biochem Biophys 258(2):465–469

Taylor JR (1997) An introduction to error analysis: the study of uncertainties in physical measurements, 2nd edn. University Science Books, Sausalito

Taylor SA, Spence J (1983) Particles in small volume injections. J Pharm Pharmacol 35(12):769–773

TQ Analyst Algorithm (2007–2010) Thermo OMNIC software manual, Thermo Electron Scientific Instruments, LLC., Madison, WI, USA (part of Thermo Fisher Scientific Inc.)

United States Pharmacopeial Convention, Inc (2009) Particular matter in injections, general chapter <788>, USP 32, NF 27. United States Pharmacopeial Convention, Inc., Rockville

Van Stokkum IHM, Spoelder HJW, Bloemendal M, Van Grondelle R, Groen FCA (1990) Estimation of protein secondary structure and error analysis from CD spectra. Anal Biochem 191:110–118

Vedantham G, Sparks G, Sane SU, Tzannis S, Przybycien TM (2000) A holistic approach for protein secondary structure estimation from infrared spectra in H_2O solutions. Anal Biochem 285:33–49

Venyaminov SW, Baikalov I, Chuen-Shang C, Yang JT (1991) Some problems of CD analysis of protein conformation. Anal Biochem 198:250–255

Vonhoffa S, Condliffeb J, Schiffterb H (2010) Implementation of an FTIR calibration curve for fast and objective determination of changes in protein secondary structure during formulation development. J Pharm Biomed Anal 51(1):39–45

Wallace BA (2005) Shining new light on protein structure and function thru synchrotron radiation circular dichroism (SRCD) spectroscopy. Australian Biochemist 36:47–50

Wallace BA, Janes RW (2010) Synchrotron radiation circular dichroism (SRCD) spectroscopy: an enhanced method for examining protein conformations and protein interactions. Biochem Soc Trans 38:861–873

Wallace BA, Whitmore L, Janes RW (2006) The protein circular dichroism data bank (PCDDB): a bioinformatics and spectroscopic resource. Proteins 62:1–3

Weissburg RP, Burris FE, Ferrell TK, Montgomery ER (2002) Validation of DSC and TGA for pharmaceutical analysis under current good Manufacturing Practices (cGMP). Am Pharm Rev 5(2):76–81

Wen J, Arakawa T, Philo JS (1996) Size-exclusion chromatography with on-line light-scattering, absorbance, and refractive index detectors for studying proteins and their interactions. Anal Biochem 240(2):155–166

Wen J, Jiang Y, Narhi LO (2008) Effect of carbohydrate on thermal stability of antibodies. Am Pharm Rev 11:98–104

Wen J, Arthur K, Chemmalil L, Muzammil S, Gabrielson JP, Jiang Y (2011) Applications of differential scanning calorimetry for thermal stability analysis of proteins: qualification of DSC. J Pharm Sci. doi:10.1002/jps.22820

Whitmore L, Wallace BA (2004) DICHROWEB, An online server for protein secondary structure analyses from circular dichroism spectroscopic data. Nucleic Acids Res 32:668–673

Whitmore L, Wallace BA (2008) Protein secondary structure analyses from circular dichroism spectroscopy: methods and reference databases. Biopolymers 89:392–400

Woody RW (1994) Contributions of tryptophan side chains to the far-ultraviolet circular dichroism of proteins. Eur Biophys J 23:253–262

Woody RW (1996) Theory of circular dichroism of proteins, circular dichroism and the conformational analysis of biomolecules. In: Fasman GD (ed) Circular dichroism and the conformational analysis of biomolecules. Plenum, New York

Wuchner K, Büchler J, Spycher R, Dalmonte P, Volkin DB (2010) Development of a microflow digital imaging assay to characterize protein particulates during storage of a high concentration IgG1 monoclonal antibody formulation. J Pharm Sci 99(8):3343–3361

Wyatt PJ (1993) Light scattering and the absolute characterization of macromolecules. Anal Chim Acta 272:1–40

Yang JT, Wu CSC, Martinez HM (1986) Calculation of protein conformation from circular dichroism. Methods Enzymol 130:208–269

Yoshii K (1997) Application of differential scanning calorimetry to the estimation of drug purity: various problems and their solutions in purity analysis. Chem Pharm Bull 45(2):338–343

Zölls S, Tantipolphan R, Wiggenhorn M, Winter G, Jiskoot W, Friess W, Hawe A (2011) Particles in therapeutic protein formulations, Part 1: overview of analytical methods. J Pharm Sci. doi:10.1002/jps.23001

Chapter 6
Application of Biophysics to the Early Developability Assessment of Therapeutic Candidates and Its Application to Enhance Developability Properties

Hasige Sathish, Nicolas Angell, David Lowe, Ambarish Shah, and Steven Bishop

6.1 Introduction and Scope

For the past few decades, both the cost and the complexity of therapeutic protein development have soared to levels that are becoming unsustainable (Adams and Brantner 2010). Additionally, there is a significant increase in the attrition rate of programs that can be attributed to, for example, complex disease pathways, an increased safety bar, and a greater number of highly complex molecules in development (Kola and Landis 2004). According to a recent analysis, the success rate for IND to market is approximately 25–30% for biotech companies (DiMasi and Grabowski 2007). Both the spiraling cost associated with drug development and the demands from end users for low-cost prescription drugs are forcing the industry to look for ways to increase efficiencies. One approach is cycle time reduction, a strategy that translates to a reduced and more efficient development timeline so that a drug can move faster to phase I and faster to market. This chapter focuses on current and future challenges in developing therapeutic biological drugs and how the use of novel biophysical tools to screen for potential developability and manufacturability risks can improve product quality, streamline development, increase manufacturing efficiency,

H. Sathish (✉) • A. Shah • S. Bishop
Formulation Sciences, Biopharmaceutical Development, MedImmune, One Medimmune Way, Gaithersburg, MD 20878, USA
e-mail: sathishh@medimmune.com; shaha@medimmune.com; bishops@medimmune.com

N. Angell
Process and Product Development, Amgen Inc., One Amgen Center Drive, MS 30E-1-C, Thousand Oaks, CA 91320, USA
e-mail: nicolasa@amgen.com

D. Lowe
Antibody Discovery and Protein Engineering, Medimmune Ltd, The Aaron Klug Building, Granta Park, Cambridge, Cambridgeshire CB21 6GH, UK
e-mail: lowed@medimmune.com

L.O. Narhi (ed.), *Biophysics for Therapeutic Protein Development*, Biophysics for the Life Sciences 4, DOI 10.1007/978-1-4614-4316-2_6,

and increase end-user convenience. Apart from selecting low-risk molecules, early developability screening can also aid in identification of platform compatible leads, which can significantly reduce both development cost and time.

6.2 Current Challenges and Future Trends in Therapeutic Drug Development

Monoclonal antibodies (mAbs) represent the fastest growing class of therapeutics in the pharmaceutical marketplace, with 147 human and 167 humanized mAbs having entered clinical development between 1985 and 2008 (Nelson et al. 2010). Because mAbs cannot readily enter cells or cross the blood–brain barrier, mAbs are typically developed against particular target classes, such as soluble proteins (e.g., cytokines and growth factors) or extracellular domains of receptors (Carter 2011). These biological limitations have resulted in a large number of similar mAbs targeting the same therapeutic indications. Hence, much of the discovery work is focused upon identifying novel therapeutic molecular targets and choosing or engineering properties that will potentially confer a clinical or commercial advantage. A third industry challenge is that of optimizing development/manufacturing design spaces to achieve a competitive advantage in terms of efficiency and timeline.

6.2.1 Novel Targets

More novel target classes, such as G-protein-coupled receptors (GPCRs), ion channels, and intracellular targets may not be evolutionarily ideally suited for antibody therapy. Therefore, these targets may require the development of therapeutic molecules that are inherently more novel than the prototypical IgG, such as alternative protein fold scaffolds or engineered receptor ligands. Recent advances in protein expression and production techniques, as well as improved in vitro libraries and selection strategies, have started to expand the coverage of these targets that are implicated in a range of different diseases (Lundstrom 2005; Ackerman and Clapham 1997; Naylor and Beech 2009). However, the types of therapeutic molecules (discussed in Sect. 6.2.2) that need to be engineered to bind these novel targets are far less well characterized in terms of large-scale development and manufacture than prototypical mAbs and are trending towards being highly complex, resulting in a greater attrition rate during discovery and development (Ma and Zemmel 2002).

6.2.2 Engineering Desired Properties

The field of antibody engineering has matured rapidly over the past decade so that properties such as affinity, potency, specificity, pharmacokinetics, and effector function are now readily tunable (Igawa et al. 2011). Also, the pharmaceutical industry is increasingly focusing on engineering molecules with novel mechanisms of action (Ma and Zemmel 2002).

For molecules with an established mechanism of action, the focus is on enhanced safety, higher affinity to target, and increased half-life. The first generation of mAbs to reach the market typically exhibited dissociation constants (Kds) in the micromolar range, in line with the in vivo ceiling of the B-cell response (Foote and Eisen 1995; Batista and Neuberger 1998). In vitro affinity maturation methods and technologies [reviewed in Igawa et al. (2011)] have allowed researchers to go far beyond these limits, reaching femtomolar Kds in many cases (Boder et al. 2000; Finch et al. 2011). Increasing the affinity of an antibody to a given antigen is achieved by targeting the variable domains, as these are regions of the molecule that have evolved to interact directly with the antigen, particularly the complementarity determining regions (CDRs). Affinity maturation typically results in several mutations in the CDR loops. As the CDR loops are surface exposed, engineering changes may result in changes to biophysical behavior, such as an increase in aggregation propensity and loss of stability (Chennamsetty et al. 2009a; Pepinsky et al. 2010).

Over the last decade, there has been an increased effort in glycoengineering and the engineering of Fc variants in order to modulate effector functions [reviewed in Kubota et al. (2009)] and improve pharmacokinetic properties through modified interaction with the neonatal Fc receptor (Presta 2008; Stewart et al. 2011; Dall'Acqua et al. 2006a, b). Glycosylation in IgG molecules is essential for structural integrity, solubility, and stability (Krapp et al. 2003; Mimura et al. 2000; Kayser et al. 2011). Any modulation of glycosylation can, in turn, have significant impact on solubility, manufacturability, and long-term storage. Likewise, engineering the Fc region of the IgG molecule to modulate effector function or pharmacokinetics may have unintended effects on the biophysical properties of the molecule, such as stability and aggregation propensity. The unintended effects of glycoengineering and the engineering of Fc variants may not become apparent until larger-scale manufacture and formulation to support clinical studies.

The search for molecules with novel mechanisms of action has resulted in an increased development of antibody-based alternative formats, such as multivalent antibodies (Carter 2011). Recently, a new set of protein therapeutics based on alternative protein scaffolds to immunoglobulin domains has entered clinical development (Gebauer and Skerra 2009). Each of these novel therapeutics is likely to present their own challenges for development, which will doubtless become apparent over the coming decade.

6.2.3 CMC Challenges

Chemistry, manufacturing, and controls (CMC) design spaces are another area of challenge in the commercialization of protein therapeutics. For molecules with a precedented mechanism of action, enhanced product performance can be achieved with superior drug product dosing and presentation. While IV administration remains an option to achieve the necessary high dose, the goal of providing greater patience convenience, lower dosing frequency, and self-administration limits the administration route to subcutaneous (SC) injection. SC injection typically requires the development of a high-concentration liquid formulation, which imposes an additional burden on the formulation and device design space and often limits the types of approaches that can be used to provide an acceptable stability and delivery profile.

Medical devices designed to enhance patient convenience are often a key aspect of product development. Typical devices include prefilled syringes (PFS) and auto injectors (AI). While the range of available devices is growing to include reconstitution and delivery of lyophilized protein formulations, the most common delivery system for self-administration is a high-concentration drug product in a PFS contained within an AI. These liquid-delivery devices pose limitations that must be considered during product development. Often certain parameters can be adjusted within the scope of the device, such as spring force and needle gauge to allow a formulation to "fit" a device, but many more limitations arise with high-concentration formulations. Injection time, force, and associated pain are all patient-related factors, but additional aspects related to stability and material compatibility must also be addressed.

Finally, from a commercial standpoint, the logistics of therapeutic protein manufacturing, fill/finish operation and storage must be considered a global activity. Exposure of a therapeutic protein with inherent instabilities to manufacturing stresses during these processes can trigger instabilities. Maintaining consistency and the stability of the therapeutic protein throughout bioprocess, fill/finish operation, transportation, shelf life, and compatibility with the administration units will be extremely challenging for unstable molecules. Typically, a detailed understanding of the sensitivity of a protein to various unit operations is determined, and a formulation control is designed or developed accordingly to maintain stability of the protein during various process steps. In general, a stable molecule will have a larger optimization window, whereas a protein with several inherent instabilities will have a very narrow optimization window, thus restricting the development of a cost-effective and commercial-friendly process and product that can be implemented on a global basis.

6.3 Rationale for Early Developability Screening

As explained in Sect. 6.2.2, there is greater focus at the discovery stage on obtaining attributes such as higher affinity, enhanced safety, and increased half-life, which translates to highly complex, often novel molecules. From the development perspective, there is an increased demand to enhance product performance and patient comfort by use of a high-concentration liquid formulation in specialty devices, with an additional burden of decreasing the overall development timeline. Taken together, these demands translate into attempts to develop novel molecules in a challenging configuration and presentation under a reduced timeline. One approach that is gaining widespread acceptance across the industry to tackle the complex challenge of increasing development efficiency and reducing the attrition rate is the early identification of liabilities and other molecular properties important for smooth and efficient CMC development. This approach is termed *developability*. Conceptually, the ideal therapeutic protein is easily manufactured, expressed at a high titer, easily purified with acceptable purity at a given yield, amenable to formulation at high concentration, and capable of presentation in a format or medical device that offers the best for patient and clinical convenience. A judicious candidate selection process that incorporates both biological activity, affinity, and safety along with developability properties such as stability, high-concentration solution properties like viscosity and opalescence, and manufacturability is critical in selecting such an ideal protein. This approach can use either a screening strategy where high-risk molecules are flagged or incorporate some of the developability properties at the earlier design phase. In the small-molecule world, assessment of "drug-like" properties is an integral part of drug discovery. In fact, the pharmaceutical industry as such is evolving from a "gatekeeper or screening type" strategy to a "drug-like properties by design" approach (Hageman 2006).

Protein therapeutics are typically highly complex large molecules that can have several instabilities. The reduction or elimination of these instabilities requires simultaneous optimization of multiple parameters. While early screening of molecule properties may lead to early identification of developability risks and aid in streamlining manufacturing and resource allotment, it is not practical or efficient to address all of these liabilities at the early discovery stage, when both resources and material availability are often limited. Hence identification of the critical instabilities and evaluation and development of novel high-throughput biophysical methods or tools to screen for and reduce or eliminate such instabilities is a prerequisite for successful developability studies.

The frontloading of developability properties assessment requires combining computational and experimental screening approaches. Application of *in silico* analysis is an emerging area and, over the past years, significant advancements have been made, especially in the area of *in silico* protein aggregation prediction (Chennamsetty et al. 2009a, b; Lauer et al. 2012; Cellmer et al. 2007). Such a predictive tool can be used to identify and eliminate, or reengineer structural and sequence liabilities to generate proteins with improved stability and manufacturability (Fowler et al. 2005). Unfortunately most of the available *in silico* tools can only be applied to predict native aggregation behavior. For prediction of the most common nonnative aggregation, as well as additional chemical instabilities and solution properties at high concentration (e.g., phase separation and viscosity), the field still relies on experimental approaches. Apart from the issue with material and resource availability, the strategies and methods that can be used during an early developability assessment should target the screening and optimization of multiple parameters in an extremely dynamic environment. In short, early developability assessment methods should allow low cost and rapid exploration of multiple parameters with

Table 6.1 Summary of common protein instabilities

Instabilities		
1.	Physical	• Conformational stability Thermal denaturation Cold denaturation Pressure-induced denaturation Chemical denaturation • Colloidal stability • Surface/interfacial-induced denaturation
2.	Chemical	• Deamidation • Asp isomerization • Fragmentation • Glycation • Oxidation • Disulfide scrambling

the objective of quick and cheap screening. Also key is better predictive strategies, which should have the ability to predict the behavior of the protein during manufacturing and storage.

6.4 Typical Protein Liabilities

Macromolecules in general and proteins in particular are highly complex in their structure and have various instabilities. These protein instabilities can be broadly classified as physical or chemical instabilities (Table 6.1). Typical physical liabilities that should be screened for in an early assessment strategy are those intrinsic stability parameters that can be used to predict overall stability and aggregation propensity and undesirable solution properties, such as solubility, high viscosity, phase separation, and opalescence. Chemical instabilities that should be screened for include deamidation, fragmentation, and oxidation propensity (Chan and Carter 2010).

6.5 Use and Description of Biophysical Tools in Molecule Selection and Development

A typical early screen for a mAb might initially focus on a property, such as specificity or strength (affinity) of binding. Subsequent secondary screens and assays typically include specificity against related molecules to the target and binding to species homologues (to support preclinical safety studies). In this way a "screening cascade" of different assays is performed, reducing the number of potential antibody variants for subsequent development. As the number of individual variants is reduced, the amount and purity of sample is typically concomitantly increased. Ideally, screens for developability characteristics, such as solubility, expression, chemical stability, and reduced propensity for aggregation would be placed as early in the screening cascade as possible in order to maximize the potential sequence diversity available in a large population of variants. For example, laboratories have reported the use of modified phage-display selections to preferentially enrich for those variants that have increased thermal stability (Jespers et al. 2004; Famm et al. 2008). Taking this approach even further, antibody-fragment libraries have been designed to maximize stability, expression levels, and aggregation resistance (Prassler et al. 2011). High-throughput (HT) assays are ideally suited for performance early in the screening cascade, allowing biophysical properties to be preferentially selected for, alongside other properties, such as affinity and specificity.

There are a wide array of biophysical techniques and strategies available to determine protein instabilities (Table 6.2). Each of these techniques has inherent advantages and disadvantages with respect to its ability to detect a meaningful change in the structure of a protein and to translate this change into information that can be used to rank order, or exclude or include, a candidate in a preclinical development program. The sensitivity of the technique combined with the predictive nature of the analysis can help determine the suitability of a method for screening preclinical candidates. Often the techniques are employed in specific studies designed to elicit a behavior that can then be directly measured. The criteria used for identifying the ideal analytical methods and testing strategies are typically those associated with the phase of project development. For the screening of large-molecule libraries, methods that offer advantages in speed and throughput while using little material are preferred. At this point, the types of methodologies that should be employed become limited to those that are robust and sensitive towards specific behaviors that enable selection of a lead candidate for CMC development.

For the biophysical assessment of a large number of candidate molecules, HT approaches are preferable. In general, available techniques should support miniaturization, simple data visualization, and interpretation to help obtain more information in well-focused studies. The techniques that are most attractive are those that can robustly predict or measure specific molecular attributes that indicate if a candidate meets a defined developability criteria that are specific to its target indication or is within an acceptable developability design space either though mechanistic rules or phenomological correlation. Such a technique should allow for exploration of a larger number of instabilities or parameters more quickly with a lower cost. In general the methods should be flexible, miniature (consume minimal amounts of material), and employable in a HT mode such that a battery of different screens can be performed on the same sample simultaneously or in a highly parallel array.

Table 6.2 List of most commonly used biophysical tools

Instabilities	Method	High-throughput amenable
Conformational stability	Differential scanning calorimetry	Yes
	Fluorescence spectroscopy	Yes
	• Extrinsic	
	• Intrinsic	
	Absorption spectroscopy	Yes
	Circular dichroism	Yes
	Fourier transform infrared spectroscopy	No
	Raman spectroscopy	No
	NMR	No
	X-ray	No
	Dynamic light scattering	Yes
Colloidal stability	Light scattering	Yes
	• SLS	
	• DLS	
	Membrane osmometer	No
	Self-interaction chromatography	Yes
	Analytical ultracentrifugation	No
	• Sedimentation velocity	
	• Sedimentation equilibrium	
Interfacial stability	Tensiometry	No
	Extensional rheology	No
Chemical stability	RPHPLC	No
	Bioanalyzer	Yes
	LC/MS	No
Solution properties		
Viscosity	Rheometry	No
	DLS	Yes
Phase separation	Nephleometry	Yes
	Right angle light scattering	No
	Absorption spectroscopy (transmission)	No

6.6 Molecular Attributes and Instabilities that Should Be Screened Early

6.6.1 Aggregation

The propensity for a molecule to aggregate is one of the single most influential and challenging factors for formulation development. Candidates that are inherently resistant towards aggregation are considered superior. Nonnative aggregation is a multistep process and can be explained by the Lumry–Eyring framework. Each state has a different molecular configuration and energy, and each reaction step proceeds through an energy barrier (ΔG^*, difference in energy or activation free energy). The step that has highest ΔG^* is the rate-limiting step (Chi et al. 2003a). The first step in this process, which is a reversible process, gives rise to an aggregation-competent partially unfolded intermediate (I) state. In the second step, the partially unfolded intermediate self-associates to from irreversible higher-order aggregates (A). The final undesired A state can be controlled either by increasing conformational stability of the native (N) state (ΔG) which results in a decreased population of aggregation-competent partially unfolded intermediates (I state) and/or by increasing the colloidal stability of the partially unfolded I state such that self-association of this state is minimized (Chi et al. 2003b). During development, controlling aggregation is achieved by screening for stabilizing excipients and solution conditions to develop appropriate formulation controls. Alternatively, such stabilization can be achieved by either selecting or designing conformationally and/or colloidally stable candidates. Mutations that increase conformational stability increase the free energies of the unfolded and transition states relative to the native state. Increasing the free energy of transition state relative to native results in a larger ΔG^*, which shifts the equilibrium towards native state. The net result is a decreased population of aggregation-competent partially unfolded state.

Both conformational and colloidal stability have an indirect but significant effect on surface- and interface-mediated properties. For example, adsorption of protein to various surfaces and interfaces during process, fill/finish, and storage can not only lead to significant reduction in yield but quite often can also induce structural perturbation in a protein resulting in reduced binding and activity as well as aggregation and particle formation (Sethuraman and Belfort 2005). Similarly shear stress on a protein during process and fill/finish operation along with mixing of the hydrophobic air–liquid interface with the bulk solution is known to induce structural perturbations and subsequent aggregations (Bekard et al. 2011; Thomas and Geer 2011; Biddlecombe et al. 2007; Bee et al. 2009). Typically, both surface- and interface-mediated aggregation and particle formation is controlled during development either by avoiding the surfaces that can have deleterious effect on a particular protein or by designing and developing appropriate formulation controls such that the adsorption process is avoided or approximately controlled. Alternatively, such surface and interface effects can be reduced by increasing the conformational stability of a protein (Karlsson et al. 2005; Norde 1986; Wendorf et al. 2004). In a study with human carbonic anhydrase mutants, Karlsson et al. (2005) showed that adsorption and desorption characteristics can be controlled by a protein-engineering approach. An early investment to either design or select molecules with better conformational stability can potentially reduce some of the process and fill/finish related surface- and interface-mediated effects. Such an approach (stability by design) will not only result in a wider process and formulation development window but can also result in development of a platform process and fill/finish conditions.

Thermal stability using differential scanning calorimetry (DSC) is a widely acceptable measure and predictor of a protein's conformational stability and can be used to form a basis for screening candidate molecules under varying buffer conditions (Remmele et al. 1998). New techniques, strategies, and instrumentation are now available that provide thermal stability in a fraction of the time typically associated with DSC and use very low amounts of material (He et al. 2010; Goldberg et al. 2011). These methodologies include spectroscopy-based techniques coupled with thermal ramping (reviewed in Chap. 2). These techniques can provide throughput on the order of 96–384 samples per hour and typically only require 10s of micrograms. Aggregation and particle formation can also be directly monitored by techniques such as dynamic light scattering (DLS), HIAC light obscuration, microflow imaging (MFI), and electrical charge-based methods (e.g., Coulter counter). Of these methods, DLS offers the greatest flexibility for early-stage use based on throughput (plate-based DLS) and material required. However, recent advances have resulted in improvements in particle detection including scanning ion occlusion spectroscopy (qNano), chip-based mass measurements (Archimedes), and nanoparticle tracking analysis (NanoSight). Each of these methods offers the possibility of increased sensitivity and throughput with a decrease in sample consumption, resulting in more screening possibilities.

6.6.2 Chemical Stability

As shown in Table 6.1, proteins have multiple liabilities that are prone to chemical degradation during manufacture and storage. The rate of degradation depends on both the solution condition and the structure of the molecule (Sinha et al. 2009; Vlasak and Ionescu 2011). Typically some of the most common liabilities like deamidation, nonenzymatic hydrolysis, and oxidation are highlighted very early and easily by *in silico* primary sequence liability analysis. Theoretically, molecules with the fewest chemical liabilities will allow one to develop a wider process and formulation design space. For molecules with identified liabilities, there will be an additional burden of determining the effect of the liabilities on product quality and designing appropriate process and formulation controls. Unlike physical stability studies, the slow rates of reaction associated with accessing chemical stability results in time-consuming forced degradation studies. For example, evaluation of the effect of deamidation on function and stability requires forced degradation studies at elevated temperature and pH, followed by liquid chromatography–mass spectrometry (LCMS) and subsequent selected ion monitoring (SIM). Similarly fragmentation propensity can be determined either using size-based separation techniques or side-chain chemistry-based analytical methods like reversed-phase high-performance liquid chromatography (RPHPLC). Even though HT analytics are available to follow fragmentation, the slow rate of reaction of forced degradation studies that can accurately predict nonenzymatic fragmentation under storage conditions typically requires time-consuming incubation studies in the range of 4–6 weeks.

6.6.3 Solution Properties at High Protein Concentrations

Other important developability criteria that are often neglected during early screening are the solution properties of a molecule at high concentration. Solution properties such as solubility, phase separation, and viscosity can often become the rate-limiting step during processing, filtration, drug delivery, etc., as solution properties of a molecule are governed by various attractive and repulsive nonspecific intermolecular interactions [reviewed in Saluja and Kalonia (2008) and Tadros (2011)]. Under concentrated solution conditions, the intermolecular distance becomes less than the molecule's diameter. This reduction in intermolecular distance will not only increase the collision frequency but will also increase the duration of such an encounter (Scherer et al. 2010; Minton 2005). Thus at high concentrations, the nature of, and relative contributions from, various nonspecific interactions is dictated by both solution conditions and the surface properties of a molecule, including net charge, charge distribution, and surface hydrophobicity. From a development perspective, the consequences of such unfavorable interactions are a change in the protein's solution behavior potentially resulting in issues such as poor solubility, phase separation, and/or increased viscosity (Saluja and Kalonia 2008; Chari et al. 2009). Typically, such unfavorable interactions can be controlled during development by modulating solution conditions like ionic strength and/or pH to develop appropriate formulation controls (Salinas et al. 2010; Yadav et al. 2012; Liu et al. 2005), but in some cases, depending on strength of intermolecular interactions, it may not be possible to achieve this control. Quite often the solution conditions might have an unintended opposite effect on various solution properties and overall stability (Salinas et al. 2010).

The rheological properties of a molecule under a given buffer system have typically been ascertained using a cone- and plate-type instrument to determine viscosity. The dimensions of the cone and plate instruments have been dramatically reduced over the years, resulting in a correlating decrease in the volume of material required. However, the time associated with these measurements has not fallen to the same degree. New techniques such as viscosity–DLS (see He et al., Chap. 2, this volume) and chip-based measurements provide an attractive alternative to the traditional cone and plate approach. Alternatively, solution properties can also be predicted by accurate measurements of net protein–protein interactions through colloidal stability (Yadav et al. 2012; George and Wilson 1994; Mehta et al. 2012). The osmotic second virial coefficient (B_{22}), which is a thermodynamic parameter, is often used to directly quantify overall protein–protein interactions (Chi et al. 2003a, b; Valente et al. 2005; Payne et al. 2006). A positive B_{22} value indicates an overall dominance of repulsive interactions, where as a negative B_{22} value indicates an overall attractive intermolecular interactions. The B_{22} of a colloidally active system can be accurately measured by various techniques like DLS (Yadav et al. 2011), static light scattering (SLS), analytical ultracentrifugation (AUC), osmometry, and self-interaction chromatography (SIC) (Le Brun et al. 2010).

One of the main reasons for limited screening of these undesirable solution properties during early development is the limited availability of HT techniques. Most of the available techniques mentioned above are low throughput in nature and also require large amounts of material. But recently significant progress has been made in this area, and several HT methods are available to either directly measure some of the solution properties (reviewed in Chap. 2) or accurately predict solution properties (Johnson et al. 2009; Bajaj et al. 2007; Sule et al. 2011; Tessier et al. 2008). Availability of such HT techniques allows one to screen and select candidate molecules with desired surface properties or interaction potential, which is a molecule that has a B_{22} value close to 0 under the desired solution conditions. Applying an approach of selecting or engineering a molecule with desired surface properties, or interaction potential, will enable the formulation scientist to focus mostly on aggregation or other chemical degradations, thus widening the formulation development design space.

One consequence of such an approach is that there is a potential for conflict between the different parameters that are being simultaneously screened. For example, if antibody affinity is found to correlate with an unfavorable biophysical characteristic, such as aggregation propensity, careful consideration would need to be given as to which characteristic would be most economic to engineer/mitigate against the aggregation propensity. Examples of candidate therapeutic antibodies that have been extensively reengineered to improve biophysical properties and as such improve developability have begun to be described (Pepinsky et al. 2010).

6.7 Case Studies

6.7.1 Case Study 1

As stated in Sect. 6.2, for molecules with a precedented mechanism of action, the focus is on enhanced safety, higher affinity to target, and increased half-life. The efficacy and potency of a molecule can be fine-tuned by efficient mediation of effector functions (Anderson et al. 1997; Weng and Levy 2003). This case study demonstrates how engineering of Fc variants to modulate these activities can have an unintentional effect on a molecule's developability and how this unintentional effect can be screened for and detected early by applying appropriate biophysical tools. In this case a triple mutant S239D/A330L/I332E (resulting in mAbA-3M) was introduced into the CH_2 domain of a human mAb (mAbA). Compared to its wild type, these mutations resulted in an approximately 10–100 fold increase in mAbA-3M binding to its receptor, but at a price of reduced stability (Fig. 6.1) (Oganesyan et al. 2008; Anandakumar et al. 2008). Subsequent developability-accelerated stability studies indicated a significant increase in the aggregation rate at 40 °C (Fig. 6.2). Based on various biophysical studies like DSC and denaturant unfolding, it was speculated that the reduced conformational stability was due to minor structural perturbations in the region around the CH_2 domain. By solving the structure of the mutant, Oganesyan et al. determined that the mutations resulted in opening of the cleft in the CH_2 domain of the Fc fragment (Oganesyan et al. 2008). They concluded that the enhanced interactions with human Fcγ(gamma)RIIIA could be due to enhanced Fc openness as well as the introduction of hydrophobic and electrostatic interactions at the corresponding interface (Oganesyan et al. 2008). Even though

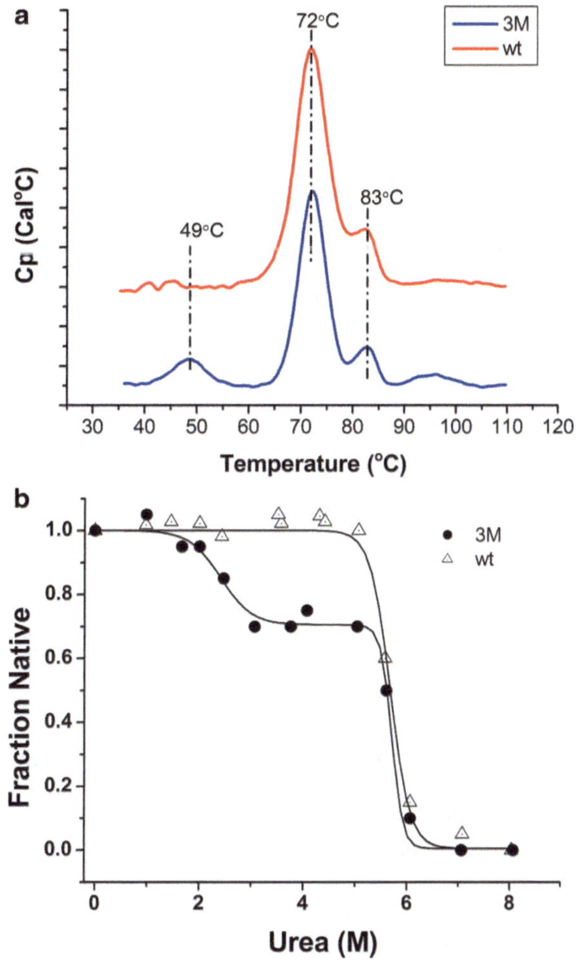

Fig. 6.1 Comparison of conformational stability of mAbA and its 3M mutant (**a**) Overlay of the DSC thermogram; figure adapted from Oganesyan et al. (2008) (**b**) Overlay of urea-induced unfolding obtained by monitoring changes in intrinsic tryptophan fluorescence of the protein at a concentration of 1 mg/mL

developability studies flagged the mAbA-3M as having a problematic aggregation potential, it was selected as the lead candidate because of its increased potency. Investigators were able to move mAbA-3M forward because appropriate formulation controls and resources were allotted based on the information obtained during developability studies.

6.7.2 Case Study 2

Case Study 2 examines three protein candidates to ascertain if any have potential liabilities with respect to manufacturing or formulation. Three IgG2 monoclonal

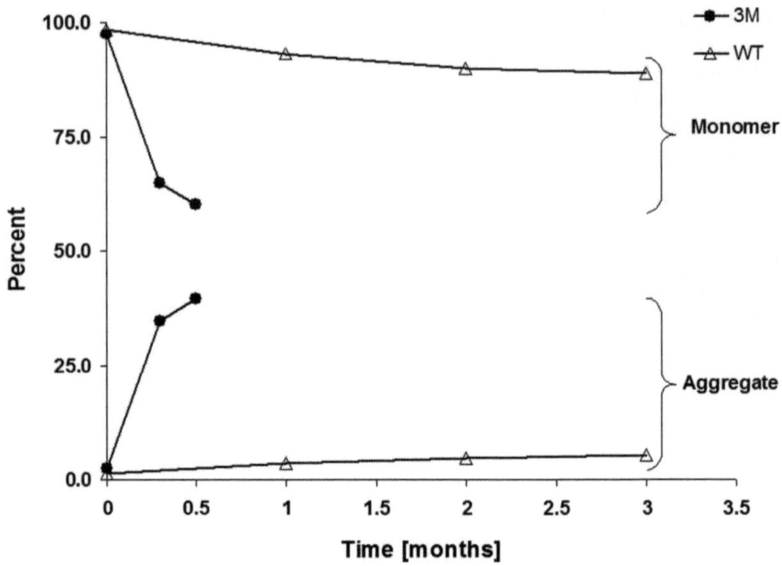

Fig. 6.2 Comparison of stability of mAbA and its 3M mutant obtained under accelerated conditions. Stability of both proteins was monitored by incubating the protein at 50 mg/mL at 40 °C and monitoring the changes in purity and aggregation by HPSEC

antibody (mAb) candidates against the same target, molecule H, molecule J, and molecule K, were provided for assessment. Using biophysical and accelerated incubation studies, these molecules were screened to see if they could adhere to a formulation and processing platform.

6.7.2.1 Formulation Stability

DSC was used to monitor the thermal stability of the candidates in the formulation at 1 mg/mL. At pH 5.2, all the candidates exhibited an apparent Fab Tm between 77 °C and 80 °C. The observation of relatively higher Fab Tm value indicated lower stability/developability risk. This was further confirmed by an incubation study performed at 70 mg/mL mAb concentration in an acetate pH 5.2 buffer. Samples were incubated in the buffer and stored at 4 °C, 25 °C, or 37 °C for 2–4 weeks. For samples stored at 37 °C, all candidates exhibited similar increases in HMW (high molecular weight) species (between 0.3% and 0.4% increase from $t=0$) over a period of 4 weeks. As expected, no significant change in the SE-HPLC % main peak (MP) was observed for any of the three candidates upon storage at 4 °C for 4 weeks or at 25 °C for 2 weeks. To simulate the pH jump upon physiological administration, candidates were subjected to pH change from pH 5.2 to ~7.0 and held for 24 h at 37 °C. An increase in % HMW species was noted for all the candidates after 24 h incubation at pH 7 at 37 °C as detected by SE-HPLC: for molecule *H* (0.8%) followed by molecule *J* (0.68%) and molecule *K* (0.67%)

6.7.2.2 Process Stability

To test the compatibility with the possible solution conditions encountered during purification, the candidates were exposed to representative buffers at pH values between 3.0 and 7.4. Potential changes in thermal stability, higher-order structure, and resulting degradation were monitored by applying various biophysical tools and SE-HPLC analysis. At pH 7.4 as expected the DSC thermogram showed three endotherms corresponding to the unfolding of the CH_2, Fab, and CH_3 domains. However, at pH 3.0, only a single endotherm was observed for all candidates indicating either increased inter-domain interactions (cooperative unfolding) or unfolding of some of the domains (even at temperatures below the starting point of the DSC analysis) at low pH. The conformation of molecule *K* appeared to be the most stable, as indicated by the higher T_m of the Fab domain at pH 3.5 and 5.0, followed by molecule J and molecule H. When reversed into PBS by dialysis, no significant irreversible change was observed for any of the three candidates, and the DSC thermograms were superimposable.

The tertiary structure of the candidates was studied by near-UV CD. The spectra of the three candidates exhibited similar features, with varying intensities of mean residue ellipticity (MRE). The spectra for all candidates at pH 3.5 and 5.0 were similar suggesting that no conformational changes occurred at pH 3.5. At pH 3.0, the near-UV CD spectrum of all candidates showed a marked loss in MRE that indicated the protein molecules had lost a significant amount of tertiary structure at this pH. Most of the loss in the near-UV CD signal occurred in the region between 250 nm and 280 nm suggesting a marked change in the disulfide environment at pH 3.0. The near-UV CD spectra of the reversibility samples in PBS overlay very well with the protein in pH 5 reversed into PBS at pH 7.2 indicating that there were no irreversible changes in the tertiary structure of any of the candidates induced by exposure to lower pH. The effect of low pH on the protein conformation and its reversibility were further evaluated by ANS-binding studies, which revealed an increase in protein surface hydrophobicity with a decrease in pH. Upon reversing the solution condition to PBS, all candidates exhibited full reversibility in surface hydrophobicity.

Implications of low pH-induced structural perturbation on candidate stability were evaluated by incubating the candidates at different pH. None of the candidates exhibited a decrease in % main peak by SE-HPLC after 6 h of exposure to actual solution pH values (pH 3.7, 5.0 and 7.4). Whereas, when exposed to pH 3.0, all candidates exhibited aggregation and a loss in % main peak determined by SE-HPLC. Molecule K exhibited the most aggregation (84%) followed by molecule H (78%) and molecule J (77%). The results obtained by SEC were further supported by DLS studies, which showed the presence of HMW species at pH 3.0 for all three candidates. At pH 3.5 and 5.0, the size distribution was superimposable and the hydrodynamic diameter (d_H) was close to that of a monomer (approximately

10–11 nm), suggesting a more monomer-like species at higher solution pH. The polydispersity index (PdI, an indicator of size heterogeneity in the sample with lower values <0.1 suggesting a homogeneous population) also showed a decreasing trend from pH 3.0 to 5.0. For samples reversed into PBS, a decrease in size distribution was not observed in the pH 3.0 samples indicating that some irreversible changes were produced. However, full reversibility was observed for samples exposed to pH values of 3.5 and 5 and then dialyzed into PBS, consistent with observations made from the other biophysical techniques tested during this study. This result was expected since the samples at pH 3.5 and 5.0 exhibited a homogeneous size distribution.

Biophysical and stability data obtained during early developability screening indicated all three candidates were low-risk molecules from a developability/manufacturability perspective and adhered to the formulation and processing platforms.

6.8 Conclusion

Recent advances in the understanding of protein structure, stability, and function properties combined with advances in equipment technology are enabling more comprehensive and accurate developability and manufacturability screening during early research stages. This early screening is starting to result in the selection of molecules more closely fitting the desired target product profile (i.e., CMC properties that maintain desired biological properties), with the promise of reduced resources and faster progression through the development phase. This process does not entirely eliminate the risk associated with taking a selected candidate forward (as biological properties are typically of higher priority than CMC properties); apart from minimizing the risk of taking forward a candidate that is difficult to develop, such an approach of selecting low-risk molecule can also result in enhanced product quality and increase end-user convenience.

Depending upon the target product profile, screening tools for CMC properties should be selected and used based on the value added from a cost-benefit and risk-benefit analysis. In an increasingly competitive global landscape, a wiser and more focused application of novel, cutting-edge biophysical technologies can provide sustainable benefits, such as early identification of developability fingerprints to allot appropriate resources and streamlining of development and manufacturing activities. Realizing these benefits can result in a significant increase in overall R&D output and productivity. Similar to the strategy employed in the small-molecule world, the biotech industry, in general, is starting a gatekeeper approach by making the screening of developability properties an integral part of preclinical drug discovery. With a better understanding of structure, stability, and activity relationships and successful integration of *in silico* and high-throughput experimental tools, the industry should evolve from a "gatekeeper" (or screening) strategy to "a stability by design" approach.

Acknowledgements Authors wish to thank Nancy Craighead at MedImmune for critical review of this chapter.

References

Ackerman MJ, Clapham DE (1997) Ion channels–basic science and clinical disease. N Engl J Med 336(22):1575–1586

Adams CP, Brantner VV (2010) Spending on new drug development1. Health Econ 19(2):130–141

Anandakumar R, Harn NR, Sathish HA, LeachWL, Oliver CN, Bishop SM (2008) American Association of Pharmaceutical Scientist, National Biotech Conference, 2008

Anderson DR, Grillo-Lopez A, Varns C, Chambers KS, Hanna N (1997) Targeted anti-cancer therapy using rituximab, a chimaeric anti-CD20 antibody (IDEC-C2B8) in the treatment of non-Hodgkin's B-cell lymphoma. Biochem Soc Trans 25(2):705–708

Bajaj H, Sharma VK, Kalonia DS (2007) A high-throughput method for detection of protein self-association and second virial coefficient using size-exclusion chromatography through simultaneous measurement of concentration and scattered light intensity. Pharm Res 24(11):2071–2083

Batista FD, Neuberger MS (1998) Affinity dependence of the B cell response to antigen: a threshold, a ceiling, and the importance of off-rate. Immunity 8(6):751–759

Bee JS, Stevenson JL, Mehta B, Svitel J, Pollastrini J, Platz R, Freund E, Carpenter JF, Randolph TW (2009) Response of a concentrated monoclonal antibody formulation to high shear. Biotechnol Bioeng 103(5):936–943

Bekard IB, Asimakis P, Bertolini J, Dunstan DE (2011) The effects of shear flow on protein structure and function. Biopolymers 95(11):733–745

Biddlecombe JG, Craig AV, Zhang H, Uddin S, Mulot S, Fish BC, Bracewell DG (2007) Determining antibody stability: creation of solid–liquid interfacial effects within a high shear environment. Biotechnol Prog 23(5):1218–1222

Boder ET, Midelfort KS, Wittrup KD (2000) Directed evolution of antibody fragments with monovalent femtomolar antigen-binding affinity. Proc Natl Acad Sci USA 97(20):10701–10705

Carter PJ (2011) Introduction to current and future protein therapeutics: a protein engineering perspective. Exp Cell Res 317(9):1261–1269

Cellmer T, Bratko D, Prausnitz JM, Blanch HW (2007) Protein aggregation in silico. Trends Biotechnol 25(6):254–261

Chan AC, Carter PJ (2010) Therapeutic antibodies for autoimmunity and inflammation. Nat Rev Immunol 10(5):301–316

Chari R, Jerath K, Badkar AV, Kalonia DS (2009) Long- and short-range electrostatic interactions affect the rheology of highly concentrated antibody solutions. Pharm Res 26(12):2607–2618

Chennamsetty N, Voynov V, Kayser V, Helk B, Trout BL (2009a) Design of therapeutic proteins with enhanced stability. Proc Natl Acad Sci USA 106(29):11937–11942

Chennamsetty N, Helk B, Voynov V, Kayser V, Trout BL (2009b) Aggregation-prone motifs in human immunoglobulin G. J Mol Biol 391(2):404–413

Chi EY, Krishnan S, Randolph TW, Carpenter JF (2003a) Physical stability of proteins in aqueous solution: mechanism and driving forces in nonnative protein aggregation. Pharm Res 20(9):1325–1336

Chi EY, Krishnan S, Kendrick BS, Chang BS, Carpenter JF, Randolph TW (2003b) Roles of conformational stability and colloidal stability in the aggregation of recombinant human granulocyte colony-stimulating factor. Protein Sci 12(5):903–913

Dall'Acqua WF, Cook KE, Damschroder MM, Woods RM, Wu H (2006a) Modulation of the effector functions of a human IgG1 through engineering of its hinge region. J Immunol 177(2):1129–1138

Dall'Acqua WF, Kiener PA, Wu H (2006b) Properties of human IgG1s engineered for enhanced binding to the neonatal Fc receptor (FcRn). J Biol Chem 281(33):23514–23524

DiMasi JA, Grabowski HG (2007) The cost of biopharmaceutical R&D: is biotech different? Manag Decis Econ 28(4–5):469–479

Famm K, Hansen L, Christ D, Winter G (2008) Thermodynamically stable aggregation-resistant antibody domains through directed evolution. J Mol Biol 376(4):926–931

Finch DK, Sleeman MA, Moisan J, Ferraro F, Botterell S, Campbell J, Cochrane D, Cruwys S, England E, Lane S, Rendall E, Sinha M, Walker C, Rees G, Bowen MA, Schneider A, Liang M, Faggioni R, Fung M, Mallinder PR, Wilkinson T, Kolbeck R, Vaughan T, Lowe DC (2011) Whole-molecule antibody engineering: generation of a high-affinity anti-IL-6 antibody with extended pharmacokinetics. J Mol Biol 411(4):791–807

Foote J, Eisen HN (1995) Kinetic and affinity limits on antibodies produced during immune responses. Proc Natl Acad Sci USA 92(5):1254–1256

Fowler SB, Poon S, Muff R, Chiti F, Dobson CM, Zurdo J (2005) Rational design of aggregation-resistant bioactive peptides: reengineering human calcitonin. Proc Natl Acad Sci USA 102(29):10105–10110

Gebauer M, Skerra A (2009) Engineered protein scaffolds as next-generation antibody therapeutics. Curr Opin Chem Biol 13(3):245–255

George A, Wilson WW (1994) Predicting protein crystallization from a dilute solution property. Acta Crystallogr D Biol Crystallogr 50(Pt 4):361–365

Goldberg DS, Bishop SM, Shah AU, Sathish HA (2011) Formulation development of therapeutic monoclonal antibodies using high-throughput fluorescence and static light scattering techniques: role of conformational and colloidal stability. J Pharm Sci 100:1306–1315

Hageman MJ (2006) Solubility, solubilization and dissolution in drug delivery during lead optimization. Springer, New York

He F, Becker GW, Litowski JR, Narhi LO, Brems DN, Razinkov VI (2010) High-throughput dynamic light scattering method for measuring viscosity of concentrated protein solutions. Anal Biochem 399(1):141–143

Igawa T, Tsunoda H, Kuramochi T, Sampei Z, Ishii S, Hattori K (2011) Engineering the variable region of therapeutic IgG antibodies. MAbs 3(3):243–252

Jespers L, Schon O, Famm K, Winter G (2004) Aggregation-resistant domain antibodies selected on phage by heat denaturation. Nat Biotechnol 22(9):1161–1165

Johnson DH, Parupudi A, Wilson WW, DeLucas LJ (2009) High-throughput self-interaction chromatography: applications in protein formulation prediction. Pharm Res 26(2):296–305

Karlsson M, Ekeroth J, Elwing H, Carlsson U (2005) Reduction of irreversible protein adsorption on solid surfaces by protein engineering for increased stability. J Biol Chem 280(27):25558–25564

Kayser V, Chennamsetty N, Voynov V, Forrer K, Helk B, Trout BL (2011) Glycosylation influences on the aggregation propensity of therapeutic monoclonal antibodies. Biotechnol J 6(1):38–44

Kola I, Landis J (2004) Can the pharmaceutical industry reduce attrition rates? Nat Rev Drug Discov 3(8):711–715

Krapp S, Mimura Y, Jefferis R, Huber R, Sondermann P (2003) Structural analysis of human IgG-Fc glycoforms reveals a correlation between glycosylation and structural integrity. J Mol Biol 325(5):979–989

Kubota T, Niwa R, Satoh M, Akinaga S, Shitara K, Hanai N (2009) Engineered therapeutic antibodies with improved effector functions. Cancer Sci 100(9):1566–1572

Lauer TM, Agrawal NJ, Chennamsetty N, Egodage K, Helk B, Trout BL (2012) Developability index: a rapid in silico tool for the screening of antibody aggregation propensity. J Pharm Sci 101(1):102–115

Le Brun V, Friess W, Bassarab S, Muhlau S, Garidel P (2010) A critical evaluation of self-interaction chromatography as a predictive tool for the assessment of protein-protein interactions in protein formulation development: a case study of a therapeutic monoclonal antibody. Eur J Pharm Biopharm 75(1):16–25

Liu J, Nguyen MD, Andya JD, Shire SJ (2005) Reversible self-association increases the viscosity of a concentrated monoclonal antibody in aqueous solution. J Pharm Sci 94(9):1928–1940
Lundstrom K (2005) Structural genomics of GPCRs. Trends Biotechnol 23(2):103–108
Ma P, Zemmel R (2002) Value of novelty? Nat Rev Drug Discov 1(8):571–572
Mehta CM, White ET, Litster JD (2012) Correlation of second virial coefficient with solubility for proteins in salt solutions. Biotechnol Prog 28(1):163–170
Mimura Y, Church S, Ghirlando R, Ashton PR, Dong S, Goodall M, Lund J, Jefferis R (2000) The influence of glycosylation on the thermal stability and effector function expression of human IgG1-Fc: properties of a series of truncated glycoforms. Mol Immunol 37(12–13):697–706
Minton AP (2005) Influence of macromolecular crowding upon the stability and state of association of proteins: predictions and observations. J Pharm Sci 94(8):1668–1675
Naylor J, Beech DJ (2009) Extracellular ion channel inhibitor antibodies. Open Drug Discov J 1:36–42
Nelson AL, Dhimolea E, Reichert JM (2010) Development trends for human monoclonal antibody therapeutics. Nat Rev Drug Discov 9(10):767–774
Norde W (1986) Adsorption of proteins from solution at the solid–liquid interface. Adv Colloid Interface Sci 25(4):267–340
Oganesyan V, Damschroder MM, Leach W, Wu H, Dall'Acqua WF (2008) Structural characterization of a mutated, ADCC-enhanced human Fc fragment. Mol Immunol 45(7):1872–1882
Payne RW, Nayar R, Tarantino R, Del Terzo S, Moschera J, Di J, Heilman D, Bray B, Manning MC, Henry CS (2006) Second virial coefficient determination of a therapeutic peptide by self-interaction chromatography. Biopolymers 84(5):527–533
Pepinsky RB, Silvian L, Berkowitz SA, Farrington G, Lugovskoy A, Walus L, Eldredge J, Capili A, Mi S, Graff C, Garber E (2010) Improving the solubility of anti-LINGO-1 monoclonal antibody Li33 by isotype switching and targeted mutagenesis. Protein Sci 19(5):954–966
Prassler J, Thiel S, Pracht C, Polzer A, Peters S, Bauer M, Norenberg S, Stark Y, Kolln J, Popp A, Urlinger S, Enzelberger M (2011) HuCAL PLATINUM, a synthetic Fab library optimized for sequence diversity and superior performance in mammalian expression systems. J Mol Biol 413(1):261–278
Presta LG (2008) Molecular engineering and design of therapeutic antibodies. Curr Opin Immunol 20(4):460–470
Remmele RL Jr, Nightlinger NS, Srinivasan S, Gombotz WR (1998) Interleukin-1 receptor (IL-1R) liquid formulation development using differential scanning calorimetry. Pharm Res 15(2):200–208
Salinas BA, Sathish HA, Bishop SM, Harn N, Carpenter JF, Randolph TW (2010) Understanding and modulating opalescence and viscosity in a monoclonal antibody formulation. J Pharm Sci 99(1):82–93
Saluja A, Kalonia DS (2008) Nature and consequences of protein-protein interactions in high protein concentration solutions. Int J Pharm 358(1–2):1–15
Scherer TM, Liu J, Shire SJ, Minton AP (2010) Intermolecular interactions of IgG1 monoclonal antibodies at high concentrations characterized by light scattering. J Phys Chem B 114(40):12948–12957
Sethuraman A, Belfort G (2005) Protein structural perturbation and aggregation on homogeneous surfaces. Biophys J 88(2):1322–1333
Sinha S, Zhang L, Duan S, Williams TD, Vlasak J, Ionescu R, Topp EM (2009) Effect of protein structure on deamidation rate in the Fc fragment of an IgG1 monoclonal antibody. Protein Sci 18(8):1573–1584
Stewart R, Thom G, Levens M, Guler-Gane G, Holgate R, Rudd PM, Webster C, Jermutus L, Lund J (2011) A variant human IgG1-Fc mediates improved ADCC. Protein Eng Des Sel 24(9):671–678
Sule SV, Sukumar M, Weiss WF, Marcelino-Cruz AM, Sample T, Tessier PM (2011) High-throughput analysis of concentration-dependent antibody self-association. Biophys J 101(7):1749–1757

Tadros T (2011) Interparticle interactions in concentrated suspensions and their bulk (rheological) properties. Adv Colloid Interface Sci 168(1–2):263–277

Tessier PM, Jinkoji J, Cheng YC, Prentice JL, Lenhoff AM (2008) Self-interaction nanoparticle spectroscopy: a nanoparticle-based protein interaction assay. J Am Chem Soc 130(10): 3106–3112

Thomas CR, Geer D (2011) Effects of shear on proteins in solution. Biotechnol Lett 33(3):443–456

Valente JJ, Payne RW, Manning MC, Wilson WW, Henry CS (2005) Colloidal behavior of proteins: effects of the second virial coefficient on solubility, crystallization and aggregation of proteins in aqueous solution. Curr Pharm Biotechnol 6(6):427–436

Vlasak J, Ionescu R (2011) Fragmentation of monoclonal antibodies. MAbs 3(3):253–263

Wendorf JR, Radke CJ, Blanch HW (2004) Reduced protein adsorption at solid interfaces by sugar excipients. Biotechnol Bioeng 87(5):565–573

Weng WK, Levy R (2003) Two immunoglobulin G fragment C receptor polymorphisms independently predict response to rituximab in patients with follicular lymphoma. J Clin Oncol 21(21):3940–3947

Yadav S, Scherer TM, Shire SJ, Kalonia DS (2011) Use of dynamic light scattering to determine second virial coefficient in a semidilute concentration regime. Anal Biochem 411(2):292–296

Yadav S, Shire SJ, Kalonia DS (2012) Viscosity behavior of high-concentration monoclonal antibody solutions: correlation with interaction parameter and electroviscous effects. J Pharm Sci 101(3):998–1011

Chapter 7
Application of Biophysics in Formulation, Process, and Product Characterization: Selected Case Studies

Satish K. Singh, Qin Zou, Min Huang, and Muralidhara Bilikallahalli

7.1 Introduction

Biophysical techniques are essential tools in the development and characterization of biotherapeutics. The information gleaned from these techniques about the (higher-order) structure of the molecule provide insight into its behavior as process and formulation conditions are changed, thus helping to identify the optimum conditions for the molecule. Earlier chapters in this book have described the most commonly used techniques such as circular dichroism (CD), fluorescence, FTIR spectroscopy, analytical ultracentrifugation (AUC), and asymmetric flow field-flow fractionation (AFFFF). In this chapter, we illustrate the utilization of these and some other novel (variations of) techniques in various aspects of biotherapeutics development.

7.2 Case Study 1: Biophysical Analysis for Comparability Exercise

The program for development of a biotherapeutic from discovery to commercialization, more often than not, involves changes to various parts of the manufacturing process and/or formulation. Changes to the process and product are also made post approval. In order to bridge quality, safety, and efficacy between product

S.K. Singh (✉) • Q. Zou • M. Bilikallahalli
Pfizer Inc., 700 Chesterfield Parkway West, Chesterfield, MO 63017, USA
e-mail: satish.singh@pfizer.com; qin.zou@pfizer.com

M. Huang
Pfizer Inc., One Burtt Road, Andover, MA 01810, USA
e-mail: min.huang@pfizer.com

L.O. Narhi (ed.), *Biophysics for Therapeutic Protein Development*, Biophysics for the Life Sciences 4, DOI 10.1007/978-1-4614-4316-2_7,

manufactured before and after such changes, comparability studies are performed (ICH Q5E 2005). In order to show that a product is comparable before and after the changes, they are required to have a high degree of similarity in their physicochemical and biological properties. Biophysical characterization is an important component of comparability studies to demonstrate similarity in higher-order structure of the product. For this purpose, low-resolution structural information is obtained through spectroscopic techniques such as far- and near-UV circular dichroism (CD) spectroscopy, fluorescence spectroscopy, and Fourier-transform infrared (FTIR) spectroscopy. These methods provide a signature for the secondary and tertiary structure of the molecule. Hydrodynamic analyses such as light-scattering spectroscopy, AUC, and AFFFF can provide information about quaternary structure. The case study below on an IgG2 antibody is presented wherein product characterization by various biophysical methods was used to explain the observed differences before and after a process-scale and formulation change and to establish criteria for comparability.

7.2.1 Methods

Far- and near-UV CD spectra were collected on two drug substance lots with pilot-scale formulation (Lots 1 and 2) and eight other lots with full-scale formulation (Lots 3–10). For heat denaturation experiments, CD spectra were taken every 5°C from 25 to 75°C. Three separate denaturation experiments were done for the assessment of experimental uncertainty. All spectra were corrected with matched buffer blanks. The spectral similarity was assessed by Similarity Match analysis in TQ Analyst (Thermo Scientific). The DSC thermogram was generated at a scan rate of 1°C/min, and the molar excess heat capacity was calculated and deconvoluted to a non-two-state model. Tryptophan fluorescence spectra were collected using an excitation wavelength of 290 nm. All spectra were corrected with matched buffer blank. For the Analytical ultracentrifugation sedimentation velocity (AUC-SV) experiments, the samples were diluted to 0.5 mg/mL with the matched buffer and spun at 45,000 rpm. The absorbance data at 280 nm was collected and analyzed using Sedfit (Schuck 2000). For details on the methods used, including the underlying theory, please see Chap. 3 in this volume.

7.2.2 Results

Simple visual examination of both the far- and near-UV CD spectra in Fig. 7.1 shows differences between the pilot-scale and full-scale materials. Differences in far-UV CD spectra (Fig. 7.1a) are mainly at 215–218 nm, suggesting alteration in β (beta)-sheet content. The near-UV CD spectra (Fig. 7.1a) differ largely at around 290 nm, mainly attributable to the change in chiral environment around tryptophan

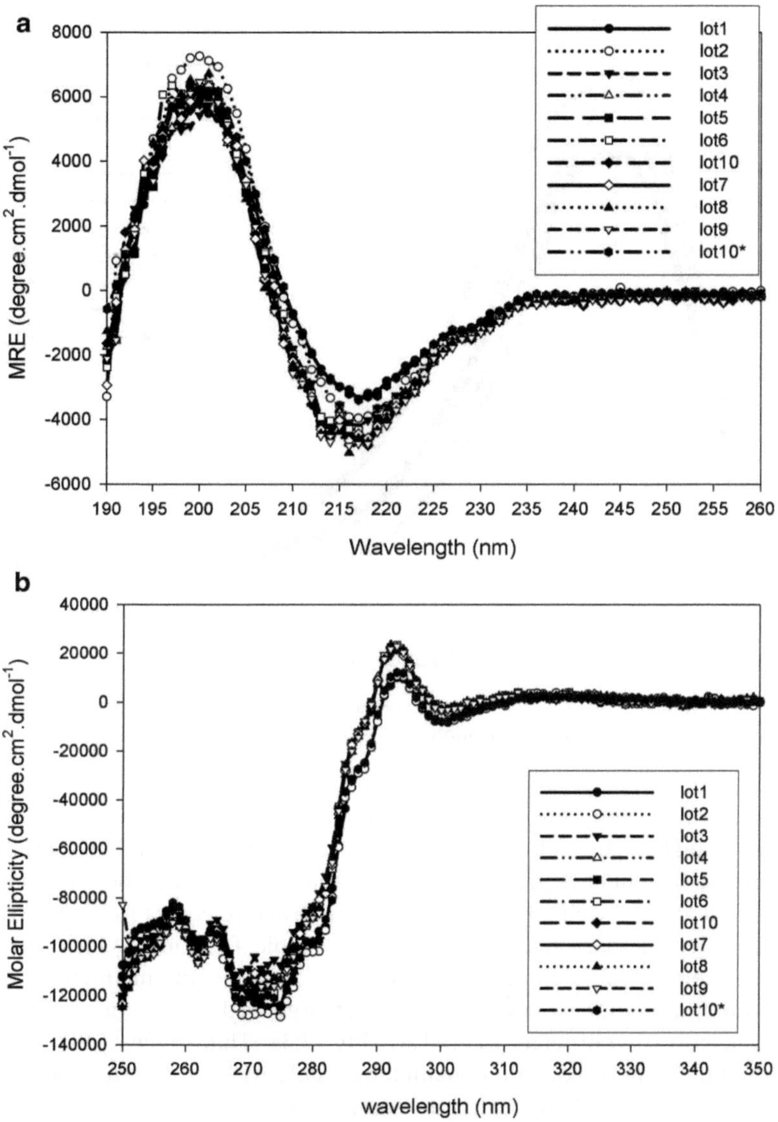

Fig. 7.1 Far-UV (**a**) and near-UV (**b**) CD spectra of pilot-scale and full-scale drug substance lots in clinical and commercial formulations, respectively. Lots 1 and 2 are pilot-scale lots; Lots 3–10 are full-scale lots. Lots 10 and 10* are the reference material in full-scale/commercial and pilot-scale/clinical formulation, respectively (see Table 7.1)

residues (Strickland 1974; Kelly and Price 2000). Intrinsic tryptophan fluorescence spectroscopy analysis also shows difference between pilot-scale and full-scale materials (Fig. 7.2), which is consistent with the difference seen in the near-UV CD spectra. However, the change in fluorescence is not easily interpreted. The

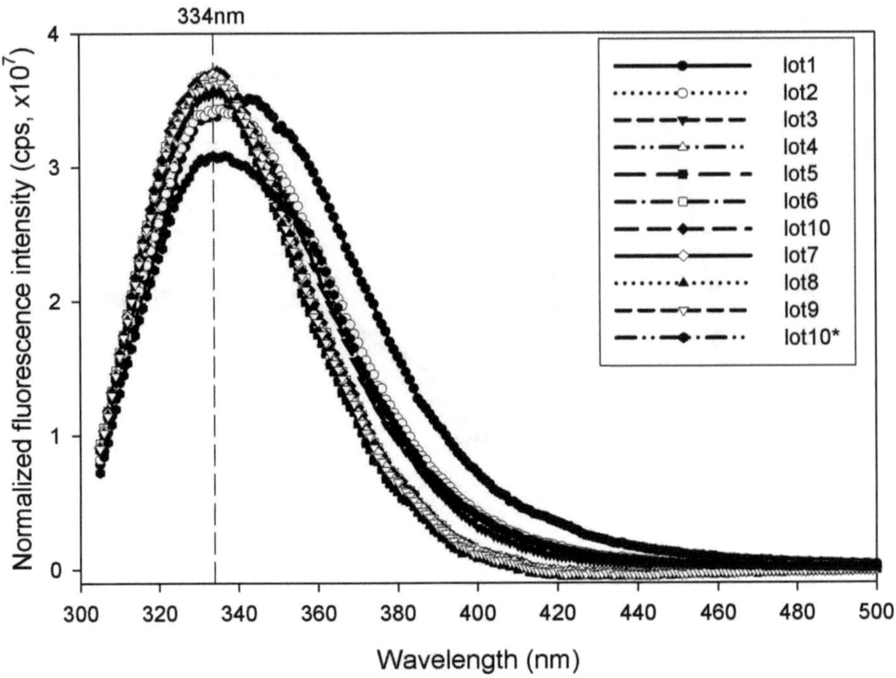

Fig. 7.2 Tryptophan emission fluorescence spectra of all different lots excited at 295 nm. See Fig. 7.1 or Table 7.1 for lot information

bandwidth of the emission spectrum for the pilot-scale formulation is relatively wider than that for the full-scale formulation, which suggests that some tryptophan residues might have greater exposure to the solvent in the former. Additional information about the changes in higher-order structure comes from the AUC-SV analysis where a sharp main peak at a reasonable frictional coefficient of f/f_0 (=1.6) is seen for the commercial formulation, whereas the main species in the pilot-scale formulation shows a wider peak at larger sedimentation coefficient and a lower frictional coefficient f/f_0 (=1.3) (Fig. 7.3). This suggests that the pilot-scale product main peak could contain a molecule with a different conformation, presumably less elongated, or that the sample has a heterogeneous population of different molecular species due to rapid self-association that could decrease the weight-averaged frictional coefficient (Dam and Schuck 2000). This can be further confirmed by a concentration-dependent AUC-SV analysis (not discussed here; see Case Study 2).

To quantitatively evaluate the similarity between the CD spectra, the Similarity Match analysis implemented in TQ Analyst™ software (Thermo Scientific) was applied to both far-UV and near-UV CD spectra using the spectra of the reference material as the standards. The algorithm uses Gram–Schmidt orthogonalism to extract the spectral information from the standard spectra and subtracts that from the test spectra. The residual spectra are then evaluated for any spectral information

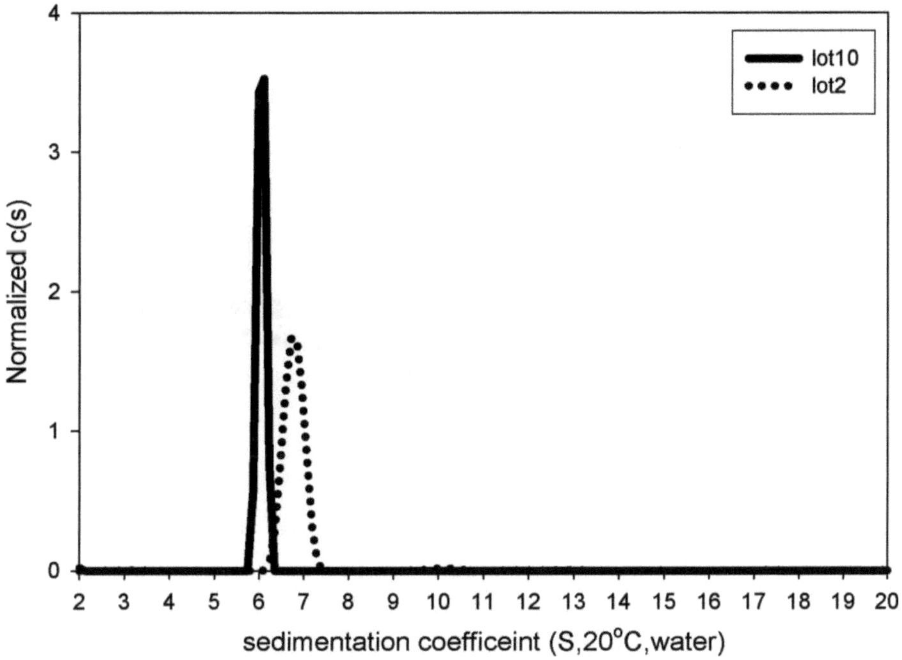

Fig. 7.3 AUC-SV profiles of a pilot-scale/clinical formulation lot (Lot 2) and a full-scale/commercial formulation lot (Lot 10)

for comparison. This application provides a single number for spectral comparison and supplements the qualitative visual assessment. The use of this commercial algorithm also eliminates tedious and error-prone spreadsheet calculations presented previously (Zou and Luo 2010). The calculated match values for the ten different lots versus the reference standard (Lot 10) are listed in Table 7.1; the difference between pilot-scale and full-scale lots is in agreement with the visual inspection. To establish a possible cut-off value for the spectral difference between "similar" and "dissimilar," comparison among a series of spectra generated by heat denaturation was conducted (Fig. 7.4a, b). The spectra at the temperatures lower than the onset denaturation temperature (60°C) as determined by DSC (Fig. 7.5) are compared pair-wise using the similarity analysis. The resulting similarity match values are subject to the *t*-test at 95% confidence level to deduce the lower limit of the similarity match value used as the spectral similarity cut-off values (87.27% and 96.50% for far- and near-UV CD spectra, respectively). As shown in Table 7.1, only one lot (Lot 6) in the full-scale material is not similar to the reference standard in the far-UV CD analysis. However, there is no difference in its near-UV CD spectrum. Careful examination of the far-UV CD spectrum of this lot reveals a high level of noise in this data, and therefore, the spectrum does not have as good a signal to noise ratio, confounding the comparison to the other samples. However, the similarity analysis is sensitive enough to pick up small differences in

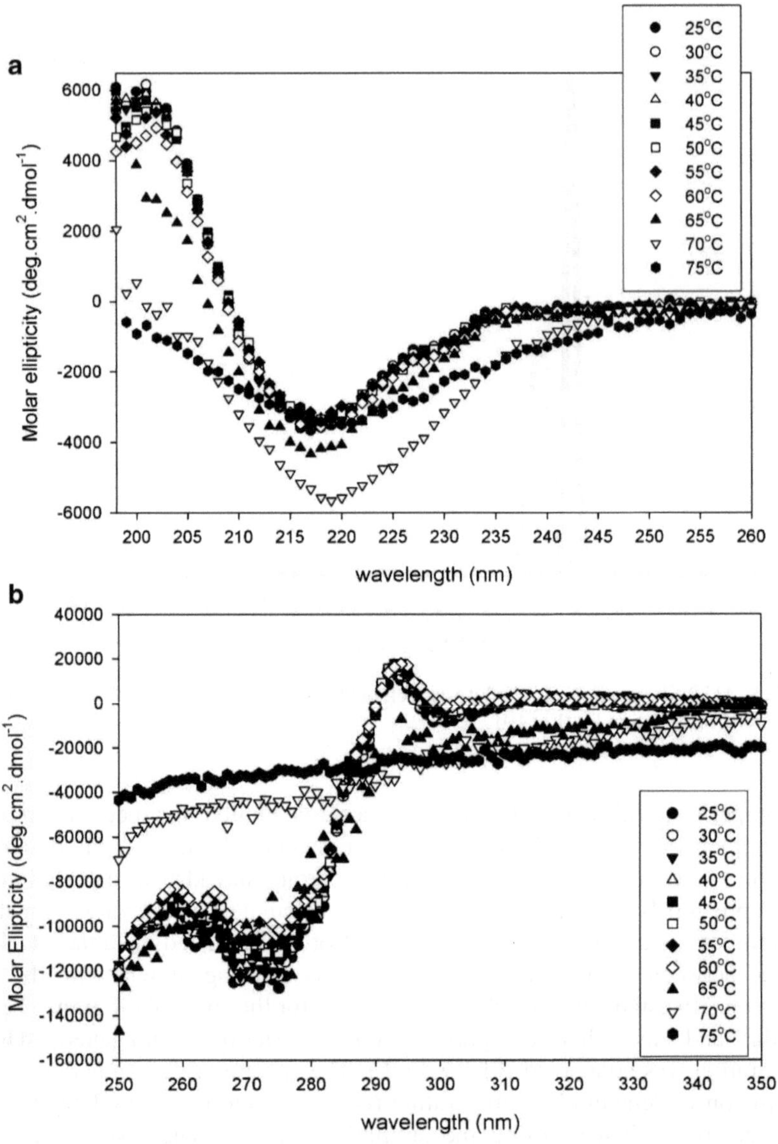

Fig. 7.4 Far- (**a**) and near-UV (**b**) CD spectra of the IgG2 antibody at different temperatures

the spectra and should serve as a good tool for comparability studies. Other type of data analyses such as data deconvolution for far-UV CD spectra can provide the estimate of relative abundance of different secondary structures (Sreerama and Woody 2000). However, it is judged to be not sufficiently sensitive to be used in comparability studies (Zou and Luo 2010).

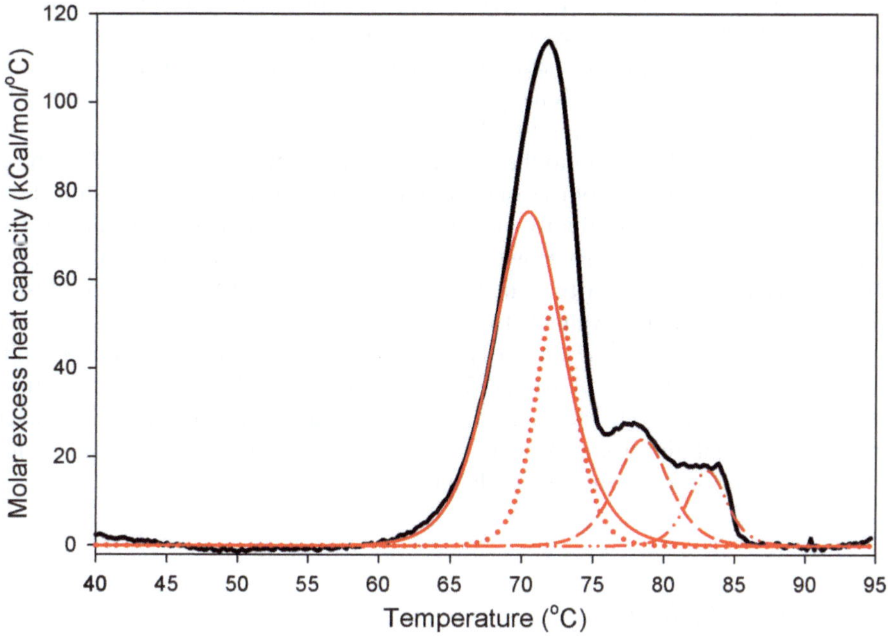

Fig. 7.5 The DSC thermogram of the reference standard (Lot 10). The traces are the fits from the non-two-state model are shown

Table 7.1 Quantitative similarity analysis for pilot-scale and full-scale drug substance lots in clinical and commercial formulations, respectively

Lot #	Drug substance /formulation	Similarity match value for far UV CD (%)	Similarity match value for near UV CD (%)
10 (reference standard)	Full-scale/commercial	100	100
1	Pilot-scale/clinical	***64.16***	***86.12***
2	Pilot-scale/clinical	***61.51***	***86.31***
3	Full-scale/commercial	93.06	99.47
4	Full-scale/commercial	90.80	99.60
5	Full-scale/commercial	90.06	99.57
6	Full-scale/commercial	85.98[a]	99.60
7	Full-scale/commercial	94.78	99.41
8	Full-scale/commercial	89.71	99.45
9	Full-scale/commercial	92.52	99.58
10* (reference standard in clinical formulation)	Pilot-scale/clinical	***58.88***	***87.30***

The values in bold are below the limits obtained from the heat denaturation study (87.27% and 96.50% for far- and near-UV CD, respectively)

[a]Difference attributable to poor quality of spectrum collected

A change in the higher-order structure of a biotherapeutic has potential to impact activity and stability. However, the structural perturbation could be reversible, in which case the risk to potency and safety would be greatly reduced. The reversibility of higher-order structural changes was demonstrated by transferring the full-scale reference material to pilot-scale formulation (Lot 10*) and comparing the far- and near-UV CD spectra to the original full-scale formulation (Fig. 7.1a, b). It is evident that changes in both secondary and tertiary structures are reversible, both by spectral overlay and similarity match analysis. The similarity values of Lot 10* are very different from Lot 10. However, a direct comparison between Lot 1 and Lot 10* showed a higher similarity value for far-UV CD spectra (92.22%) in contrast to the much lower similarity value between Lot 10 and Lot 10* (58.88%) (Table 7.1). For near-UV CD spectra, the similarity between Lot 1 and Lot 10* is 99.00% while that between Lot 10 and Lot 10* is 87.30%, much lower than the similarity limit (96.50%) obtained from the heat denaturation study. This formulation-related structural reversibility was consistent with the observation that the activity is not affected by the formulation change (data not shown).

7.2.3 *Conclusions*

An in-depth biophysical characterization with quantitative criteria enhances the utility of the comparability exercise. This not only provides information needed for product comparability but also helps to assess the impact of differences should they be observed during the comparability exercise.

7.3 Case Study 2: Self-Association and Aggregation of an Fc-Fusion Protein

Comprehensive characterization of a biotherapeutic before it goes into clinical trials and later for commercialization is an expectation from both the industry and regulatory bodies (Kozlowski and Swann 2006). The characterization of biotherapeutics can be generally divided into three areas: (1) biochemical characterization for primary sequence, chemical composition, and post-translational modification; (2) biophysical characterization for higher-order structure, association state in solution, and ligand-binding mechanism; and (3) biological characterization for in vitro activity, ex vivo potency, and mode of action. A thorough characterization effort provides essential knowledge needed for the development of analytical methods that are used to evaluate the critical quality attributes during manufacture. One of the critical quality attributes for biotherapeutics is the level and type of aggregates, which arise from intermolecular interactions and are potentially linked to immunogenicity. Size-exclusion chromatography (SEC) is the most common method to

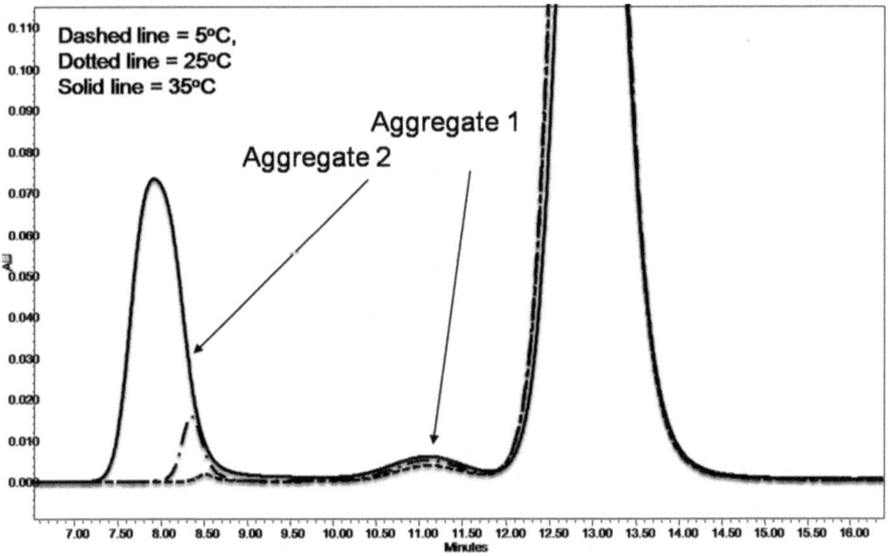

Fig. 7.6 SEC chromatogram of representative stability samples of an Fc-fusion protein

quantify aggregates, but the upper limit of size range is limited to approximately 50 nm. Thus, this technique can assess oligomers but misses the larger aggregates in the submicron and micron range (den Engelsman et al. 2011). Limitations of SEC in terms of accuracy and specificity, when evaluating aggregates, have been discussed elsewhere (Philo 2009; Carpenter et al. 2010). Orthogonal methods such as AUC and AFFFF are therefore recommended to supplement the SEC results.

The second case study presented in this chapter involves an Fc-fusion protein which showed significant aggregation during storage at accelerated and stressed conditions as measured by SEC. Application of AUC in two modes (sedimentation velocity and gravitational sweep) provided some interesting insights into the self-association behavior and impact of salt in the mobile phase for SEC.

7.3.1 *Methods*

The SEC analysis was performed using a TSKgel G3000SW column (Tosoh Bioscience) with a mobile phase buffer of 200 mM sodium phosphate, 450 mM NaCl, and pH 6.5. The elution profile was monitored at 214 nm. A sample volume of 20 μL at 0.5 mg/mL was injected. AUC sedimentation velocity was run at 20,000 rpm and absorbance data at 280 nm was collected. The data were analyzed using Sedfit (Schuck 2000). For the concentration-dependent study, samples at three different concentrations were analyzed at 50,000 rpm, and the data were fitted using Sedphat for reversible association (Schuck 2003). Gravitational sweep AUC

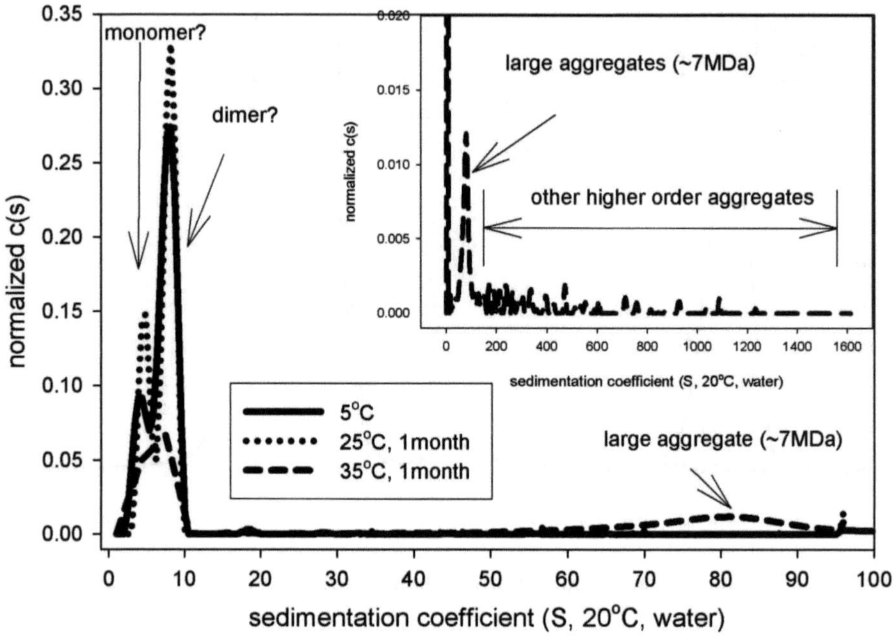

Fig. 7.7 AUC-SV sedimentation coefficient distribution for the stability samples of an Fc-fusion protein

(AUC-gs) was performed between 3,000 rpm and 50,000 rpm, and the absorbance data at 280 nm was analyzed using SedAnal (Stafford and Braswell 2004). For more detailed discussion of the techniques, please see Chap. 3.

7.3.2 Results

Representative SEC chromatograms from the stability samples are shown in Fig. 7.6. AUC-SV was used to verify the SEC results. AUC-SV was run at relativly low speeds (20,000 rpm) on the same stability samples in an effort to capture the larger aggregates (Fig. 7.7) that would otherwise sediment too quickly. AUC-SV was performed in a matching formulation buffer that has a lower salt content (compared to the SEC mobile phase). Comparison between AUC-SV and SEC results showed significant differences in relative abundance and size profiles of the aggregates (Tables 7.2 and 7.3). The SEC method underestimated the level of aggregates, especially for the samples stressed at 35°C. The peak area recovery in SEC was assessed to be reasonable and would not account for the loss of aggregates (data not shown). The possibility that some of the aggregates are salt dependent and could

Table 7.2 Relative abundance of aggregates by SEC analysis for the Fc-fusion protein stability samples

Sample treatment (1 month)	Total aggregates	Aggregate 1	Aggregate 2
5 °C	1.6%	1.4%	0.2%
25 °C	3.7%	1.8%	1.8%
35 °C	20.9%	2.4%	18.5%

dissociate at high salt concentrations in the mobile phase was considered. The discrepancy prompted further development of a more suitable SEC method. To further characterize the aggregates profile, gravitational sweep AUC (AUC-gs) with speeds between 3,000 and 45,000 rpm was employed. This technique expands the dynamic range and allows detection of even larger aggregates if present, although quantitation is difficult (Stafford and Braswell 2004). The resultant profile shown in Fig. 7.8 demonstrates that some very large aggregates are present in the stressed sample. Comparison of the size distribution from the single speed at 20,000 rpm to that from the variable speeds showed some interesting similarities as well as differences. Both methods captured the large aggregate at the apparent sedimentation coefficient of 55 S (78 S in standard condition, Fig. 7.8), but the gravitational sweep method detected even larger aggregates at 1,800 S. Due to the changing speed during the sedimentation, it is not possible to determine the true size and mass of aggregates in Fig. 7.8. However, if treated as compact spheres, the minimal size and molar masses of the aggregates in an AUC-gs run can be estimated, and this data is provided in Table 7.4. The results show that a broad size range of aggregates are formed in this biotherapeutic after being subject to stress conditions.

In addition to the broad distribution of aggregates, the AUC profile at 20,000 rpm shows two overlapping main peaks (Fig. 7.7). This suggests the existence of a dynamic equilibrium during the sedimentation process. To further understand the potential for reversible self-association, a concentration-dependent AUC-SV analysis was applied to the sample at 5°C. Figure 7.9 shows the size distributions at three different concentrations. The concentration-dependent shift in sedimentation coefficient of the second peak suggests that there is rapid association/dissociation equilibrium at the timescale of the sedimentation experiment (Schuck 2003). Global data fitting to the Gilbert–Jenkins theory implemented in Sedphat gives a dimer dissociation constant of 40 μM and the free energy of the association as −5.9 kcal/mol, suggesting a moderate strength protein–protein interaction (Horton and Lewis 1992). It is likely that this association is dependent on salt concentration so that this equilibrium does not occur in the SEC mobile phase, and the SEC analysis therefore does not detect the presence of protein self-association. It may be possible that the non-Fc part of the fusion protein has pH- and salt-dependent self-association behavior too.

Table 7.3 AUC-SV results of the stability samples for the Fc-fusion protein

Sample treatment (1 month)	Monomer			Dimer			Large aggregates			Other higher order aggregates
	Mol. wt. (kDa)	Sed. coeff. ($S_{20^{\circ}C,w}$)	Fraction %	Mol. wt. (kDa)	Sed. Coeff. ($S_{20^{\circ}C,w}$)	Fraction %	Mol. wt. (kDa)	Sed. coeff. ($S_{20^{\circ}C,w}$)	Fraction %	Fraction %
5 °C	85	4.1	21.5	222	7.7	78.0				1.2[a]
25 °C	86	4.7	28.5	202	8.0	68.7				3.0[a]
35 °C							7	78.2	33.9	27.8[b]

[a]species that are larger than the “dimer” peak
[b]species that are larger than the 7 MDa peak

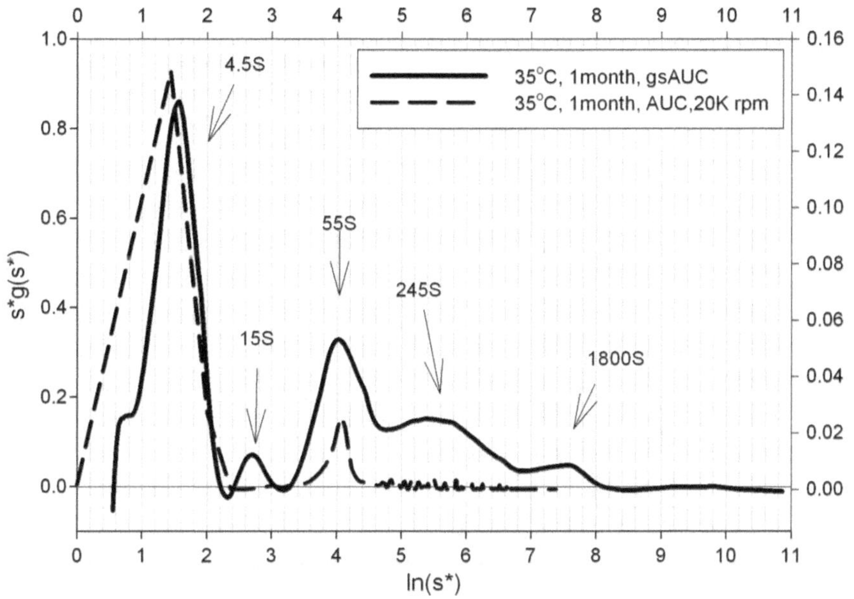

Fig. 7.8 Gravitational sweep AUC analysis of the 35°C 1-month stability sample of an Fc-fusion protein showing the wide range of aggregate sizes present

Table 7.4 Size and molar mass of the hypothetical compact spheres for the species in the gravitational sweep AUC (also see Fig. 7.8)

Species	Minimum molecular weight (MDa) for compact spheres	Minimum hydrodynamic radius (nm) for compact spheres
1800 S	565.6	54.7
245 S	28.4	20.2
55 S	3.0	9.6
15 S	0.4	5
4.5 S	0.1	2.7

7.3.3 *Conclusions*

Orthogonal characterization based upon AUC in different modes, allowed the nature of aggregation behavior of the Fc-fusion protein to be comprehensively elucidated in contrast to the SEC analysis. The observed discrepancy prompted an improvement in the SEC method, while detailed analysis of the AUC runs showed the existence of a reversible self-association behavior in the system at the timescale of the AUC experiment.

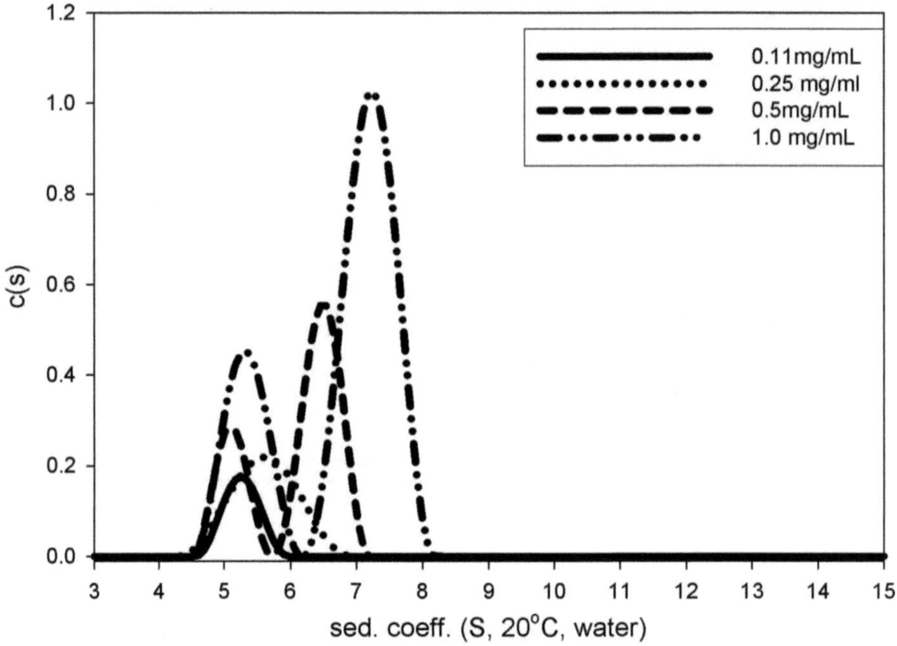

Fig. 7.9 Concentration-dependent AUC-SV analysis of the 5°C 1-month sample showing the concentration-dependent shift in position of the second peak

7.4 Case Study 3: Formation of Subvisible and Visible Particles in a Monoclonal Antibody Drug Substance Solution

Aggregation and more specifically proteinaceous (subvisible and visible) particles have come to be regarded as critical quality attributes of biotherapeutics since they have the potential to impact the safety and efficacy of these products (Carpenter et al. 2009). The mechanism for generation of (non-native) proteinaceous particles is generally considered in terms of the nucleation and growth model where small aggregates eventually grow large enough to become subvisible (greater than 1 μm in apparent diameter) or larger (Weiss et al. 2009; Roberts 2006). The model envisions a continuum of aggregate sizes, ranging from dimer to multimers larger than 100 μm, present in solution. However, the kinetics of the step-wise growth model allows for certain size ranges to not be populated to any significant extent.

Visible particles are defined as those that can be seen using a regular, consistent, visual inspection method, usually larger than 100 μm with the actual size dependent on the method developed by the individual manufacturer. The mechanism of visible particle formation for an IgG4 antibody was explored using dynamic light scattering and is described below.

7.4.1 Methods

Dynamic light-scattering (DLS) experiments were performed on the drug substance lot without further dilution. The data were collected and analyzed without any correction for solvent viscosity, so only apparent hydrodynamic size was calculated. The correlation of scattering intensity was deconvoluted using the CONTIN algorithm to obtain the size distribution. For more details on the technique, please see Chap. 3.

7.4.2 Results

Visible particles were observed in an IgG4 monoclonal antibody drug substance after storage under the preferred conditions. However, SEC analysis did not show any changes in soluble aggregate levels before the emergence of the visible particles. Concern was raised as to the suitability of the SEC method being stability indicating since it seemed to have missed any early signs of physical instability. AUC analysis on the drug substance agreed with the SEC results showing similar abundance and distribution of soluble aggregates (data not shown). To understand the mechanism of formation of the visible particles and potential subvisible particles, dynamic light scattering was employed. A DLS size distribution profile of the drug substance stability sample that contained visible particles is shown in Fig. 7.10. There is a peak around 8 μm, suggesting the presence of very large

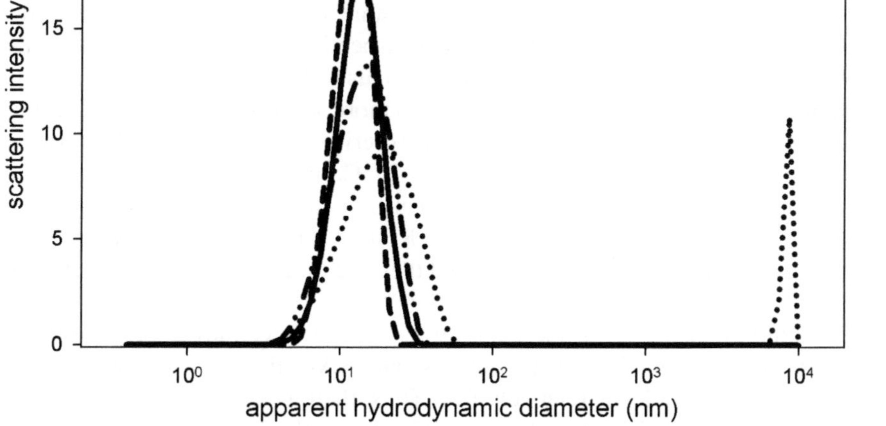

Fig. 7.10 Dynamic light-scattering size distribution of the particles in drug substance stability samples. Both backscattering (173°) and forward-scattering (13°) results from unfiltered and filtered samples are shown

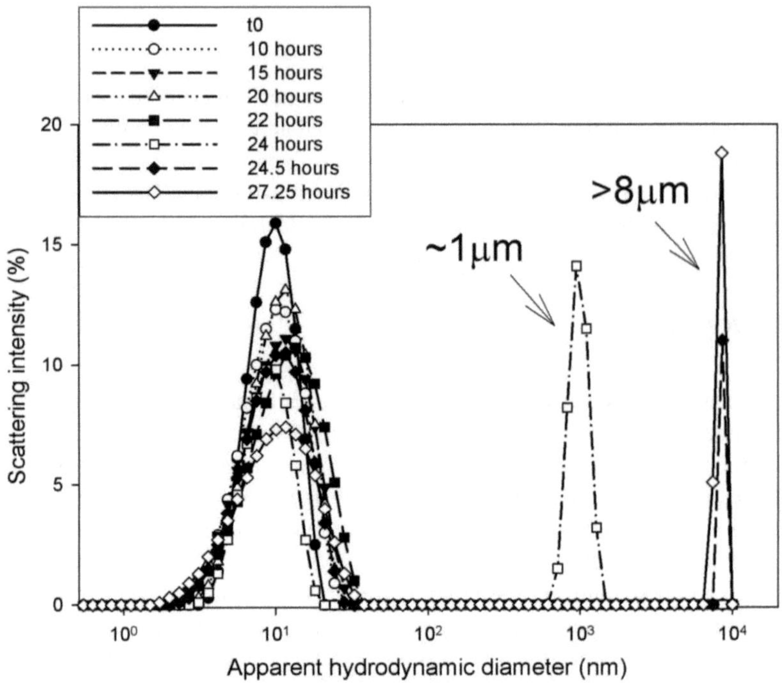

Fig. 7.11 Real-time monitoring of particle formation in the filtered samples by DLS at 37°C

aggregates/subvisible particles. However, the actual size information is not reliable since there is the possibility of sedimentation of the larger particles during the measurement itself, so that they are not detected. The amount of these aggregates could also be low enough that they are below the level of detection. Even though data on the exact sizes is not available, the results still provide valuable information about the presence of very large aggregates. This was demonstrated by the absence of the peak at around 8 μm in the same stability sample after filtration through a 0.2 μm filter (Fig. 7.10). Visual inspection of the filtered sample also showed no visible particles. Data in Fig. 7.10 also demonstrated that the forward scattering at a scattering angle of about 13° is more sensitive to large particles than the backscattering at 173°. A time-course experiment with DLS in the forward-scattering mode was then carried out at 37°C, monitoring the formation of submicron and subvisible particles in real-time at that temperature. Filtered stability samples were used for this study. Results show the generation of submicron particles after 24 h of incubation and then the rapid appearance of larger particles (~8 μm) within another half hour (Fig. 7.11).

The simplest scenario for the inability of SEC to detect any aggregation trend before the appearance of visible particles is that the kinetics of formation of subvisible and visible particles is relatively fast. Thus, there is no accumulation of small aggregates that would be detectable by SEC analysis (Fig. 7.12).

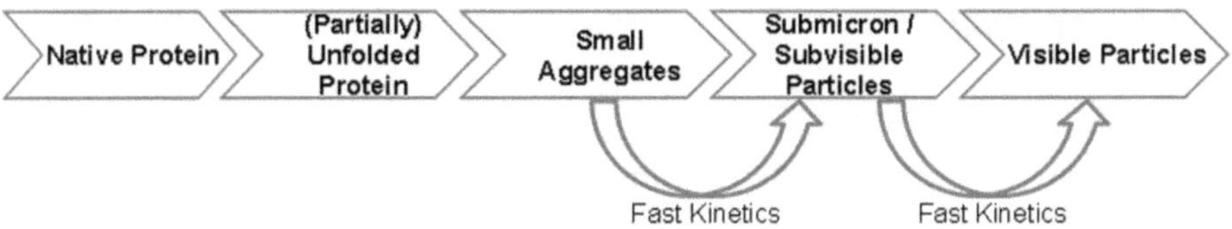

Fig. 7.12 A schematic diagram illustrating the mechanism for visible particle formation in the IgG4 antibody

7.4.3 Conclusions

A real-time DLS study was able to help elucidate the mechanism of visible particle formation and explain the lack of detection of aggregation by SEC. Once aggregates reach a certain size range, they rapidly coalesce to form larger visible particles, thus effectively eliminating all but a small amount of aggregates, detectable by SEC, from the solution. No aggregates in intermediate size ranges are detectable, and therefore, no changes are seen in the SEC results, despite the appearance of visible particles.

7.5 Case Study 4: Probing Aggregation Propensity of Different Batches of Biotherapeutics Using Nile Red Fluorescence Binding Kinetics

The process of bringing a biotherapeutic to the patient (expression, purification, formulation, distribution) subjects it to numerous stresses (Cromwell et al. 2006; Singh 2011). The recombinant protein undergoes steps with high concentration, high ionic strengths, pH extremes, shear stresses, air–liquid and solid–liquid interfaces, freeze/thaw, and long-term storage. Even as it is purified, it is possible for a batch of material to accumulate small amounts of misfolded or partially unfolded species that are sufficiently similar to the native desired molecule to be indistinguishable in any standard assays but which may carry the seeds for long-term instability through propensity for higher rates of aggregation or even chemical degradation. The ability to predict susceptibility to aggregation upon long-term stability between various batches of material by detecting such aggregation precursor species would be a very useful tool for the characterization of biotherapeutics.

Nile red is a hydrophobic dye that is sensitive to the polarity of its environment and binds to hydrophobic regions of proteins (Sackett and Wolff 1987; Demeule et al. 2007). It fluoresces only upon binding to hydrophobic surfaces such as those that are created by protein unfolding/aggregation. Its utility in detecting small amounts of aggregation prone species was explored in this study. It was found that utilizing Nile red fluorescence in the conventional manner was not sufficiently sensitive, but the sensitivity could be enhanced by thermal incubation and monitoring the kinetics of change in peak fluorescence intensity.

7.5.1 Methods

Nile red extrinsic dye-based fluorescence was carried out according to the procedure described by (Demeule et al. 2007), with some modifications. Nile red was dissolved in DMSO to obtain a 10 mM stock solution. Three microliter of the dye stock solution was added to 2,997 μL of protein at 20 mg/mL for the fluorescence experiments. Thermal incubation was conducted by preheating the cuvette to the target temperature and then adding 3 mL of dye + protein solution into the cuvette at time zero.

An IgG2 mAb was prepared at 20 mg/mL in 20 mM histidine buffer pH 5.5, without any other excipients or stabilizers. Samples were purposely stressed to induce aggregation including thermal treatment at 40°C (7 weeks) or 50°C (10 days), or 20 cycles of freeze/thaw stress. The level of aggregation was measured by SEC, and the samples were also studied by CD, intrinsic fluorescence, and bis-ANS fluorescence, apart from the Nile red fluorescence described above. An aggregate sample was also created by mixing aliquots of 40°C 7 weeks and 5°C control at 1:1 (v/v) ratio.

7.5.2 Results

SEC analysis of the test solutions showed fairly low and similar levels of aggregates (labeled as high molecular mass species, HMMS) (see Table 7.5). Characterization of these samples by circular dichroism (Fig. 7.13), intrinsic fluorescence (not shown), bis-ANS fluorescence (Fig. 7.14), and Nile red fluorescence (Fig. 7.15) showed little differences between them. DLS analysis did not pick up any significant particles beyond the main peak at around 9 nm (data not shown). However, when samples containing the Nile red dye were incubated at 63°C (just below the onset of CH_2 domain-melting peak by DSC), and monitored for peak fluorescence, an interesting trend was seen. Once at temperature, the peak Nile red fluorescence intensity of the samples increased in a linear fashion that was distinct for the different stresses (Fig. 7.16). The freeze/thaw cycled sample and the control sample were very similar in the evolution of the peak fluorescence while the thermally stressed samples followed a distinctly different pattern with a more rapid increase in the peak intensity. The results of

Table 7.5 Aggregates (% HMMS) in stressed samples of IgG2 mAb measured by size-exclusion chromatography

Sample treatments	HMMS (%)
5°C control	0.9
Freeze/thaw cycles FT20	2.0
Heat treated 40 °C 7 weeks + 5 °C control 1 + 1 v/v	1.0
Heat treated 40 °C 7 weeks	1.1
Heat treated 50 °C 10 days	1.9

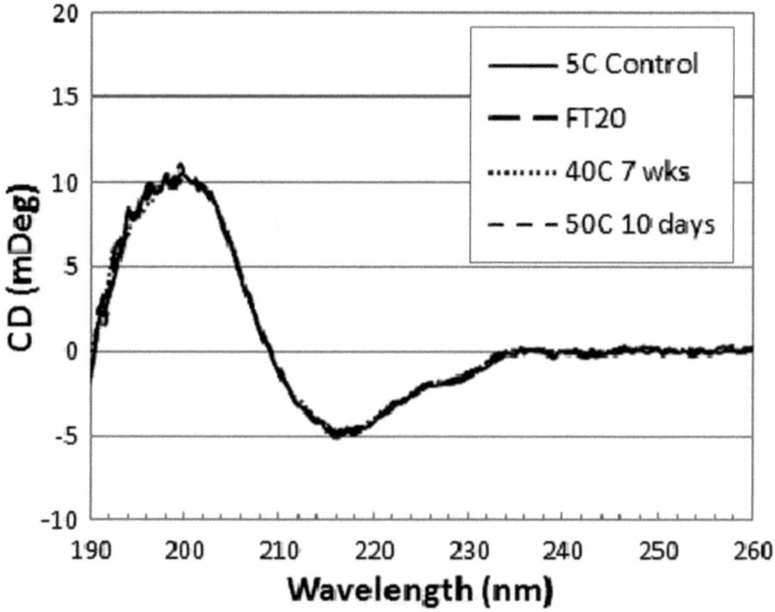

Fig. 7.13 Circular dichroism spectra of stressed samples of IgG2 mAb. Sample treatments are as follows: *5°C control* control unstressed sample, *FT20* 20 cycles of freeze/thaw stress, *40°C 7 weeks* thermal stress at 40°C for 7 weeks, *50°C 10 days* thermal stress at 50°C for 10 days

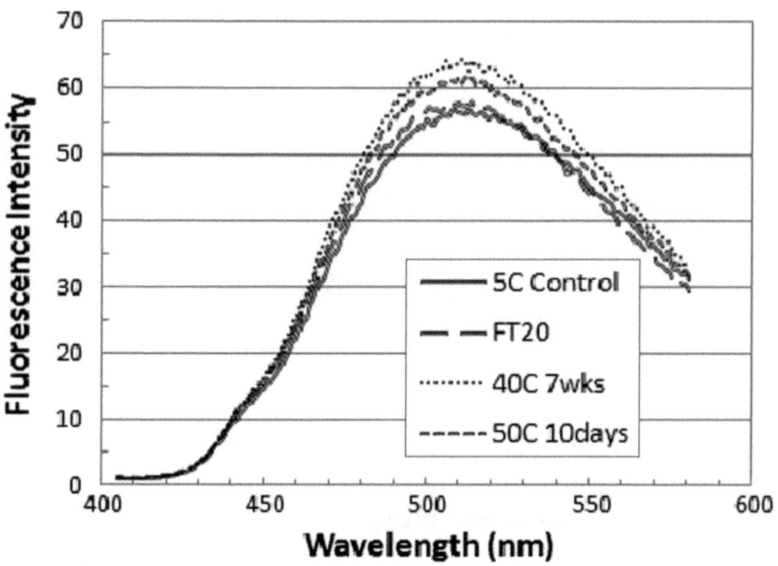

Fig. 7.14 Extrinsic (bis-ANS) fluorescence spectra of stressed samples of IgG2 mAb. Sample treatments are as follows: *5°C control* control unstressed sample, *FT20* 20 cycles of freeze/thaw stress, *40°C 7 weeks* thermal stress at 40°C for 7 weeks, *50°C 10 days* thermal stress at 50°C for 10 days

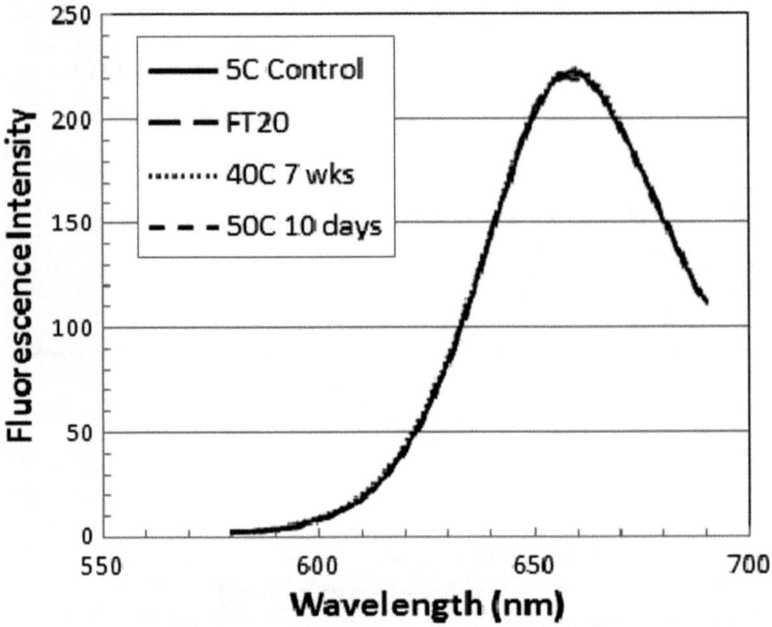

Fig. 7.15 Extrinsic (Nile red) fluorescence spectra of stressed samples of IgG2 mAb. Sample treatments are as follows: *5°C control* control unstressed sample, *FT20* 20 cycles of freeze/thaw stress, *40°C 7 weeks* thermal stress at 40°C for 7 weeks, *50°C 10 days* thermal stress at 50°C for 10 days

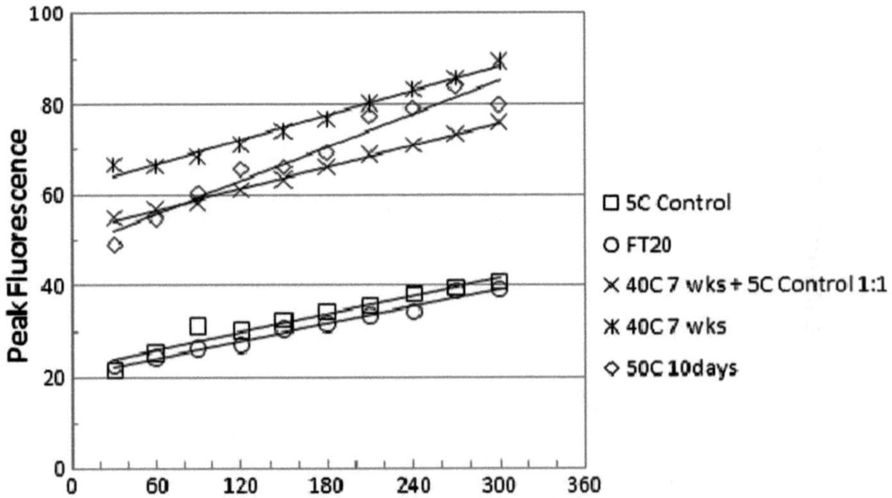

Fig. 7.16 Time course of Nile red peak fluorescence for stressed samples of IgG2 mAb under incubation at 63°C. Sample treatments are as follows: *5°C control* control unstressed sample, *FT20* 20 cycles of freeze/thaw stress, *40°C 7 weeks* thermal stress at 40°C for 7 weeks, *50°C 10 days* thermal stress at 50°C for 10 days, *40°C 7 weeks* 5°C control 1:1 is a 1:1 by volume mixture of the thermally stressed and unstressed control samples. Linear regression plots are shown (see Table 7.6)

Table 7.6 Linear regression analysis for time course of Nile red peak fluorescence from stressed samples of IgG2 mAb shown in Fig. 7.16

Sample treatments	HMMS (%)	Slope (h^{-1})	Intercept	R^2
5°C control	0.9	0.0662	22.062	0.9465
Freeze/thaw cycles FT20	2.0	0.0633	20.377	0.9824
Heat treated 40°C 7 weeks + 5°C control 1 + 1 v/v	1.0	0.0795	51.973	0.9963
Heat treated 40°C 7 weeks	1.1	0.0897	61.477	0.9797
Heat treated 50°C 10 days	1.9	0.1231	48.408	0.9983

a linear model fit are summarized in Table 7.6. The fit parameters do not correlate with the measured HMMS values, but they form two distinct groups. It is clear from the fluorescence data in Fig. 7.15 that the aggregate (levels) or the applied stress per se cannot be distinguished by their ability to bind the hydrophobic dye Nile red. However, when incubated at 63°C, it is likely that molecules that have small structural perturbations and misfolds as a consequence of the stress history of the samples, begin to unfold in the cuvette. As this occurs, the Nile red binding and fluorescence response increases. The data in Fig. 7.16 therefore suggests that molecules that underwent freeze/thaw stress, either aggregate or refold back to the native state when thawed. Therefore, few partially unfolded or misfolded species are created by this stress, and the response with incubation (in Fig. 7.16) is similar to that of the control sample. However, when the mAb is thermally stressed, a variety of such species are created which are not detectable by SEC since they are not aggregates, and are not distinguishable by conventional spectroscopic techniques since they are either too few and/or too similar to the native molecule.But the structural susceptibility of these molecules is amplified by incubating the sample at 63°C, where the presence of the Nile red dye is able to capture this phenomenon, which would otherwise be too small energetically to be detected by DSC (data not shown).

7.5.3 Conclusions

This fluorescent dye-based thermal-scanning method seems to be able to probe and amplify the presence of small fractions of structurally perturbed species or species with small amount of structural perturbations in a pool of properly folded molecules. Since it is likely that the presence of perturbed species could lead to further increases in aggregation over long-term storage at normal (e.g., 2–8°C) temperatures, this technique offers a valuable tool to characterize various batches of biotherapeutics for their long-term viability.

A significant limitation of dye-based techniques is that they cannot be employed on formulated material containing surfactants, since the hydrophobic dye will interact with surfactant (micelles) and fluoresce.

7.6 Case Study 5: Probing Aggregation Propensity of Biotherapeutics Using Bis-ANS Binding Thermodynamics and Kinetics

Detection and control of aggregation is a key aspect for the successful development of a biotherapeutic product. Exposed hydrophobic patches either by design or through some degree of misfolding are generally understood to act as seeds for the nucleation step of aggregation. Such regions may be present in only a small fraction of the population and would therefore be difficult to detect. An ability to detect small amounts of such precursors would therefore be useful as a predictor of long-term behavior of a particular batch of the molecule. Ability to detect small differences in the exposed hydrophobic regions can also be of value in selecting among candidate molecules, as an indicator of their aggregation propensity.

Following upon the above concept of exposed hydrophobic regions, it was decided to use bis-ANS binding as a probe. The proposed hypothesis is that bis-ANS would bind more efficiently (tighter and faster) to such hydrophobic regions.

7.6.1 *Methods*

Thermodynamic binding parameters were obtained using an isothermal titration calorimeter interfaced with a computer for data acquisition and analysis using Origin 7 software (VP-ITC, GE-Healthcare). Protein and bis-ANS samples were degassed at 25°C before loading into the ITC cell and syringe, respectively. A typical injection schedule included the addition of 10 μL of 2 mM bis-ANS solution to 20–30 μM of protein (mAb or HSA) in the calorimetric cell with 25–28 injections spaced at 3-min intervals. For the first injection, 3 μL of bis-ANS was added and the corresponding data point was deleted from the analysis. The binding isotherms were fit to a two-class binding-site model using Origin 7.

The kinetics of change in fluorescence intensity at 25°C was monitored using stopped-flow instruments (Hi-Tech and Chirascan, Applied Photophysics). The dead time of mixing to monitoring was about 5 ms. Proteins were mixed at 1:1 volume ratio with bis-ANS to a final stoichiometric concentration of tenfold molar excess of ligand, to optimize the timescales. The kinetic profiles were monitored for different timescales ranging from 200 ms to 200 s, all yielding biphasic kinetics of bis-ANS binding. The time base was optimized for 20 s, and the data were fit to biphasic kinetics equation using a nonlinear, least-squares ProK algorithm (Applied Photophysics) to derive first-order rate constants.

7.6.2 *Results*

Three monoclonal antibodies were evaluated for their interaction with the hydrophobic dye bis-ANS. The titration curves are fit to the model as shown in Table 7.7.

Table 7.7 Thermodynamic parameters of bis-ANS binding to different IgG mAbs and HSA derived from ITC data at 25°C fit to a two-class binding-site model

Protein	p*I*	*N*1	*K*1	Δ*G*1	Δ*H*1	*T*Δ*S*1	*N*2	*K*2	Δ*G*2	Δ*H*2	*T*Δ*S*2
IgG2 Mol A	8.6	13	11	−9.6	0.18	−9.77	22	0.12	−6.93	−0.20	6.26
IgG2 Mol B	8.0	10	20	−9.9	0.11	−9.98	18	0.16	−7.16	−1.38	5.72
IgG2 Mol C	8.3	3	10	−9.6	0.10	−9.66	17	0.19	−7.19	−1.67	5.54
HSA	4.9	11	82	−10.8	0.29	−11.0	9	0.46	−7.72	−1.91	5.81

The p*I* of each molecule is also listed. Note the opposite signs of Δ*H* and −*T*Δ*S* for the two classes of binding sites for all proteins
$N = \pm 1$ (SD); $K = 1 \times 10^6$ M (5% SD); ΔG = kcal/mol (5% SD); ΔH = kcal/mol (5% SD); $-T\Delta S$ = kcal/mol (5% SD)

HSA was also included in the study as a reference. The results show that bis-ANS binds to the molecules with varying stoichiometry and affinity. The data was analyzed using a two-class binding-site model. Analysis shows that the first class of binding sites are specific and entropically driven (K_{D1} = 0.0–0.2 μM; Δ*H* positive; *T*Δ*S* negative; K_{D1} is inverse of K_1, the association constant). This is likely a result of the interaction of the naphthalene ring structure of bis-ANS molecules with hydrophobic surfaces/pockets of the protein. The second class of binding sites are nonspecific, low-affinity sites (K_{D2} = 10–40 μM; Δ*H* negative; *T*Δ*S* positive; K_{D2} is inverse of K_2, the association constant), probably a result of hydrophilic/ionic interactions through the sulfonic acid groups. For reference, although the HSA–bis-ANS N1 value is similar to the mAbs (despite being a smaller protein), the equilibrium constant (K_{D1}) is significantly larger, indicating a stronger affinity of bis-ANS for the more hydrophobic HSA protein. Based on the affinity and stoichiometry of the first binding site (N1), the aggregation propensity driven by hydrophobicity is rank-ordered as HSA > IgG2 MolA > IgG2 MolB > IgG2 MolC.

For aggregation to occur as a consequence of the presence of hydrophobic patches, the patches must be accessible. It was hypothesized that the patches that are accessible would be the first to which bis-ANS would bind, and thus the molecule with more of these accessible patches would have a faster initial binding of bis-ANS. To distinguish between the molecules on the basis of the (extent of their) highly accessible hydrophobic patches, binding kinetics was studied by the stopped-flow technique. Stopped-flow measurements on the four test molecules are shown in Fig. 7.17 with the model fit data in Table 7.8. A biphasic-binding model was used for the data analysis. Hydrophobic HSA has the highest k_1, indicating the fastest binding rate, and also correlates with the high aggregate level (% HMMS) measured by SEC. Among the mAbs, the low general level of aggregation did not correlate with the kinetics. However, based on k_1, the rank order of aggregation propensity is as follows HSA > IgG2 MolA > IgG2 MolC > IgG2 MolB.

The rank ordering based on stopped-flow kinetics was found to correlate with the long-term stability behavior of the mAbs.

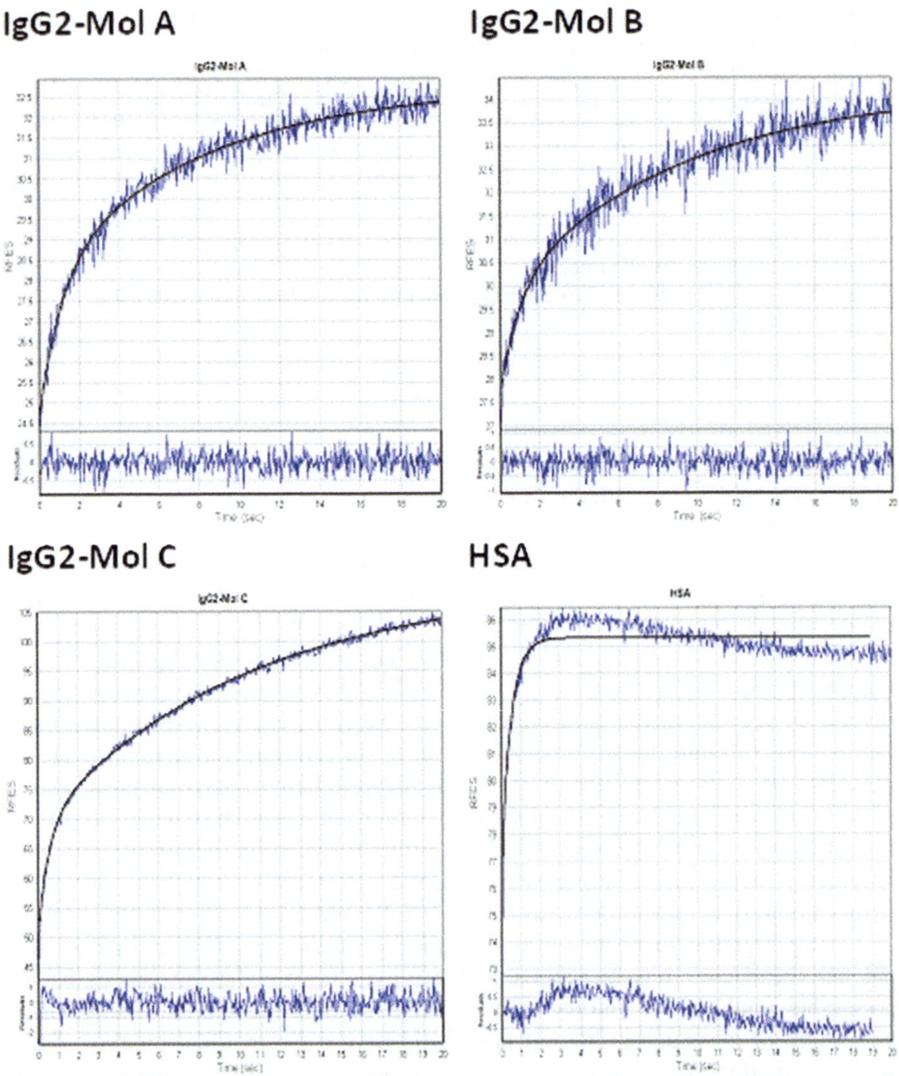

Fig. 7.17 Stopped-flow binding kinetics for mAbs and HSA with bis-ANS. Raw data is compared at 20s time scale.

Table 7.8 Bis-ANS binding kinetic parameters derived from a biphasic kinetic model fit to stopped-flow data

Protein	HMMS (%) (SEC)	k_1 (s^{-1})	k_2 (s^{-1})	$A1$ (RFES)[a]	$A2$ (RFES)[a]	R^2
IgG2-Mol A	0.3	2.44±0.02	0.048±0.003	3.34±0.23	15.15±0.48	0.99
IgG2-Mol B	1.0	1.41±0.24	0.122±0.010	2.18±0.11	4.49±0.10	0.94
IgG2-Mol C	0.7	1.78±0.09	0.093±0.002	1.71±0.34	40.7±0.18	0.99
HSA	17.3	11.82±0.35	1.32±0.05	3.16±0.54	7.01±0.27	0.98

[a]Relative fluorescence emission signal

7.6.3 Conclusions

Novel approaches such as isothermal titration calorimetry and stopped-flow to extract more information out of extrinsic fluorescence dye-binding studies may provide newer insight into the aggregation propensity of biotherapeutics.

7.7 Conclusions

Biophysical methods such as differential scanning calorimetry, circular dichroism, (intrinsic) fluorescence spectroscopy, and AUC are widely applied to assess the higher-order structure of biotherapeutics and the impact of process and formulation stresses. The case studies in this chapter illustrate the use of a variety of biophysical methods in comparability studies as well as in troubleshooting product or process impact. Some novel applications of dynamic light scattering and fluorescence spectroscopy as well as isothermal titration calorimetry and stopped-flow are also provided. Combinations of the methods are often required to answer any given question due to the complementary information the methods provide.

References

Carpenter JF, Randolph TW, Jiskoot W et al (2009) Overlooking subvisible particles in therapeutic protein products: gaps that may compromise product quality. J Pharm Sci 98:1202–1205

Carpenter JF, Randolph TW, Jiskoot W, Crommelin DJA, Middaugh CR, Winter G (2010) Potential inaccurate quantitation and sizing of protein aggregates by size exclusion chromatography: essential need to use orthogonal methods to assure the quality of therapeutic protein products. J Pharm Sci 99:2200–2208

Cromwell MEM, Hilario E, Jacobsen F (2006) Protein aggregation and bioprocessing. AAPS J 8:E572–E579

Dam J, Schuck P (2000) Calculating sedimentation coefficient distributions by direct modeling of sedimentation velocity concentration profiles. In: Johnson ML, Brand L (eds) Methods in enzymology, vol 384. Part E: numerical computer methods. Elsevier Academic, New York, pp 185–212

Demeule B, Gurny R, Arvinte T (2007) Detection and characterization of protein aggregates by fluorescence microscopy. Int J Pharm 329:37–45

den Engelsman J, Garidel P, Smulders R, Koll H, Smith B, Bassarab S, Seidl A, Hainzl O, Jiskoot W (2011) Strategies for the assessment of protein aggregates in pharmaceutical biotech product development. Pharm Res 28:920–933

Horton N, Lewis M (1992) Calculation of the free energy of association for protein complexes. Protein Sci 1:169–181

ICH Q5E (2005) Comparability of biotechnological/biological products. Available from www.ICH.org

Kelly SM, Price NC (2000) The use of circular dichroism in the investigation of protein structure and function. Curr Protein Pept Sci 1:349–384

Kozlowski S, Swann P (2006) Current and future issues in the manufacturing and development of monoclonal antibodies. Adv Drug Del Rev 58:707–722

Philo JS (2009) A critical review of methods for size characterization of non-particulate protein aggregates. Curr Pharm Biotechnol 10:359–372

Roberts CJ (2006) Non-native protein aggregation: Pathways, kinetics, and shelf-life prediction. In: Murphy RM, Tsai AM (eds) Misbehaving proteins: protein misfolding, aggregation, and stability. Springer, New York, pp 17–46

Sackett DL, Wolff J (1987) Nile red as a polarity-sensitive fluorescent probe for hydrophobic protein surfaces. Anal Biochem 167:228–234

Schuck P (2000) Size distribution analysis of macromolecules by sedimentation velocity ultracentrifugation and Lamm equation modeling. Biophys J 78:1606–1619

Schuck P (2003) On the analysis of protein self-association by sedimentation velocity analytical ultracentrifugation. Anal Biochem 320:104–124

Singh SK (2011) Impact of product-related factors on immunogenicity of biotherapeutics. J Pharm Sci 100:354–387

Sreerama N, Woody RW (2000) Estimation of protein secondary structure from circular dichroism spectra: comparison of CONTIN, SELCON, and CDSSTR methods with an expanded reference set. Anal Biochem 287:252–260

Stafford WF, Braswell EH (2004) Sedimentation velocity, multi-speed method for analyzing polydisperse solutions. Biophys Chem 108:273–279

Strickland HE (1974) Aromatic contributions to circular dichroism spectra of proteins. CRC Crit Rev Biochem 2:113–175

Weiss WFI, Young TM, Roberts CJ (2009) Principles, approaches, and challenges for predicting protein aggregation rates and shelf life. J Pharm Sci 98:1246–1277

Zou Q, Luo Y (2010) Biophysical characterization for product comparability. BioPharm Int Suppl 12–17

Chapter 8
Biophysical Analysis in Support of Development of Protein Pharmaceuticals

Sreedhara Alavattam, Barthelemy Demeule, Jun Liu, Sandeep Yadav, Mary Cromwell, and Steven J. Shire

8.1 Introduction

Proteins as pharmaceuticals, especially monoclonal antibodies, have dominated recent new drug development and are projected to have a market value in excess of $140 billion worldwide by 2015 (RNCOS 2012). This development of proteins as pharmaceuticals has led to increased challenges compared to traditional small molecule drugs because of the increased complexity of the chemical and physical stability of protein pharmaceuticals. Stability of small molecule drugs often involves a limited number of chemically distinct functional groups which can be assayed using perhaps one or two methods. Proteins are biopolymers made up of 20 common amino acids and thus consist of a variety of functional groups susceptible to different chemical degradation pathways. In addition to this increased variety of chemical degradation pathways, the folding of the polypeptide chain is critical for maintaining efficacy and safety of the protein drug. Thus, the overall stability of a protein involves chemical degradation as well as physical degradation pathways. Although it is well recognized that proteins can undergo aggregation, it is important to consider that chemical alterations can induce conformational changes as well as promote formation of aggregates, i.e., the physical stability may be impacted by the chemistry. The overall stability of a protein is depicted in Fig. 8.1 where the native properly folded conformer is in equilibrium with the unfolded protein, which is actually an ensemble of different conformers. The aggregates that form may arise from an assembly of folded or unfolded forms, and all these forms may or may not be chemically altered. An important point to consider is that even in the native state,

S. Alavattam • B. Demeule • J. Liu • S. Yadav • M. Cromwell • S.J. Shire (✉)
Late Stage Pharmaceutical Development, Genentech, 1 DNA Way,
South San Francisco, CA 94080, USA
e-mail: alavattam.sreedhara@gene.com; demeule.barthelemy@gene.com; liu.jun@gene.com; yadav.sandeep@gene.com; cromwell.mary@gene.com; shire.steve@gene.com

L.O. Narhi (ed.), *Biophysics for Therapeutic Protein Development*, Biophysics for the Life Sciences 4, DOI 10.1007/978-1-4614-4316-2_8,

Labile Amino Acid Residues	Primary Reactions
Asn, Gln	Deamidation Racemization Isomerization
Asp	Hydrolysis Racemization Isomerization
Met, Cys, His,Trp,Tyr	Oxidation
Ser, Thr, Cys	Beta-Elimination Racemization
Cys	Disulfide Exchange

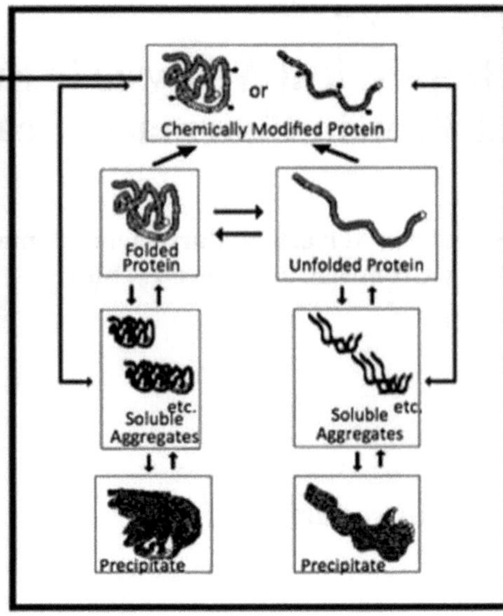

Fig. 8.1 Schematic of overall protein stability

amino acid residues, especially those accessible to solvent, are susceptible to chemical alteration. Thus, development of a stable protein formulation requires several assays to monitor the different chemical degradation routes as well as appropriate analytics to assess the physical state including conformation and aggregation of a protein. This necessitates the use of biophysical methods to investigate the physical properties of proteins, but this should not be done without the corresponding analysis of chemical degradation routes since generally the chemical and physical properties are interlinked and ultimately dictate the stability of the molecule. Often the optimal formulation conditions for physical stability are not optimal for chemical stability, and thus the challenge is to create a formulation where the rates for chemical and physical degradation are optimized to get the desired shelf life for a commercial drug product.

The choice of which biophysical methods to use is dictated by the types of studies and specific requirements that support protein drug development. The type of studies that are often done can be grouped into three general areas: (1) early screening assessments, (2) intense characterization, and (3) confirmatory studies. Early screening studies are used to select a molecule from several versions that are being considered for further development to initiate clinical trials. Once a molecule is chosen, several formulations need to be analyzed to screen impact on physical stability of the protein in these conditions. These screening assessments done early in development require high-throughput as well as small amounts of protein since there are a large number of samples with limited amount of material. Thus, biophysical methods that use small amounts of protein, which are rapid and amenable to automation technology would be considered. Studies that provide in-depth characterization can use more labor-intensive biophysical methods such as analytical ultracentrifugation (AUC). These studies may guide development work or provide a greater understanding of the physical properties of the protein to ensure long-term stability. The confirmatory studies provide alternative methods to support more rapid conventional analysis. There are several reviews that summarize the use of biophysical analysis in the development of protein biopharmaceuticals (Samra and He 2012; Kamerzell et al. 2011; Manta et al. 2011), and this chapter is not intended to be an all-encompassing review of the work that has been done. In this chapter we cite some key examples from the literature and present some examples on how biophysical analysis has been used at Genentech supporting protein pharmaceutical development.

8.2 Early Screening Assessments

8.2.1 Early Molecule Assessments

Early in development there may be several choices of molecule, and it would be prohibitively expensive to move forward with all the choices. In addition, although the molecules designed by research may be acceptable in early testing for activity, there are many attributes such as long-term stability, bioavailability, and manufacturability that may dictate the final choice of molecule. Generally as a first pass the primary structure of the molecule is used to identify particular "hot spots" for potential chemical degradation. Amino acid residues such as methionine and tryptophan are labeled as potential sites for oxidation (Ji et al. 2009; Wang 1999), and potential deamidation and aspartic acid isomerization sites are identified based on N and C terminal flanking residues (Robinson and Robinson 2004). Assessment of physical stability can be a more daunting task since it is not obvious how to link physical stability to primary sequence information. Attempts to assess propensity for aggregation and solubility (Agostini et al. 2012; Tartaglia et al. 2005, 2008) may be useful biophysical computational tools for choice of molecules, but the accuracy of such methods still needs to be verified for protein molecules in development. A nice example of the use of computational algorithms is the recent work of Chennamsetty et al. (2009a, b) where they used spatial aggregation propensity (SAP) using molecular dynamics simulation to predict aggregation prone areas in antibodies and to make mutants with enhanced stability. Stability of the antibodies was confirmed using turbidity, size exclusion chromatography (SEC), and differential scanning calorimetry (DSC) measurements. Recently Lauer et al. (2012) presented a developability index which is an *in silico* method for mAbs that uses SAP and the mAbs net charge to predict aggregation. Wu et al. (2010) have shown that structure-based engineering identified an aggregation promoting hydrophobic patch that could be modified with a carbohydrate moiety to provide a more soluble and stable analog of human recombinant IL13 antibody. All the above-published literature clearly indicates that computational methods can be used in understanding the physical behavior of mAbs, especially during candidate selection prior to entering clinical development. Although computational methods may be useful, the actual screening of protein molecule choices using biophysical analysis is highly desirable. As stated earlier since the amount of protein available is limited during such early assessments, assays are required that use small amounts of material. Thus, assays such as those SEC can be used to assess levels of aggregate in solution. However, assays that require small amounts of protein to assess particulate formation and physical properties such as viscosity may need to be developed. This can be very challenging especially for the development of high-protein concentration formulations such as those being developed for therapeutic monoclonal antibodies for subcutaneous (SC) delivery (Shire et al. 2004). As an example, viscosity measurements using standard rheometers generally require high volumes, but alternative techniques have been developed such as high-frequency quartz impedance methods (Saluja and Kalonia 2004, 2005; Patel et al. 2009), and several instrument companies are also offering commercial viscometers that can handle small volumes of proteins. Although these alternative techniques require smaller volumes, they may not be readily adaptable for high throughput. Dynamic light scattering (DLS) has been suggested as a way to predict the viscoelastic properties of proteins by monitoring protein–protein interactions (Saluja et al. 2007; Yadav et al. 2012). In this method the diffusion constant is determined as a function of protein concentration resulting in an interaction parameter, k_d. A negative value for k_d is indicative of attractive protein–protein interactions whereas a positive value represents repulsive interactions. It was shown that net attractive interactions determined using the k_d interaction parameter correlated with high viscosities measurements using quartz impedance viscometry (Yadav et al. 2012). Recently this method was adapted for high-throughput screening (HTS) using a DLS plate reader (Connolly et al. 2012). The results for 29 monoclonal antibodies showed overall a very good correlation between negative k_d values and viscosity measurements at high concentrations (Fig. 8.2) and demonstrate the utility of this approach.

8.2.2 Formulation Screening

Gibson et al. (2011) recently have reported the use of HTS methods for protein solubility using an IgG1 mAb. The authors were able to modify a PEG-induced precipitation method in a 96 well format and used ultra violet-visible (UV–Vis) spectroscopy to screen various buffer compositions and pH that help to maintain mAb solubility. Relative solubility profiles of both chimeric and human IgG1 mAbs were determined using this HTS method. He et al. (2010a) reported the use of extrinsic fluorescence in a 96-well format to detect and quantify IgG aggregation. It can be envisioned that several other analytical techniques, such as turbidity, DLS, and various chromatographic methods, can be utilized in the future in an HTS format to help

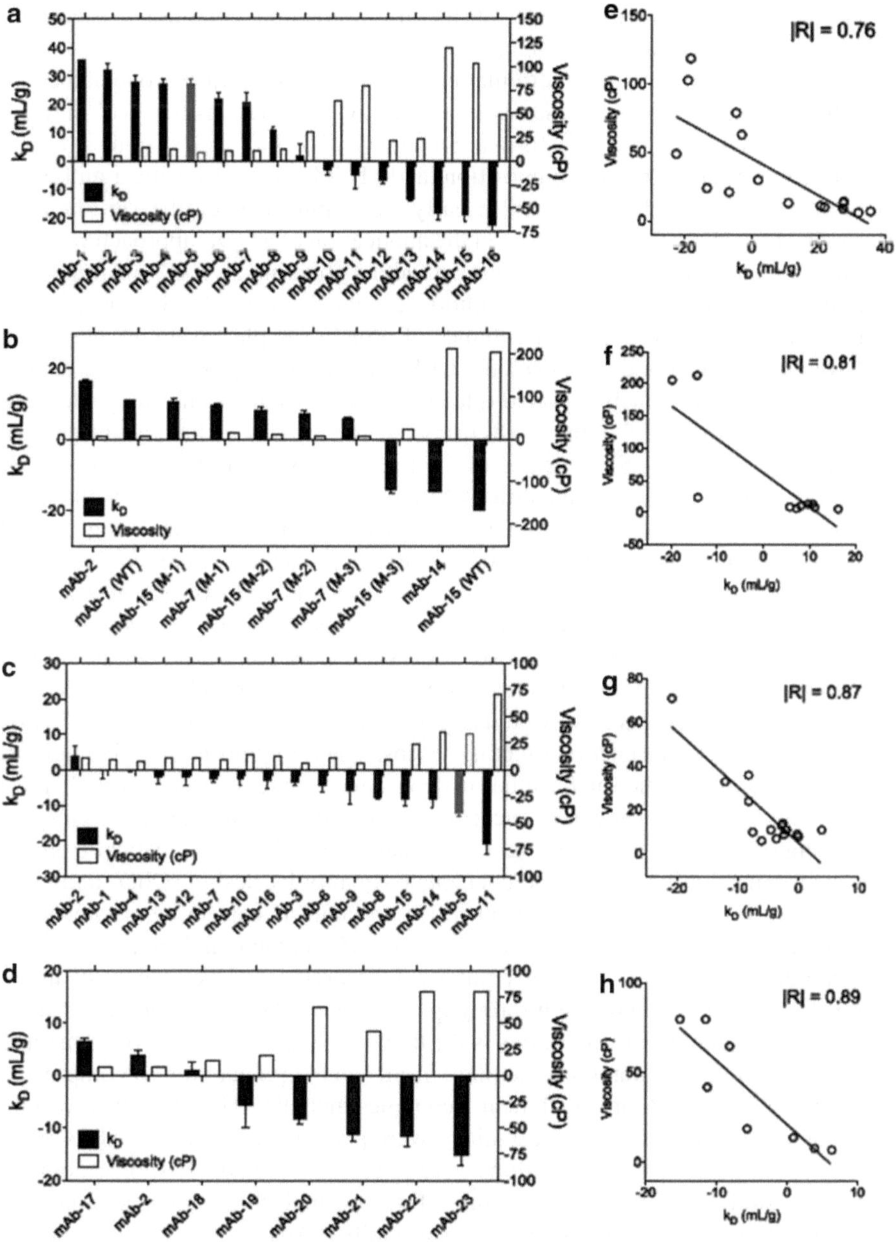

Fig. 8.2 Bar plots comparing kD (*filled square*) and mAb solution viscosity (*open square*) in (**a**) 20 mM His-OAc, pH 5.5, (**b**) 30 mM His-Cl, pH 6.0, (**c**) 200 mM Arg-Cl, pH 5.0, and (**d**) 200 mM Arg-Succ, pH 5.5. Scatter plots display correlation between k_d and viscosity for the corresponding mAbs in (**e**) 20 mM His-OAc, pH 5.5, (**f**) 30 mM His-Cl, pH 6.0, (**g**) 200 mM Arg-Cl, pH 5.0, and (**h**) 200 mM Arg-Succ, pH 5.5. Viscosity was measured at 175 mg/mL by cone and plate rheometry (published previously in Connolly et al. 2012)

determine protein solubility under different formulation conditions, especially using design of experiments (DOE) principles.

As high concentration mAb formulations become more prevalent, one of the factors that will need to be addressed is screening for formulations that will mitigate high solution viscosity in a HTS format. Since several factors such as ionic strength, protein concentration, buffer composition, and pH influence viscosity, a multivariate analysis for formulation screen in early development would be very helpful. Recently He et al. (2011) demonstrated the application of a DOE approach with HT formulation screening to identify the main factors affecting an IgG2 mAb thermostability and solution viscosity. The authors describe the use of DLS measurements of diffusion coefficients of small volumes of protein samples using an automated 384-well plate reader and correlated the DLS data to viscosity of concentrated protein solutions as previously reported (He et al. 2010b). In the same study, He et al. report the thermostability of the mAbs using a relatively new technique, differential scanning fluorimetry (DSF), instead of the traditionally used DSC, using 96-well plates and significantly low quantities of the mAb. Several factors as well as the interactions between factors on both the thermostability and viscosity using a 3^4 full factorial design (81 conditions) and robust statistical methods were explored. Full and reduced models considering main effects of factors and two-way interactions were appraised, and selection of optimal formulation conditions was evaluated based on data generated using the JMP® software. Conditions wherein the protein maintains thermostability and has reasonable viscosity at particular pH and protein concentrations can be easily viewed using the contour plot and profiler. In addition to estimating the main factors affecting viscosity and thermostability, the authors were able to apply prediction formulas to determine formulations that can help meet predetermined values. They have successfully explored a formulation design space consisting of different factors using a combination of HTS and statistical methods that contribute significantly to the concepts of quality by design (QbD) and DOE approach to formulation development. Such an analysis can be very valuable to determine optimized formulations for Phase III and commercial development.

Zhao et al. (2010) have recently published formulation development of mAbs using a robotic system and a high-throughput laboratory. While the individual study setup has been published before, this is the first of a kind of work that demonstrates the end to end of formulation screening—from sample preparation to comprehensive sample analysis. Many analytical techniques including UV–vis, DLS, various chromatographic methods, and turbidity—were performed for optimized formulation screens, and the best formulations were selected in a relatively short amount of time. The study demonstrates the value of such a high-throughput laboratory in preformulation development of proteins, especially mAbs.

Lehermayr et al. (2011) compared various biophysical properties of several monoclonal antibodies using a variety of techniques. Protein–protein interactions, which define mAb properties of low and high concentration solutions, were assessed via determination of the second virial coefficient (A_2) using SLS and DLS. As measuring A_2 values using SLS is tedious, the authors measured the interaction parameter k_d using DLS. A plot of k_d (from DLS) and A_2 (from SLS) for eight different mAbs provided a linear fit and showed the relation as $k_d = 1.06\, A_2 M - 8.9$, where M is the molecular weight of the mAb. Overall, this methodology using DLS could be used to analyze A_2 using low protein quantities and is amenable to HTS as demonstrated in this work.

8.3 Intense Characterization Studies

8.3.1 Biophysical Techniques in Intense Characterization

Biophysical measurements that are labor-intensive such as AUC are not amenable to a high-throughput methodology but can provide important information to guide further development of a protein therapeutic. As an example, Lu et al. (2008) used sedimentation velocity AUC to demonstrate its usefulness in monitoring long-term stability and molecular integrity of antibodies. Specifically, their data was used to support the notion that a single point mutant in the hinge region of an IgG4 (S241 to P241) led to an increase in stability of the molecule against freeze–thaw-induced aggregation.

AUC has also been used to characterize the complexes formed in vitro between an anti IgE–IgG1 antibody and IgE (Liu et al. 1995). Monoclonal antibodies that bind to free-circulating IgE can be used for the treatment of IgE-induced allergic asthma, as they prevent the loading of IgE on mast cells or basophils, which can result in release of inflammatory molecules such as leukotrienes and histamine after exposure to an allergen (Fig. 8.3). Theoretically since there are two sites on each IgE where anti IgE can bind and since each anti IgE molecule is bivalent, the complexes could become very large (Fig. 8.4). Sedimentation velocity (Fig. 8.5) and equilibrium measurements were used to determine weight-average molecular weight and size distribution (Fig. 8.6) to clearly show that this did not happen and that the complexes that formed were of limiting size. The complexes formed are dependent on the molar ratio of IgE:anti IgE as shown in a schematic diagram (Fig. 8.7). The formation of these complexes dictates the pharmacokinetics of the drug therapy since IgE has a clearance time of ~6 h whereas an IgG1, due to the binding to FcRN neonatal receptor, has a typical half-life of ~2 weeks in serum. The complexes that form between IgE and anti IgE take on the long half-life typically seen for an IgG1 (Fox et al. 1996). The amount of free IgE in plasma should be related to the clearance rate of anti IgE:IgE complexes, free IgE, unbound anti IgE, and relative binding affinities of high-affinity receptor for IgE and that of IgE with anti IgE (Fig. 8.8). Thus, the dose required for effective lowering of free IgE in plasma will be related to stability of the complexes when interacting with high-affinity receptor. Since it is possible to detect the formed IgE:anti IgE complexes using sedimentation velocity AUC, it is possible to perform competitive binding experiments using AUC. AUC experiments were performed with preformed anti IgE:IgE complexes at a molar ratio of 6:1 where there is an excess of anti IgE and a predominance of a trimeric species consisting of two anti IgE molecules bound to one IgE (Fig. 8.7). A soluble

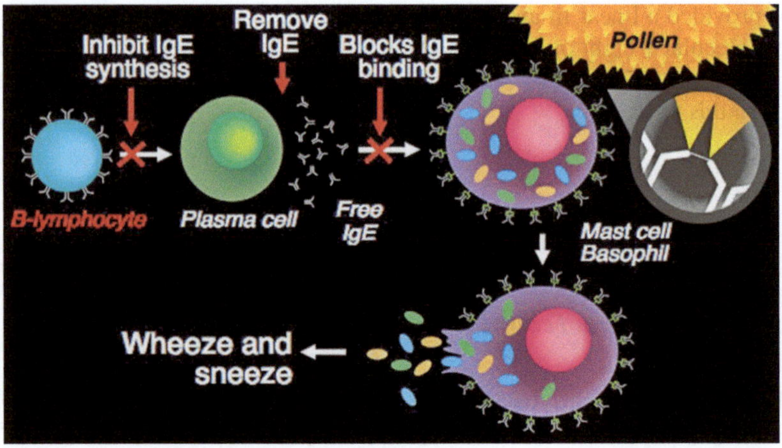

Fig. 8.3 Mechanism of action of an anti IgE mAb in the treatment of IgE allergic-mediated disease. The anti IgE mAb can inhibit IgE synthesis and binds free-circulating IgE preventing the IgE from interacting with the FcεRIα high-affinity receptors on the surface of mast cells or basophils

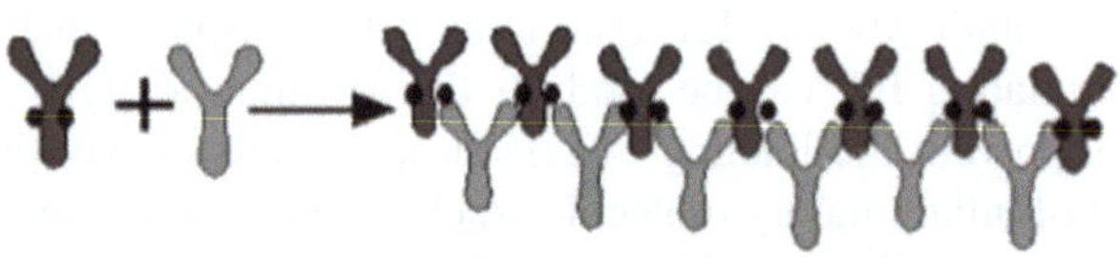

Fig. 8.4 The theoretical interaction of IgE (*dark grey*) with an anti IgE mAb (*light grey*) via binding of the two high-affinity Fc receptor sites on IgE

form of the high-affinity receptor, sFcεRIα, was then added at several molar ratios, and the results (Fig. 8.9a) clearly showed that the receptor has a greater affinity for IgE than anti IgE necessitating excess dosing of anti IgE to effectively lower free-circulating IgE. In particular, at a molar ratio of IgE:anti IgE of 1:6, all the IgE is incorporated into a trimeric complex at ~13.3 S, and upon addition of soluble receptor at 0.1:1 of complex, there is a reduction of this 13.3 S peak as well as a slight increase in the baseline between 8 and 9 S and an increase of the unbound anti IgE peak at ~7 S. As more soluble receptor is added, there are additional increases in the anti IgE peak and the appearance of a peak that is likely the receptor:IgE complex, which was characterized as a dimer (Fig. 8.7) in earlier work using sedimentation velocity and equilibrium AUC as well as static light scattering (SLS) (Liu et al. 1997). These data were generated using the differential sedimentation method of Stafford (Stafford 1992), which does not take into account diffusion resulting in a broadening of the peak and is the likely reason that at lower concentrations it is difficult to detect a single peak representative of the receptor:IgE dimer. As an extension of this technique, a competition binding AUC experiment was also done

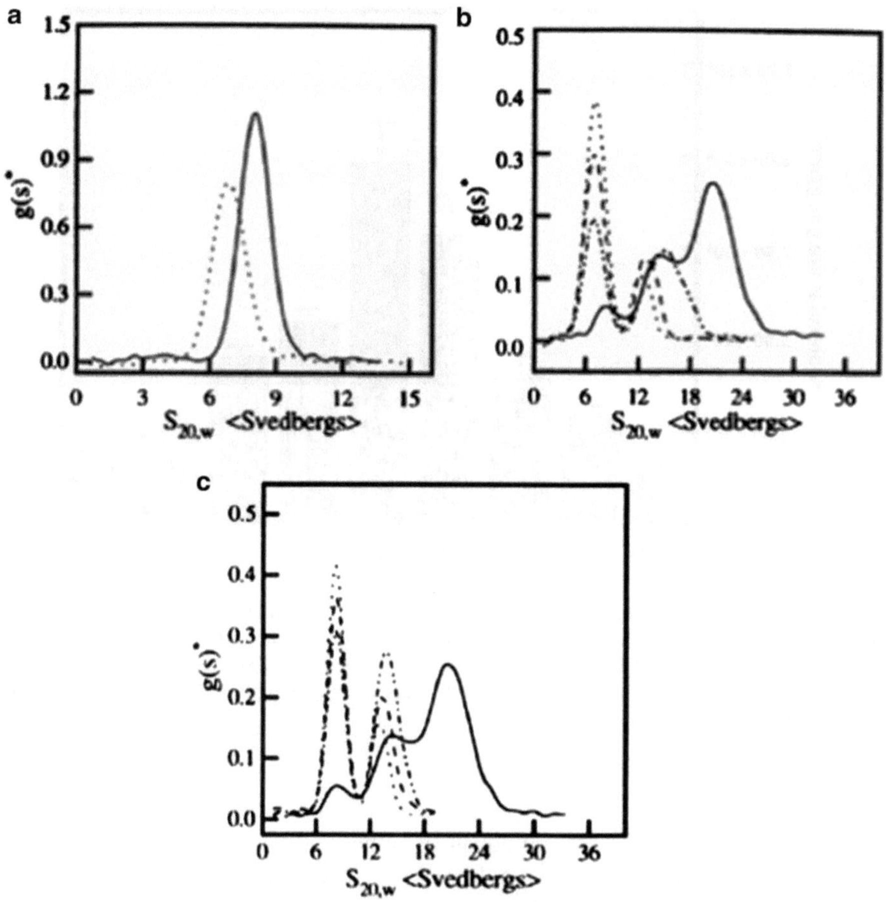

Fig. 8.5 Differential sedimentation coefficient distribution of IgE (*solid line*) and anti IgE (*dotted line*) monomers at 0.64 mg/mL (**a**); IgE and anti IgE complexes at various molar ratios (**b** and **c**) in PBS at 10°C. The molar ratios of IgE:anti IgE were as follows: (**b**) 1:1 (*solid line*), 1:3 (*dash-dotted line*), 1:6 (*dashed line*), and 1:10 (*dotted line*): (**c**) 1:1 (*solid line*), 3:1 (*dash-dotted line*), 6:1 (*dashed line*), and 10:1 (*dotted line*). The sedimentation coefficients have been corrected to the standard condition of water at 20°C. No faster moving species was observed in early scanning (previously published in Liu et al. 1995)

using a genetically engineered version of an anti IgE, referred to as anti IgEZ, which was designed to have a greater affinity for IgE. The competition binding analysis shows that anti IgEZ does indeed have higher affinity since even at a ratio of soluble receptor to IgE at 1:1 there is very little disruption of the complex (Fig. 8.9b).

These examples are typical of how biophysics can be used to study the behavior of biotherapeutics in vitro and help in determining adequate dosing. However, rather little information on chemical and physical stability is available once the drug is administered to humans, since characterization under physiological conditions

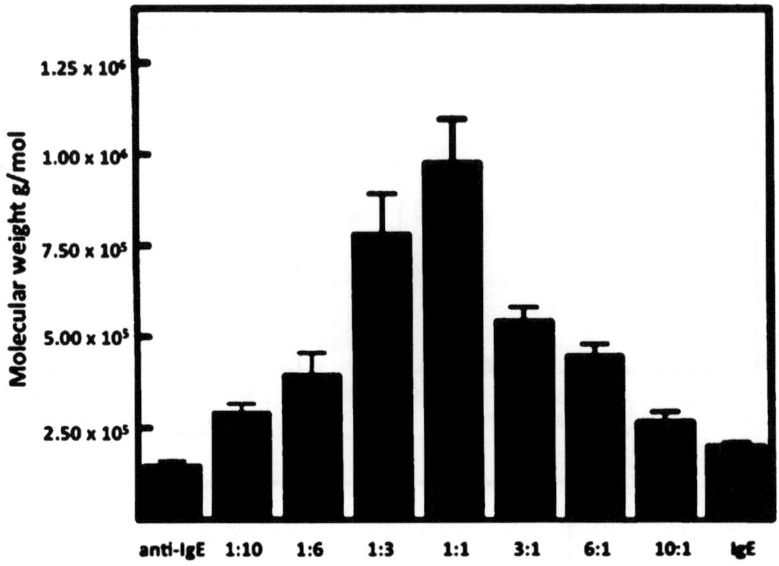

Fig. 8.6 Sedimentation equilibrium analysis of IgE and anti IgE mAb complex formation in PBS at 10°C. The weight-average molecular weights of complexes at different molar ratios were obtained by analyzing the data from three different rotor speeds (5,000, 7,000, and 10,000 rpm) as a single ideal species simultaneously. The error bars correspond to a 95% confidence interval (adapted from Liu et al. 1995)

requires specialized tools. Recent advances in hardware have resulted in the capability to use fluorescence optics in AUC and have been specifically used to detect an anti IgE molecule binding to IgE in a complex matrix such as human serum (Demeule et al. 2009a). Two main differences were noticed for the anti IgE binding to IgE in serum vs. PBS. The absence of the 21 S peak and the presence of an 8.7 S peak in serum for the anti IgE complex with IgE, instead of a 7.3 S peak in PBS, were noticeable. The absence of the 21 S peak and the different profile observed in serum compared to PBS underlines the importance to characterize the molecules under physiologically relevant conditions. Additionally, the absence of the 7.3 S peak that is replaced by a 8.7 S peak further confirms the distinct behavior in serum compared to PBS. The authors hypothesize that affinity of an anti IgE molecule towards IgE was higher in serum compared to that in PBS and that the largest anti IgE:IgE complex observed in serum was smaller than expected. AUC equipped with fluorescence optics was clearly demonstrated to be useful to characterize biopharmaceuticals under physiological conditions. Direct characterization in serum may now allow for better drug candidate selection and should be carefully considered during molecular assessment in early research stages.

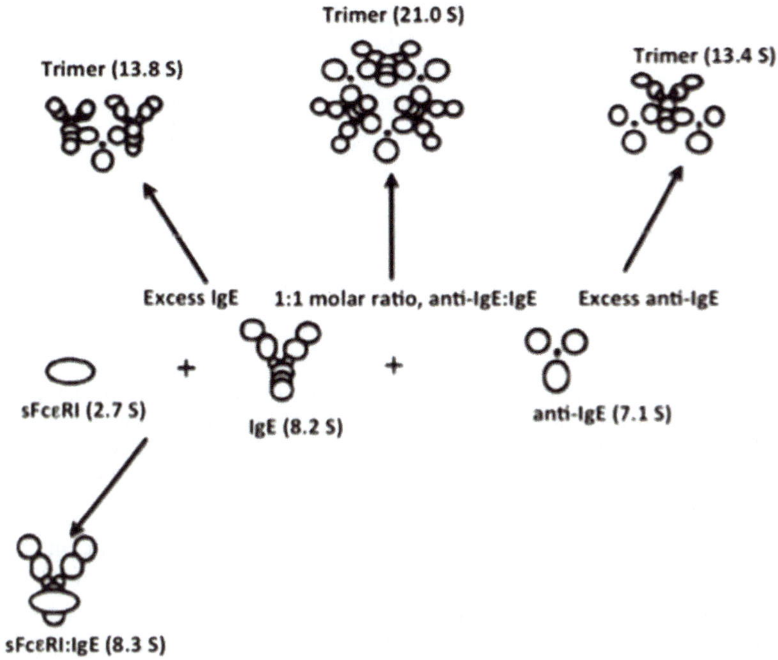

Fig. 8.7 Schematic diagram of complex formation by IgE and anti IgE and IgE and soluble high-affinity Fc receptor, sFcεRIα (adapted from Liu et al. 1997)

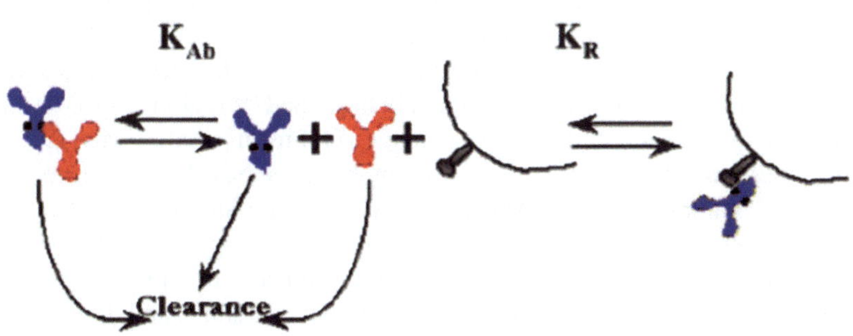

Fig. 8.8 Overall scheme for clearance of IgE (*purple*), anti IgE (*red*), and formation and clearance of complexes at excess anti IgE vs. binding of IgE with high-affinity receptor, FcεRIα (*grey*), on mast cells and basophils

8.3.2 Higher Order Structure Determinations

Alterations in conformation of proteins obviously require the use of biophysical analytical tools. Hydrodynamic measurements such as a diffusion coefficient from DLS or sedimentation coefficient from AUC provide such information as long as

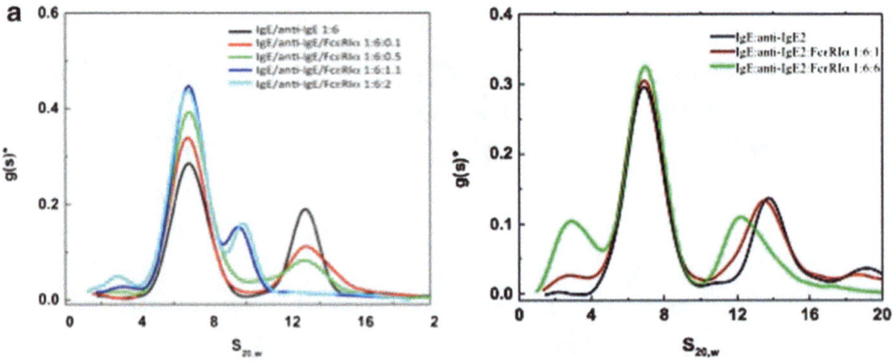

Fig. 8.9 Differential sedimentation coefficient distribution of anti IgE, FcεRIα, and IgE:anti IgE and IgE:FcεRIα complexes (**a**) and anti IgEZ, FcεRIα, and IgE:anti IgEZ and IgE:FcεRIα complexes (**b**): assessment of competition of binding of soluble high-affinity receptor, FcεRIα, with either preformed IgE:anti IgE or IgE:anti IgEZ complexes using AUC sedimentation velocity

the overall conformational changes are large enough to elicit a significant change in the determined values. SEC when used with an on-line light scattering detector to determine weight-average molecular weight can also detect mis-folded conformers as long as the shape change is large enough to allow for separation from the properly folded protein monomer (Philo 2006).

Spectrophotometric techniques such as circular dichroism (CD) have often been used to assess the folded state of a protein. Secondary structures (including α helices and β sheets) can be identified in the far-UV region, 190–240 nm, of the CD spectrum. These ellipticity changes stem from the distinct chiral positioning of the amide chromophores within different secondary structures (Johnson 1990). In addition, the local environment of aromatic chromophores such as tryptophan, tyrosine, and phenylalanine gives rise to a CD signal in the near-UV region, 240–340 nm, that can be used to infer tertiary structure changes (Johnson 1990). Thus, CD can be used to determine if the structure of a recombinant DNA-derived protein is that expected of "natural sourced" proteins or whether the structure is altered when exposed to stress conditions. As an example, the far-UV spectrum of a mAb reference material from Genentech showed a minimum at 218 nm indicative of the β-sheet character consistent with IgG1 antibodies (Brahms and Brahms 1980). Comparison of both near- and far-UV spectra between reference material and the stress panel shows no differences in structure by CD, as shown in Fig. 8.10, indicating that any subtle structural changes due to the applied stress (pH, oxidation, etc.) may not be picked up in this technique. However, it has been shown recently by Li et al. (2011) that CD can be used to monitor conformational changes in proteins during manufacturing process conditions. Li et al. (2011) describe the effect of pH or denaturants during purification and show a quantitative method to compare CD spectra using the OMNIC QC compare algorithm. Near-UV CD spectrum of several protein candidates was assessed either in sodium citrate buffer (pH 3.0) or in PBS (pH 7.4). Careful evaluation of the near-UV CD spectrum of these candidates

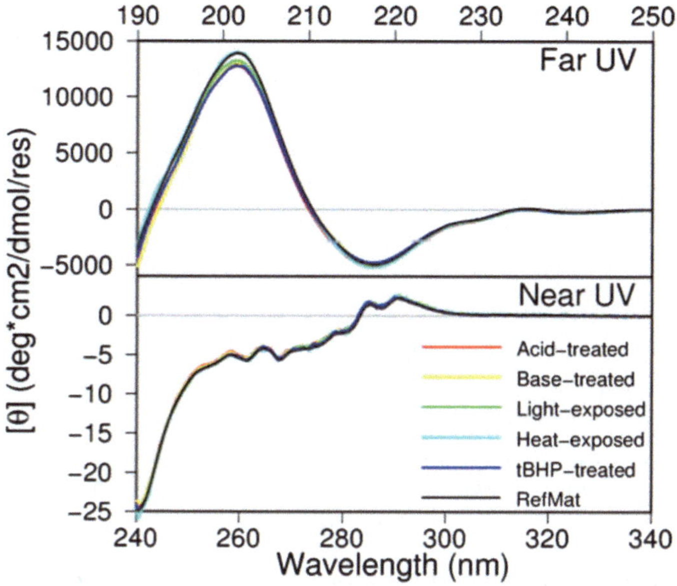

Fig. 8.10 Far- and near-UV CD spectra of a monoclonal antibody after exposure to different stress conditions

indicated that fewer changes were induced by low pH on candidate 1 over candidate 2, and hence the former may be more amenable to manufacturing processes at low pH. Far-UV CD spectrum of the same two candidates indicated that both proteins underwent the loss of the native β structure when incubated at pH 3.0; however, the secondary structure of candidate 1 was found to be relatively more stable than candidate 2.

Although CD is often used on its own to probe conformation of proteins, it can be a very valuable tool when used in conjunction with other biophysical techniques. The self-association properties and conformation of recombinant DNA-derived human relaxin, a pregnancy hormone, were studied by sedimentation equilibrium analytical ultracentrifugation and CD (Shire et al. 1991). The sedimentation equilibrium AUC data were consistent with a monomer–dimer self-association model with an association constant of ~6×10^5 M^{-1}. An approximate five fold increase in weight fraction of human relaxin monomer elicited by dilution of the protein resulted in no change in the far-UV CD spectrum at 220 nm.

In contrast, after the same increase in weight fraction of monomer, the near-UV circular dichroism spectra for human relaxin showed a significant decrease in the intensity of the CD bands near 277 and 284 nm. Although human relaxin has two tryptophan residues, the near-UV CD spectra exhibit only a broad shoulder near 295 nm rather than the strong CD bands often found for tryptophan. Moreover, there is little change in this broad band after dilution of human relaxin to concentrations that resulted in a five fold increase in the monomer weight fraction (Fig. 8.11).

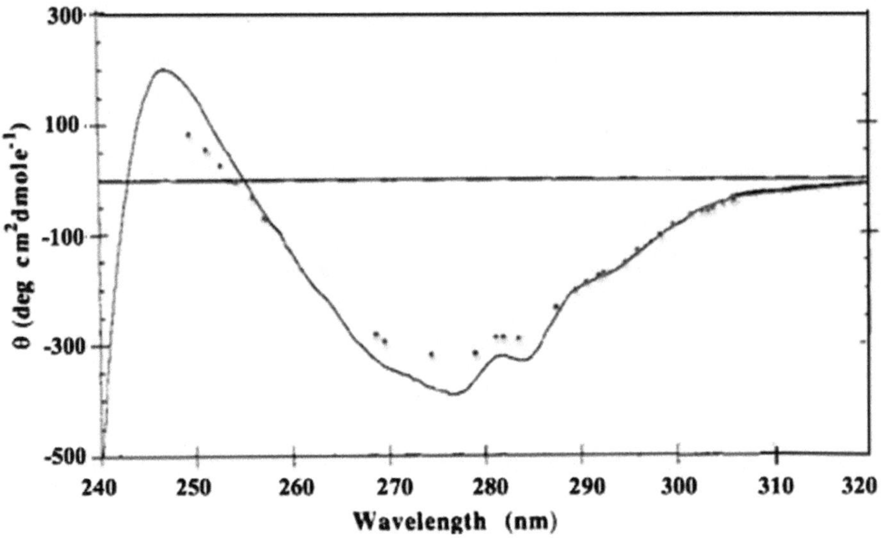

Fig. 8.11 Near-UV circular dichroism of human relaxin at 0.5 mg/mL (*solid line*) and 20 μg/mL (*dotted line*). Relaxin at 0.5 mg/mL was thermostated at 20 °C in a 1-cm cell, whereas relaxin at 20 μg/mL was in an unthermostated 10-cm cylindrical cuvette. The temperature in the sample compartment was ~27°C during the data collection process. The CD data were collected at 0.25-nm intervals at a spectral bandwidth of 0.5 nm and are the result of an average of three scans using an average time for each single data point collection of 5 s for the 0.5 mg/mL samples and the result of an average of ten scans using an average time for each single data point collection of 10 s for the 20 μg/mL sample. The weight fraction of human relaxin monomer estimated from the determined association constant by sedimentation equilibrium AUC of $100(g/L)^{-1}$ is 0.13 at 0.5 mg/mL and 0.50 at 20 μg/mL (adapted from Shire et al. 1991)

These data suggest that dissociation of the human relaxin dimer to monomer is not accompanied by large overall changes in secondary structure or alteration in the average tryptophan environment, whereas there is a significant change in the tyrosine environment. This conclusion was affirmed by the X-ray crystal structure of human relaxin, which crystallized as a dimer with the lone tyrosine from each monomer at the dimer interface (Eigenbrot et al. 1991). Thus, the solution studies were in good agreement with the crystal studies, suggesting that the determined crystal structure is very similar to the structure of the protein in solution.

Another spectroscopic technique widely used to assess conformational changes in proteins is Fourier transform infrared (FTIR) spectroscopy, which can be used to probe secondary structural elements including α helices and β sheets in solution as well as in solid-state dosage forms. These structural features appear in FTIR spectra as broad, characteristic absorption bands in the regions 1,700–1,620 cm^{-1} (Amide I) and 1,600–1,500 cm^{-1} (Amide II), among others (Byler and Susi 1986; Dong et al. 1990; Jackson and Mantsch 1995). These absorption bands are caused by a combination of bending and stretching vibrations of bonds along the peptide backbone. Since the technique can be used to assess conformation of proteins in solution as

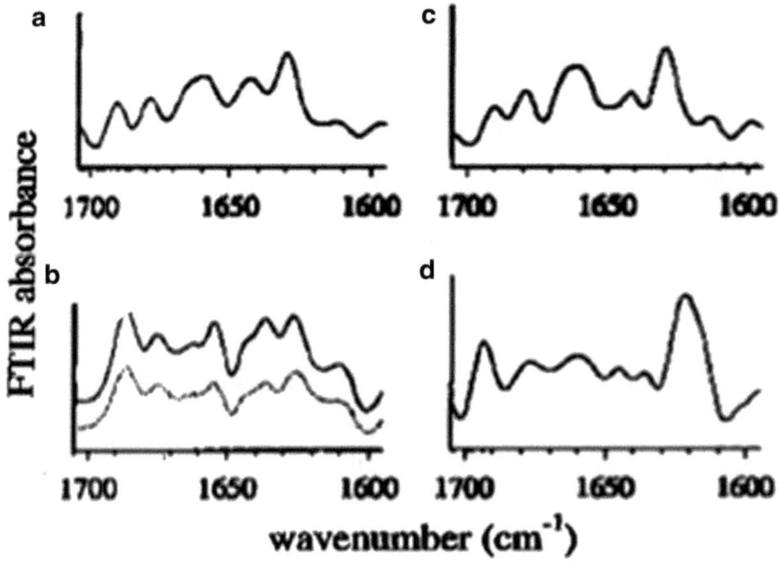

Fig. 8.12 Inverted second derivative of the amide I IR spectra of rhDNase I: (**a**) in aqueous solution (EGTA treated, no incubation), (**b**) lyophilized-untreated power (*solid line*) and lyophilized EGTA-treated power (*dotted line*) (no incubation for either sample), (**c**) deamidated form (aqueous protein incubated in 1 mM $CaCl_2$ for 120 days at 40°C), and (**d**) aggregated form (aqueous EGTA-treated protein incubated in the absence of exogenous calcium ions for 120 days at 40°C) (published previously in Chen et al. 1999)

well as in the solid state, it allows for assessments of excipients used to stabilize the protein during drying. Studies using omalizumab clearly showed that the native conformation of a monoclonal antibody could be preserved when a lyoprotectant such as sucrose was added to the formulation. Freeze drying in the absence of lyoprotectant resulted in formation of covalent aggregates, which were linked by disulfide bonds as shown by nonreducing and reducing SDS polyacrylamide gel electrophoresis (SDS PAGE) (Andya et al. 2003). FTIR has also been used to investigate the conformational stability of Pulmozyme in the aqueous and solid states (lyophilized). Pulmozyme is a recombinant DNA-derived DNase used for the treatment of cystic fibrosis and requires calcium ions for stability and activity (Chen et al. 1999). Exogenous calcium can be removed by treatment with EGTA leaving one tightly bound calcium ion per rhDNase molecule. Analysis of the FTIR spectra in the amide III region in either the aqueous or lyophilized state demonstrated that removal of exogenous Ca^{2+} by EGTA treatment had little effect on the secondary structure (Fig. 8.12, Table 8.1). This result for the aqueous state was confirmed using CD that also showed that there was no large overall change in the secondary or tertiary structure upon the removal of calcium. The primary degradation route for rhDNase in solution is deamidation. For the EGTA-treated protein, there was also severe covalent aggregation, e.g., formation of intermolecular disulfides facilitated

Table 8.1 FTIR analyses of various rHDNase I samples[a] (published previously in Chen et al. 1999)

	Secondary structure (%)		
Sample	α helix	β sheet	Other[b]
Aqueous solution, with Ca^{2+}[c,d]	21±2	23±3	56±6
Aqueous solution, EGTA treated	20±2	26±2	54±3
Lyophilized, with Ca^{2+}[d]	13±2	41±3	46±2
Lyophilized, EGTA treated	14±2	45±3	41±3
Deamidated form[e]	21±1	25±2	54±2
Aggregated form[f]	10±1	44±4	46±3

[a]The secondary structure of rhDNase was calculated by Gaussian curve fitting the original amide III spectra
[b]Other secondary structure includes random coil and turns and extended chains
[c]The aqueous solution contained 1 mM calcium chloride
[d]Data from Saluja and Kalonia (2004)
[e]Aqueous protein incubated in 1 mM $Cacl_2$ for 120 days at 40°C
[f]Aqueous EGTA-treated protein incubated in the absence of exogenous calcium ions for 120 days at 40°C

by the cleavage of Cys173-Cys209, which resulted in complete loss of activity. The aggregates had a very different secondary structure compared to the native protein. In particular, the β-sheet band observed at ~1,620 cm^{-1} wave number in the amide I second derivative spectra was increased.

For the protein lyophilized in the presence of Ca^{2+}, there was no increase in deamidated species during solid-state storage; however, some aggregation was observed. For the lyophilized EGTA-treated protein, aggregation was even more pronounced, and there was some loss in enzymatic activity upon reconstitution. Thus, the removal of calcium ions by EGTA treatment decreased the stability of rhDNase in both the aqueous and solid states even though no large overall calcium-induced structural changes could be observed by FTIR or CD. This example shows one of the problems in using spectroscopic methods such as FTIR or CD to assess structural alterations. Unless there are gross overall changes, it is unlikely that these techniques can detect the more subtle changes that may occur which still can lead to large amounts of aggregate and loss of potency. Thus, biophysical techniques that can detect subtle conformational changes are required, and recent improvement of an old technique, hydrogen–deuterium (H/D) exchange introduced by Linderstrøm–Lang in 1954 (Hvidt and Linderstrom-Lang 1954), may address those needs.

8.3.3 H/D Exchange by MS

Hydrogen/deuterium exchange mass spectrometry (HDX-MS) utilizes mass spectrometry to identify hydrogen atoms that are exchanged for deuterium and is based on the principle that the kinetics of H/D exchange is greatly affected by local structure of the peptide bond. Houde et al. (2009) used HDX-MS to study changes to IgG1mAb conformation as a result of deglycosylation by comparing the glycosylated and deglycosylated forms of the antibody. Two regions of the IgG1 (residues 236–253 and 292–308) were found to have altered exchange properties upon deglycosylation. Previous studies using X-ray crystallography and NMR have indicated that the residues in the same region are involved with Fc receptor binding confirming the HDX-MS is a useful tool to show subtle conformation changes.

Burkitt et al. (2010) recently reported the use of HDX-MS in the analysis of oxidized mAbs. Structural alterations in a number of segments of the Fc region corresponding to large changes in the hydrogen exchange profile of residues 247–253 of the heavy chain that contains the peptide LFPPKPKDTL were observed for an IgG1 mAb after oxidative stress. The peptide is located in the Fc region of the antibody adjacent to Met253 and spatially close to Met430, both of which are found to easily oxidized under various conditions.

Zhang et al. (2012) have recently demonstrated the use of HDX-MS to study the aggregation phenomenon of bevacizumab. The authors demonstrate that freeze–thaw-induced aggregation formed native aggregates that increased with F/T cycles and inversely proportional to protein concentration, whereas, the molecule underwent nonnative-like aggregation after thermal stress. The three CDRs on the molecule showed reduced H/D exchange indicating that they are involved in strong intermolecular contacts during F/T-induced aggregation. Structural comparisons of the aggregates after F/T or thermal stress using intrinsic Trp fluorescence and extrinsic ANS fluorescence confirmed that HDX-MS analysis showed that unfolded structures are predominant during thermal stress-induced aggregation.

8.3.4 *Assessing Impact of Chemical Changes on Physical Stability*

While chemical changes are examined using various analytical methods, they typically do not monitor structural changes that may accompany the chemical change. Liu et al. (2008) have recently reported the structural changes that accompany Met oxidation in the CH_2 domain of an *Escherichia coli* expressed Fc protein. The authors report that methionine oxidation led to subtle changes in the protein conformation and used various biophysical techniques such as CD, DSC, and NMR to confirm their results. Using far-UV CD, it was reported that methionine oxidation led to an ellipticity decrease of about 10% at 218 nm, indicating a small but detectable change in the secondary structure of the protein. Near-UV CD spectra showed a positive increase in ellipticity indicating that Met oxidation also leads to changes in the tertiary structure of the protein as well. The authors also used 2D, ^1H–^{15}N HSQC NMR experiments to corroborate their findings on the structural impact of Met oxidation. The resonances of many of amino acids in close proximity to the oxidized Met are affected post H_2O_2 treatment indicating that upon oxidation, the protein undergoes a dramatic conformational change. Using transgenic mice with human FcRn, Wang et al. (2011) have recently reported that oxidized Met residues (Met 252 and Met 428) have significantly shorter serum half-life.

Bertolotti-Ciarlet et al. (2009) have used surface plasmon resonance and cell binding assays to study the impact of methionine oxidation on the binding of two humanized IgG1 antibodies to Fcγ receptors and to the neonatal Fc receptor (FcRn). SPR analysis showed an increase in k_d values that was approximately proportional to level of Met oxidation, reaching a several-fold higher value in highly oxidized species. The authors conclude that while Met oxidation did not result in substantial changes to Fcγ binding (except FcγRIIa), binding to FcRn was significantly affected. Wang et al. (2011) further evaluated the impact of Met oxidation on serum half lives of two humanized IgG1 mAbs in transgenic mice with human FcRn. Results obtained from these studies corroborate the fact that Met oxidation leads to significant reduction of half-life in serum after Met oxidation in the Fc region of the mAb and correlates directly to the level of Met oxidation and binding constants as measured by SPR. The above studies indicate that techniques that help understand and prevent methionine oxidation during mAb production would be a significant addition to the arsenal used in characterizing proteins.

8.3.5 *Biophysical Techniques in High-Concentration Formulation Development*

The recent increase in development of monoclonal antibody therapeutics for indications requiring administration in a clinic, physician's office, or by self-administration at home has resulted in the need for liquid subcutaneous (SC) formulations to ensure user and patient convenience. Since monoclonal antibody therapies are generally high dosing, on the order of several mg/kg, administration via the SC route with acceptable dose volumes (<1.5 mL) necessitates development of mAb formulations over a concentration range from 100 to 200 mg/mL or even higher depending on the indication. Development of such high-concentration formulations can result in several manufacturing, stability, analytical, and delivery challenges (Shire et al. 2004). Manufacturing of high-concentration protein formulations is typically done using tangential flow filtration (TFF), which has limitations as the viscosity of the solution being concentrated increases leading to high back pressure during the TFF process that may exceed the capacity of the pumps. The high protein concentrations required for formulations of mAbs often result in a dramatic increase in the viscosity of the solution as shown by the concentration–viscosity profile for three mAbs (Fig. 8.13). These three mAbs are constructed from the same human IgG1 Fc framework. mAb2 and mAb3 have very similar behavior, and the solid line is a fit to the extended Mooney equation, which assumes only repulsive hard-sphere interactions (Ross and Minton 1977):

$$\eta = \eta_0 \exp\left[\frac{[\eta]c}{1-\frac{k}{v}[\eta]c}\right] \tag{8.1}$$

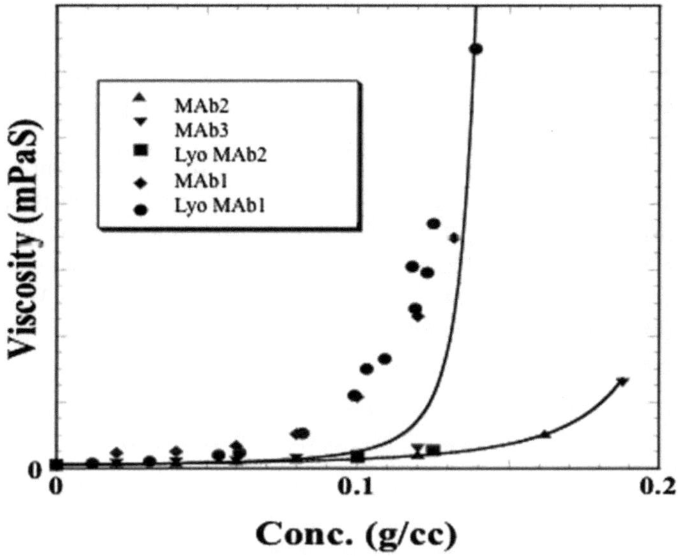

Fig. 8.13 Viscosity of mAbs1–3 as a function of protein concentration. mAb1, mAb2, and mAb3 aqueous samples have a composition of 266 mM sucrose, 16 mM histidine, and 0.03% polysorbate 20 at pH 6.0. mAb2 reconstituted lyophilized samples are at either 100 mg/mL MAb2 in 240 mM trehalose, 40 mM histidine, and 0.04% polysorbate 20 or 125 mg/mL in 300 mM trehalose, 50 mM histidine, and 0.05% polysorbate 20, pH 6. The solid curves are the result of a nonlinear regression of all the data to the modified Mooney equation using a solvent viscosity of 1.1 mPas and intrinsic viscosity of 6.3 cm^3/g (adapted from Liu et al. 1995)

where η_0 is the solution viscosity, η is the solvent viscosity, [η] is the intrinsic viscosity of the protein, *k* is a "crowding factor," and ν is the Simha parameter related to the ellipsoid of revolution used to model the protein. The viscosity–concentration profile for mAb1, on the other hand, cannot be described by the hard-sphere model (Fig. 8.13, Liu et al. 2005) suggesting that there are other interactions in addition to those involving excluded volume that mitigate the viscosity behavior of this particular mAb. It was hypothesized by Liu et al. that net-attractive interactions that result in reversible concentration-dependent self-association increase the viscosity of a concentrated monoclonal antibody in aqueous solution. At low concentrations the three monoclonal antibodies are essentially monomeric in solution as determined by SEC analysis. Thus, in order to determine if, in fact, mAb1 undergoes self-association at high concentration, the analysis needs to be done at high concentration. This is a challenging task since most analyses are done at low concentrations, typically at 1 mg/mL. AUC sedimentation equilibrium has been used to investigate protein self-association at high concentration by using thin plastic gaskets as centerpieces (Minton and Lewis 1981), but this is not an easy method due to deformations of the gasket material when torquing the AUC cells. The thin plastic gaskets that result in a narrow cell pathlength are necessary in order to deal with the high refractive index that results in a deviation of the light path away from the detecting photomultiplier in the centrifuge (Gonzalez et al. 2003). An alternative technique previously described by Minton is preparative AUC (Minton 1989). In this method a preparative centrifuge is used at low speeds to generate sedimentation equilibrium concentration gradients. After equilibrium is attained the centrifuge tubes are loaded onto a low-volume fraction collector (Brandel®, Brandel Inc, Gaithersburg, MD), and 5 μL fractions are dispensed into a 96-well plate. All the wells are then diluted to bring the absorbance reading down to within the dynamic range of the UV plate reader (Fig. 8.14). The resulting absorbance as a function of radial position is then used to determine the apparent weight-average molecular weight, $M_{\mathrm{w,\ app}}$, at each radial position using the following equation:

$$C(r) = C_0 \exp\left[\frac{M_{\mathrm{w,app}}(1-\bar{v}\,\rho)(r^2 - r_0^{\ 2})}{2RT}\right] \tag{8.2}$$

where $c(r)$ is the protein concentration at radial position r, c_0 is the initial loading protein concentration, $\bar{v}$ is the partial specific volume, ρ is the buffer density, ω is the angular velocity, and r_0 is the reference radial position. At these high concentrations there is a huge amount of non-ideality. Since mAb2 and mAb3 viscosity can be accounted for using the Mooney equation, it is assumed that mAb2 and 3 exist mainly as monomeric molecules in solution, and the non-ideality correction for the charge and excluded volume effects can be obtained in the absence of added NaCl for an IgG1 monomer with 150 kDa molecular weight. These corrections are obtained from the relationship between apparent molecular weight, M_{a}, at weight/volume concentration c and actual molecular weight, M (Chatelier and Minton 1987):

$$M_{\mathrm{a}} = M\left[1 + c\left(\frac{\mathrm{d}\ln\gamma}{\mathrm{d}c}\right)\right] \tag{8.3}$$

where γ is the activity coefficient of the monoclonal antibody. Assuming that mAb2 and 3 are essentially monomeric in solution leads to a multiplicative correction factor as a function of c when M is set equal to 150 kDa. The resulting corrected molecular weight for mAb1 (Fig. 8.14) supports the hypothesis that this monoclonal antibody undergoes a concentration-dependent reversible self-association. Another technique, SLS, at high concentrations has essentially corroborated the AUC analysis (Scherer et al. 2010). A rigorous analysis of the SLS data shows that mAb1 self-associates much more than mAb2. Thus, while literature has focused on protein–protein interactions in dilute solutions, this work has emphasized the need to develop newer techniques to understand intra-protein and inter-protein interactions in highly concentrated protein solutions, especially because self-association of proteins under such conditions appears to be crucial to appreciate the underlying solution viscosity (Yadav et al. 2010). Kanai et al. proposed that the observed self-association of mAb1, at pH 6.0, originates from multiple Fab–Fab interactions.

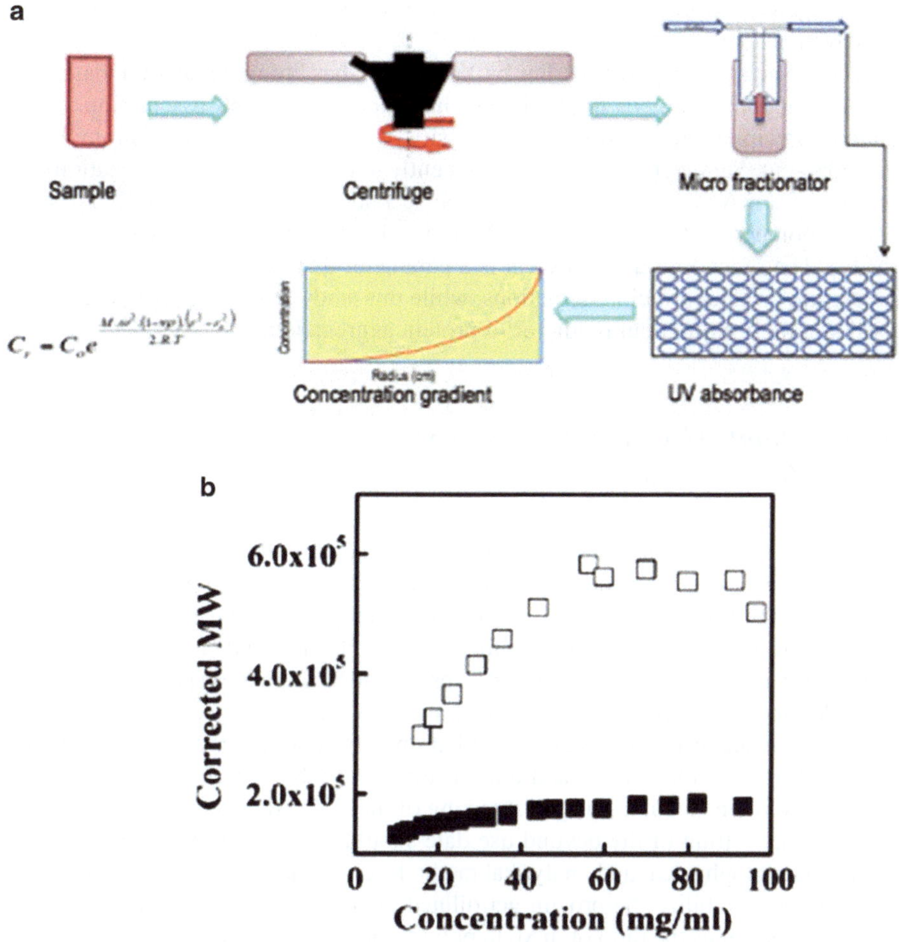

Fig. 8.14 Overview of a preparative AUC experiment. Centrifugation was carried out in an Beckman XL/A without the installed optics using a swinging bucket rotor. After attainment of sedimentation, equilibrium samples were fractionated using a Brandell micro fractionator and collected into a 96-well quartz plate and diluted to generate an absorbance vs. radial position gradient that was used to generate a concentration for analysis to determine weight-average molecular weight (overview drawing of preparative AUC provided by Sandeep Yadav, corrected MW plot previously published in Liu et al. 1995)

More recently, Yadav et al. (2011) showed that the presence of specific attractive interactions at pH 6.0 leads to the self-association and high viscosity of mAb-1 at high concentrations. They also demonstrate that exposed charged residues, especially histidyl residues, in the CDR of mAb1 are critical in determining the self-associating and highly viscous behavior observed at high concentrations. The authors used various biophysical techniques including CD, sedimentation equilibrium, high-frequency ultrasonic rheology, DLS, zeta potential, and net charge determinations to understand the physical properties of several IgG1 molecules (Yadav et al. 2012).

The recent study reported earlier by Yadav et al. has brought attention that developing orthogonal methods for mAbs that have potential for self-association, aggregation, precipitation, and/or high viscosity is crucial for successful molecule development. Saluja et al. (2010) have recently demonstrated that aggregation propensity of a mAb closely correlated to the second virial coefficient (B_2), k_d and k_s (interaction parameters and for the diffusion and sedimentation coefficients, respectively). The second virial coefficient is a parameter that is known to be valuable in predicting protein–protein interactions, while this study also suggests that k_d can be used as a high-throughput predictor of protein aggregation.

8.3.6 Biophysical Techniques to Support Clinical In-Use Studies

Pharmacists are responsible for setting a "beyond use" date based on USP 797, wherein the beyond use date for the compounded sterile preparation (CSP) is defined as the time by which the compounded preparation must be used to avoid risks for product degradation, contamination, etc. Physical and chemical stability of the CSP can be difficult to maintain over extended storage, especially since the formulation components are diluted within the IV bag contents. Recent published reports have suggested the use of extended time, beyond that recommended by the manufacturer, for the storage and administration of CSP. These recommendations were based on inadequate analytical testing of the CSP. Alavattam et al. 2012 demonstrate that setting of the beyond use date should be carefully assessed using the appropriate biophysical and analytical methods, given the fact that important excipients that help stabilize the protein get diluted upon dilution in saline IV bags. This study shows that as the ratio of mAb to polysorbate 20 increased, the levels of aggregates also increased with agitation as detected by the SEC assay. This suggests that there is less surfactant to compete with the protein for the air–liquid interface in the IV bag, thereby allowing mAb to aggregate at the interface. This study also suggests that while the ratio of mAb to polysorbate 20 is an important contributing factor to form soluble aggregates, headspace also played a key role. No clear trending was observed in the subvisible particle analysis using the light obscuration (HIAC–Royco) method. When performing particle analysis, there may be variability in the particulate measurements attributed to sample handling and/or instrument accuracy, and care must be taken to ensure the proper procedures are followed (Cao et al. 2010). However, it was evident that mAb1 and mAb2 with 0% polysorbate had a larger number of subvisible particles at the ≥10 μm size in bags that contained headspace, suggesting at least 0.0001% polysorbate may be necessary to protect the protein at the air–water interface from particulate formation. It appeared that subvisible particles increased upon agitation in IV bags with headspace in mAb1 solutions that did not contain polysorbate 20, whereas the subvisible particles were not significantly different for mAb2 under similar solution conditions. When mAb3 was diluted tenfold into PO IV bags and subjected to agitation, a significant increase in soluble aggregates was observed (13.5% over 1 h). A larger increase in soluble aggregates was observed for mAb3 than for mAb1 and mAb2 over similar time scales despite higher polysorbate 20 levels in the IV bags illustrating the high sensitivity of mAb3 to agitation. Visible precipitation of the protein was observed on further agitation of mAb3 in the IV bag, and no additional testing was performed on samples that were agitated for longer periods of time. A clear correlation between subvisible particles and soluble aggregates could not be found in this study.

Kumru et al. (2012) recently reported the physical stability of an IgG4 monoclonal antibody upon dilution into intravenous (i.v.) bags containing 0.9% saline. Soluble aggregates and subvisible particles were characterized using a variety of analytical (including SEC) and biophysical methods such as light obscuration, nanoparticle tracking analysis (NTA), microflow-digital imaging (MFI), and turbidity measurements. Characterization studies with FTIR microscopy and extrinsic fluorescence spectroscopy demonstrated that isolated particles contained native-like secondary structure with partially altered tertiary structure, compared with heat-denatured and non-stressed controls. Transmission electron microscopy (TEM) and MFI analysis showed particles had an amorphous morphology of varying sizes.

Demeule et al. (2009b) recently reported a case study on characterizing trastuzumab samples diluted in either saline or dextrose solutions using asymmetrical flow field-flow fractionation (FFF), fluorescence spectroscopy, fluorescence microscopy, and TEM. When trastuzumab samples were analyzed by FFF using a standard separation method, no difference could be seen between trastuzumab diluted in sodium chloride and trastuzumab diluted in dextrose. However, during FFF measurements made with appropriate changes in the protocol, trastuzumab aggregates were detected in 5% dextrose. This suggests that trastuzumab aggregates are overlooked if the FFF analysis of trastuzumab diluted in dextrose solution is performed using 0.9% NaCl. The authors were also able to confirm aggregates using fluorescence microscopy and TEM.

All the above studies emphasize the need to perform orthogonal biophysical tests to clearly understand different mechanisms for protein degradation in IV bags and support the proper use of protein therapeutics in clinical trials.

8.3.7 Biophysical Techniques in Antibody Drug Conjugate Formulation Development

Trastuzumab-DM1 (T-DM1) is one of the most successful antibody drug conjugates (ADCs) to enter clinical trials in the recent past. It has a non-cleavable linker using lysine chemistry to link trastuzumab (Herceptin®) to the potent antimicrotubule thiol containing a maytansine analog (DM1) via a thioether linkage. In addition to developing analytical techniques to characterize the antibody itself, ADCs need special chemical and biophysical techniques to understand various properties of the small

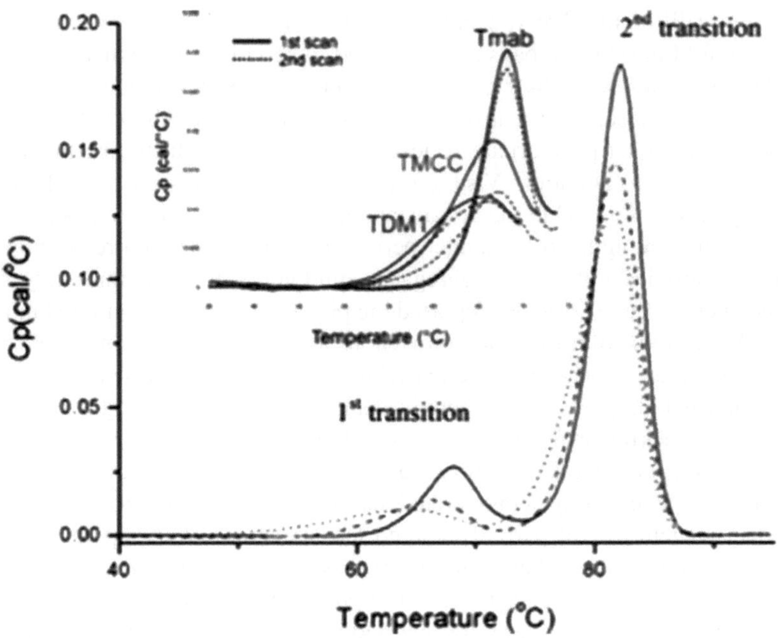

Fig. 8.15 Comparison of the DSC thermograms of Tmab, T-MCC, and T-DM1 ([antibody] = 1 mg/mL) in 10 mM sodium succinate, 6% (w/v) trehalose dihydrate, and 0.02% (w/v) polysorbate 20. Inset in the figure depicts reversibility of the first (CH_2) transition for Tmab, T-MCC, and T-DM1 (previously published in Wakankar et al. 2010)

molecule chemotherapeutic drug in order to evaluate drug-to-antibody ratio (DAR), hydrophobicity due to the small molecule, drug distribution, linker, and drug stability as well as influence on protein conformation upon drug binding. The challenges in developing in vivo assays for ADC characterization have recently been reported and will not be a focus in this chapter (Stephan et al. 2011). Due to the large variance in physical properties of the antibody itself (typically water soluble) and the small molecule drug (typically lipophilic), it is often found that aggregation is a major issue in ADCs. These aggregates seem to reflect the properties of the small molecule drug in that they are driven by lipophilic association (Hollander et al. 2008). Hollander et al. have used SEC to identify conditions during conjugation steps in order to identify reaction conditions that minimize aggregation of calicheamicin conjugated antibodies. They found that a combination of propylene glycol and octanoic acid produced the least aggregate and high protein recovery with good drug loading. However, further characterization of the ADC has not been reported in this chapter. Wakankar et al. (2010) have recently reported thorough characterization of T-DM1, including the stability of Tmab (unconjugated IgG1) and the intermediate T-MCC using a variety of techniques such as DSC, SEC, CE-SDS, and LC-MS. DSC analysis showed two transitions typical of IgG1 (Fig. 8.15). DSC data from this ADC showed that conjugation with DM1 leads to a 4.4°C drop in the melting temperature of the first transition due to CH_2 domain (t_m is in the order of Tmab>TMCC>T-DM1) whereas only a 0.8°C drop in the second transition due to CH_3 domain was noticed. These results clearly suggest that conjugation of Tmab has significant impact on the thermal stability of the CH_2 domain. Further stability characterization using SEC showed aggregate formation was in the order of Tmab<T-DM1<TMCC, an order that is not predicted based on DSC results alone. This indicates that mechanisms other than the typical hydrophobic interactions may be important, and a plethora of biophysical and analytical techniques need to be carefully considered during formulation development.

In addition to understanding the effect of conjugation on the melting temperature and potential aggregation propensity of the ADC, it is important to understand the effect of the small molecule drug on the higher order structure of the molecule. Since most of the small molecule cytotoxic drugs exhibit strong absorbance in the near UV region, analysis using CD spectroscopy may be complicated in this region. Little to no information has been published in characterizing ADCs using this technique. Any alterations in higher order structure will probably be difficult to tease out from variations because of DAR, especially in the near UV region. Far-UV CD, however, may provide information and can be employed to understand higher order structure for ADCs.

8.4 Confirmatory Studies

Given that most analytical methods have limitations in characterizing proteins, alternate methods are used to characterize protein pharmaceuticals. These orthogonal techniques provide vital information that is necessary for clinical development and are typically done during licensure activities. One area that has been heavily scrutinized by regulatory agencies is the determination of aggregates and fragments by SEC. SEC determines molecular size based on the hydrodynamic volume of the molecules and thus molecules with an altered shape, or extensive glycosylation may yield incorrect molecular weight when assessed using typical globular protein standards. This problem can be circumvented using on-line light-scattering detectors to determine weight-average molecular weight. However, interactions with the column matrix may result in loss of aggregates as well as decreased resolution. Often organic modifiers or salts are added to decrease the matrix interactions, but this may also impact the actual distribution of aggregate size. In addition, if the protein undergoes reversible self-association that is concentration dependent, the dilution on typical SEC analysis may result in detection of only lower molecular weight species. Regulatory agencies have become aware of such limitations and have been asking for confirmatory studies using orthogonal biophysical techniques to verify the accuracy of the SEC determination that are often used to evaluate protein aggregation (Carpenter et al. 2010). Although several biophysical techniques such as SLS, FFF, and AUC have been used to characterize aggregates and fragments, AUC

Table 8.2 Comparison of AUC and SEC results of a monoclonal antibody

	% HMWS (total)	
Samples	AUC	SEC
A	1.2 (31.4% RSD)[a]	0.2
B	1.9	1.3
C	4.8	5.5
D	6.4	6.6
E	7.6 (7.9% RSD)	7.2

Sample A consists of a representative mAb drug product batch, Sample B consists of a mAb batch subjected to light exposure at 1.2 mlux hours, Sample C consists of a mAb batch subjected to light exposure at 3.6 mlux hours, Sample D consists of a mAb batch subjected to acid treatment at pH 3.2, and Sample E consists of purified basic variants from IE-HPLC

AUC analytical ultracentrifugation, *HMWS* high-molecular weight species, *RSD* relative standard deviation, *SEC* size exclusion high-performance liquid chromatography

Samples A and B have HMWS levels that are below the limit of quantitation of the AUC technique (Gabrielson and Arthur 2011)

has been one of the more extensively used techniques to confirm SEC analysis(Berkowitz 2006), as described in Chap. 5.

8.4.1 AUC as an Orthogonal Technique to SEC

As an example of the use of AUC in a confirmatory study, a monoclonal antibody reference standard from Genentech was analyzed by SEC and AUC. The amount of high-molecular weight species (HMWS), generally dimer, and low-molecular weight species (LMWS) were determined by SEC to be 0.1–0.2% and 0.0–0.1%, respectively. There was no difference in HMWS upon dilution, suggesting that the aggregates are non-dissociable. Measurements were also made on the same sample at 0.55 mg/mL in formulation buffer using AUC as an orthogonal technique to measure HMWS. The samples were loaded in Spin Analytical 12-mm 2-sector cells and spun at 20°C and 40,000 rpm in an XLA/I (Beckman) analytical ultracentrifuge. The sedimentation coefficient distributions $c(s)$ were calculated with Sedfit v12.1b (Schuck 2000) using absorbance data collected at 280 nm. In addition, various levels of aggregates were generated by light stress, acid treatment, or purified basic variants using IE-HPLC (Table 8.2) and analyzed using SEC and AUC. There was good agreement between AUC and SEC results in terms of HMWS quantitation, showing no evidence of SEC missing or underestimating major HMWS. A concordance plot demonstrating the agreement between AUC and SEC is shown in Fig. 8.16.

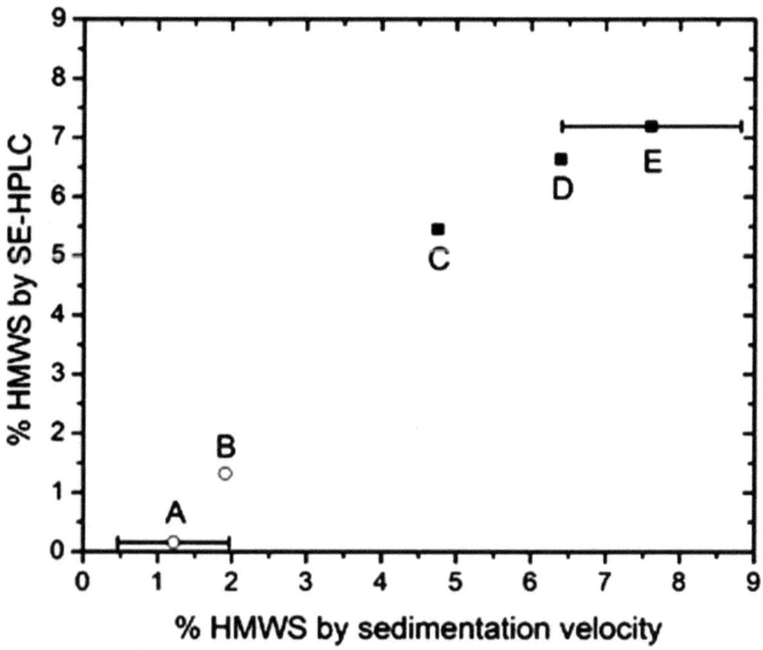

Fig. 8.16 Concordance plot of AUC sedimentation velocity and SEC analysis of a monoclonal antibody. AUC, analytical ultracentrifugation; HMWS, high-molecular weight species; SEC, size exclusion high-performance liquid chromatography. The error bars represent two standard deviations from $n=3$ determination. All other data points denote a single determination. Circles denote samples that have HMWS levels below the LOQ of the AUC technique (Gabrielson and Arthur 2011)

8.4.2 *Flow Field-Flow Fractionation as an Orthogonal Technique*

Another orthogonal technique that has been used to characterize protein aggregates is FFF (Liu et al. 2006; Rambaldi et al. 2011). This technique is a flow-based separation method, whereby separation is achieved by applying an externally generated field that is perpendicular to laminar flow within a buffer filled open channel. Although several different external fields have been used (Giddings 2000), characterization of protein aggregates has been done mainly by application of solution flow as the external perpendicular field and is termed flow FFF. This method, which covers a wide range of molecular sizes from 0.001 to 50 μm in size, provides assessments of size and quantity of protein aggregates without the use of standards or solid-state matrices. Among all the flow-based FFF methods, asymmetrical flow FFF (AF4), where only the bottom wall of the channel has a semipermeable membrane, is the one most commonly used for therapeutic proteins (Fraunhofer and

Table 8.3 Analysis of a mixture of three proteins (described in the text) using AF4, SEC, and AUC

Samples	MW (kDa)	% Species (actual)	% Species AF4 ($n=5$, mean±σ)	Accuracy[a] AF4 (%)	% Species SEC ($n=5$, mean±σ)	Accuracy[a] SEC (%)	% Species AUC ($n=6$, mean±σ)	Accuracy[a] AUC (%)
Protein I	150	32.99	31.97±0.13	1.02	32.80±0.03	0.19	31.1±0.00	1.9
Protein II	100	33.88	35.53±0.08	−1.65	34.31±0.05	−0.43	33.0±0.00	0.9
Protein III	50	33.13	32.50±0.05	0.63	32.87±0.04	0.26	35.9±0.00	−2.8

[a]Accuracy was determined by subtracting the experimental values from the actual values

Winter 2004). The applied cross flow drives macromolecules towards the membrane, and differences in diffusion coefficients create a concentration distribution in the laminar flow, resulting in different elution times. These elution times allow for determination of the translational diffusion coefficient, which can be used to compute the molecular weight assuming spherical structures for all species. Although there is no solid-state support for potential interaction of protein, there have been issues with protein interaction with the channel membrane. Liu et al. (2012) have investigated this problem and show that with correct choice of membrane and solvent conditions, it is possible to use this technology to confirm results by SEC. In particular, they compared the analysis of three mAb preparations, protein I (a full-length monoclonal antibody), protein II (a single-armed antibody, i.e., with only one Fab), and protein III (a Fab fragment) using SEC, AUC, and AF4. The results show very good agreement between all three techniques (Table 8.3) demonstrating that AF4 is a viable technique that can be used for confirmatory studies.

8.5 Concluding Remarks

This chapter has covered a variety of biophysical techniques, which have been used in the development of protein biopharmaceuticals, including some perspective of formulation development and long-term stability. It has not meant to be an all-encompassing review but rather to demonstrate in what areas of development biophysics has been of use to aid in the production of a stable drug product and to compliment the other chapters in this book. We have not discussed the use of different biophysical methods to characterize protein particulates since that was covered in Chap. 4. We also have highlighted some of the ways we have used biophysical analysis at Genentech and hope this will be useful for researchers who wish to apply such technologies.

References

Agostini F, Vendruscolo M, Tartaglia GG (2012) Sequence-based prediction of protein solubility. J Mol Biol 421(2–3):237–241

Andya JD, Hsu CC, Shire SJ (2003) Mechanisms of aggregate formation and carbohydrate excipient stabilization of lyophilized humanized monoclonal antibody formulations. AAPS PharmSci 5(2):E10

Alavattam S et al (2012) Stability of IgG1 monoclonal antibodies in intravenous infusion bags under clinical in use conditions. J Pharm Sci 101(1):21–30

Berkowitz SA (2006) Role of analytical ultracentrifugation in assessing the aggregation of protein biopharmaceuticals. AAPS J 8(3):E590–E605

Bertolotti-Ciarlet A et al (2009) Impact of methionine oxidation on the binding of human IgG1 to FcRn and Fc gamma receptors. Mol Immunol 46(8–9):1878–1882

Brahms S, Brahms J (1980) Determination of protein secondary structure in solution by vacuum ultraviolet circular dichroism. J Mol Biol 138(2):149–178

Burkitt W, Domann P, O'Connor G (2010) Conformational changes in oxidatively stressed monoclonal antibodies studied by hydrogen exchange mass spectrometry. Protein Sci 19(4):826–835

Byler DM, Susi H (1986) Examination of the secondary structure of proteins by deconvolved Ftir spectra. Biopolymers 25(3):469–487

Cao S, Jiang Y, Narhi LO (2010) A light-obscuration method specific for quantifying subvisible particles in protein therapeutics. Pharmacopeial Forum 36(3):824–834

Carpenter JF et al (2010) Potential inaccurate quantitation and sizing of protein aggregates by size exclusion chromatography: essential need to use orthogonal methods to assure the quality of therapeutic protein products. J Pharm Sci 99(5):2200–2208

Chatelier RC, Minton AP (1987) Sedimentation equilibrium in macromolecular solutions of arbitrary concentration. I. Self-associating proteins. Biopolymers 26(4):507–524

Chen B et al (1999) Influence of calcium ions on the structure and stability of recombinant human deoxyribonuclease I in the aqueous and lyophilized states. J Pharm Sci 88(4):477–482

Chennamsetty N et al (2009a) Aggregation-prone motifs in human immunoglobulin G. J Mol Biol 391(2):404–413

Chennamsetty N et al (2009b) Design of therapeutic proteins with enhanced stability. Proc Natl Acad Sci USA 106(29):11937–11942

Connolly BD et al (2012) Weak interactions govern the viscosity of concentrated antibody solutions: high-throughput analysis using the diffusion interaction parameter. Biophys J 103:69–78

Demeule B, Shire SJ, Liu J (2009a) A therapeutic antibody and its antigen form different complexes in serum than in phosphate-buffered saline: a study by analytical ultracentrifugation. Anal Biochem 388(2):279–287

Demeule B et al (2009b) New methods allowing the detection of protein aggregates: a case study on trastuzumab. MAbs 1(2):142–150

Dong A, Huang P, Caughey WS (1990) Protein secondary structures in water from second-derivative amide I infrared spectra. Biochemistry 29(13):3303–3308

Eigenbrot C et al (1991) X-ray structure of human relaxin at 1.5 A. Comparison to insulin and implications for receptor binding determinants. J Mol Biol 221(1):15–21

Fox JA et al (1996) Tissue distribution and complex formation with IgE of an anti-IgE antibody after intravenous administration in cynomolgus monkeys. J Pharmacol Exp Ther 279(2):1000–1008

Fraunhofer W, Winter G (2004) The use of asymmetrical flow field-flow fractionation in pharmaceutics and biopharmaceutics. Eur J Pharm Biopharm 58(2):369–383

Gabrielson JP, Arthur KK (2011) Measuring low levels of protein aggregation by sedimentation velocity. Methods 54(1):83–91

Gibson TJ et al (2011) Application of a high-throughput screening procedure with PEG-induced precipitation to compare relative protein solubility during formulation development with IgG1 Monoclonal antibodies. J Pharm Sci 100(3):1009–1021

Giddings JC (2000) The field-flow fractionation family: underlying principles. In: Schimpf ME, Caldwell KD, Giddings JC (eds) Field-flow fractionation handbook. Wiley, New York

Gonzalez JM, Rivas G, Minton AP (2003) Effect of large refractive index gradients on the performance of absorption optics in the Beckman XL-A/I analytical ultracentrifuge: an experimental study. Anal Biochem 313(1):133–136

He F et al (2010a) Detection of IgG aggregation by a high throughput method based on extrinsic fluorescence. J Pharm Sci 99(6):2598–2608

He F et al (2010b) High-throughput dynamic light scattering method for measuring viscosity of concentrated protein solutions. Anal Biochem 399(1):141–143

He F et al (2011) Screening of monoclonal antibody formulations based on high-throughput thermostability and viscosity measurements: design of experiment and statistical analysis. J Pharm Sci 100(4):1330–1340

Hollander I, Kunz A, Hamann PR (2008) Selection of reaction additives used in the preparation of monomeric antibody-calicheamicin conjugates. Bioconjug Chem 19(1):358–361

Houde D et al (2009) Characterization of IgG1 conformation and conformational dynamics by hydrogen/deuterium exchange mass spectrometry. Anal Chem 81(14):2644–2651

Hvidt A, Linderstrom-Lang K (1954) Exchange of hydrogen atoms in insulin with deuterium atoms in aqueous solutions. Biochim Biophys Acta 14(4):574–575

Jackson M, Mantsch HH (1995) The use and misuse of FTIR spectroscopy in the determination of protein structure. Crit Rev Biochem Mol Biol 30(2):95–120

Ji JA et al (2009) Methionine, tryptophan, and histidine oxidation in a model protein, PTH: mechanisms and stabilization. J Pharm Sci 98(12):4485–4500

Johnson WC Jr (1990) Protein secondary structure and circular dichroism: a practical guide. Proteins 7(3):205–214

Kamerzell TJ et al (2011) Protein-excipient interactions: mechanisms and biophysical characterization applied to protein formulation development. Adv Drug Deliv Rev 63(13):1118–1159

Kumru OS et al (2012) Compatibility, physical stability, and characterization of an IgG4 monoclonal antibody after dilution into different intravenous administration bags. J Pharm Sci 101(2):3636–3650

Lauer TM et al (2012) Developability index: a rapid in silico tool for the screening of antibody aggregation propensity. J Pharm Sci 101(1):102–115

Lehermayr C et al (2011) Assessment of net charge and protein-protein interactions of different monoclonal antibodies. J Pharm Sci 100(7):2551–2562

Li CH et al (2011) Applications of circular dichroism (CD) for structural analysis of proteins: qualification of near- and far-UV CD for protein higher order structural analysis. J Pharm Sci 100(11):4642–4654

Liu J et al (1995) Characterization of complex formation by humanized anti-IgE monoclonal antibody and monoclonal human IgE. Biochemistry 34(33):10474–10482

Liu J, Ruppel J, Shire SJ (1997) Interaction of human IgE with soluble forms of IgE high affinity receptors. Pharm Res 14(10):1388–1393

Liu J et al (2005) Reversible self-association increases the viscosity of a concentrated monoclonal antibody in aqueous solution. J Pharm Sci 94(9):1928–1940

Liu J, Andya JD, Shire SJ (2006) A critical review of analytical ultracentrifugation and field flow fractionation methods for measuring protein aggregation. AAPS J 8(3):E580–E589

Liu D et al (2008) Structure and stability changes of human IgG1 Fc as a consequence of methionine oxidation. Biochemistry 47(18):5088–5100

Liu J et al (2012) Assessing and improving asymmetric flow field-flow fractionation of therapeutic proteins. In: Williams KR, Caldwell KD (eds) Field flow fractionation in biopolymer analysis. Springer, New York

Lu Y et al (2008) The effect of a point mutation on the stability of IgG4 as monitored by analytical ultracentrifugation. J Pharm Sci 97(2):960–969
Manta B et al (2011) Tools to evaluate the conformation of protein products. Biotechnol J 6(6):731–741
Minton AP (1989) Analytical centrifugation with preparative ultracentrifuges. Anal Biochem 176(2):209–216
Minton AP, Lewis MS (1981) Self-association in highly concentrated solutions of myoglobin: a novel analysis of sedimentation equilibrium of highly nonideal solutions. Biophys Chem 14(4):317–324
Patel AR, Kerwin BA, Kanapuram SR (2009) Viscoelastic characterization of high concentration antibody formulations using quartz crystal microbalance with dissipation monitoring. J Pharm Sci 98(9):3108–3116
Philo JS (2006) Is any measurement method optimal for all aggregate sizes and types? AAPS J 8(3):E564–E571
Rambaldi DC, Reschiglian P, Zattoni A (2011) Flow field-flow fractionation: recent trends in protein analysis. Anal Bioanal Chem 399(4):1439–1447
RNCOS (2012) Global-protein-therapeutics-market-forecast-to-2015. http://www.pharmaceutical-market-research.info/research/PMAAAWUF-Global-Protein-Therapeutics-Market-Forecast-to-2015.shtml
Robinson NE, Robinson AB (2004) Prediction of primary structure deamidation rates of asparaginyl and glutaminyl peptides through steric and catalytic effects. J Pept Res 63(5):437–448
Ross PD, Minton AP (1977) Hard quasispherical model for the viscosity of hemoglobin solutions. Biochem Biophys Res Commun 76(4):971–976
Saluja A, Kalonia DS (2004) Measurement of fluid viscosity at microliter volumes using quartz impedance analysis. AAPS PharmSciTech 5(3):e47
Saluja A, Kalonia DS (2005) Application of ultrasonic shear rheometer to characterize rheological properties of high protein concentration solutions at microliter volume. J Pharm Sci 94(6):1161–1168
Saluja A et al (2007) Ultrasonic storage modulus as a novel parameter for analyzing protein-protein interactions in high protein concentration solutions: correlation with static and dynamic light scattering measurements. Biophys J 92(1):234–244
Saluja A et al (2010) Diffusion and sedimentation interaction parameters for measuring the second virial coefficient and their utility as predictors of protein aggregation. Biophys J 99(8):2657–2665
Samra HS, He F (2012) Advancements in high throughput biophysical technologies: applications for characterization and screening during early formulation development of monoclonal antibodies. Mol Pharm 9(4):696–707
Scherer TM et al (2010) Intermolecular interactions of IgG1 monoclonal antibodies at high concentrations characterized by light scattering. J Phys Chem B 114:12948
Schuck P (2000) Size-distribution analysis of macromolecules by sedimentation velocity ultracentrifugation and Lamm equation modeling. Biophys J 78(3):1606–1619
Shire SJ, Holladay LA, Rinderknecht E (1991) Self-association of human and porcine relaxin as assessed by analytical ultracentrifugation and circular dichroism. Biochemistry 30(31):7703–7711
Shire SJ, Shahrokh Z, Liu J (2004) Challenges in the development of high protein concentration formulations. J Pharm Sci 93(6):1390–1402
Stafford WF 3rd (1992) Boundary analysis in sedimentation transport experiments: a procedure for obtaining sedimentation coefficient distributions using the time derivative of the concentration profile. Anal Biochem 203(2):295–301
Stephan JP, Kozak KR, Wong WLT (2011) Challenges in developing bioanalytical assays for characterization of antibody-drug conjugates. Bioanalysis 3(6):677–700
Tartaglia GG et al (2005) Prediction of aggregation rate and aggregation-prone segments in polypeptide sequences. Protein Sci 14(10):2723–2734
Tartaglia GG et al (2008) Prediction of aggregation-prone regions in structured proteins. J Mol Biol 380(2):425–436

Wakankar AA et al (2010) Physicochemical stability of the antibody-drug conjugate Trastuzumab-DM1: changes due to modification and conjugation processes. Bioconjug Chem 21(9):1588–1595

Wang W (1999) Instability, stabilization, and formulation of liquid protein pharmaceuticals. Int J Pharm 185(2):129–188

Wang WR et al (2011) Impact of methionine oxidation in human IgG1 Fc on serum half-life of monoclonal antibodies. Mol Immunol 48(6–7):860–866

Wu SJ et al (2010) Structure-based engineering of a monoclonal antibody for improved solubility. Protein Eng Des Sel 23(8):643–651

Yadav S, Shire SJ, Kalonia DS (2010) Factors affecting the viscosity in high concentration solutions of different monoclonal antibodies. J Pharm Sci 99(12):4812–4829

Yadav S et al (2011) Establishing a link between amino acid sequences and self-associating and viscoelastic behavior of two closely related monoclonal antibodies. Pharm Res 28(7):1750–1764

Yadav S, Shire SJ, Kalonia DS (2012) Viscosity behavior of high-concentration monoclonal antibody solutions: correlation with interaction parameter and electroviscous effects. J Pharm Sci 101(3):998–1011

Zhang A et al (2012) Distinct aggregation mechanisms of monoclonal antibody under thermal and freeze-thaw stresses revealed by hydrogen exchange. Pharm Res 29(1):236–250

Zhao H et al (2010) Formulation development of antibodies using robotic system and high-throughput laboratory (HTL). J Pharm Sci 99(5):2279–2294

Chapter 9
Case Studies Applying Biophysical Techniques to Better Characterize Protein Aggregates and Particulates of Varying Size

Tingting Wang, Sangeeta B. Joshi, Ozan S. Kumru, Srivalli Telikepalli, C. Russell Middaugh, and David B. Volkin

9.1 Introduction

From a pharmaceutical perspective, protein storage stability is defined by the analytical techniques used to monitor chemical and physical changes in protein structure. As new analytical methods with increased sensitivity are introduced, resulting in an enhanced ability to detect chemical alterations in amino acid residues (oxidation, deamidation, etc.) or physical changes in protein higher-order structure (partial unfolding, aggregation, etc.), questions and concerns can arise about how best to set specifications to define storage stability. For example, physical changes leading to protein aggregation have been historically monitored by a combination of size-exclusion chromatography (SEC) and visual assessments of solution clarity. As awareness of SEC limitations has grown (Carpenter et al. 2010a), combined with introduction of improved biophysical methods to detect protein aggregates of varying size at lower limits of detection (see chapters in this book), it has become apparent that a "new" physical pathway of protein degradation should be monitored as part of protein stability studies: protein particle formation in the submicron (0.1–1 μm) and subvisible (1–100 μm) size ranges (Carpenter et al. 2009; Singh et al. 2010).

The challenge of detecting "new" impurities or degradants, due to development of improved analytical methods, has been an inherent part of therapeutic protein drug development for decades. For example, as the ability to detect residual host cell impurities increased with the introduction of new analytical methods, the need for new control strategies and specifications for residual host cell protein and DNA

T. Wang • S.B. Joshi • O.S. Kumru • S. Telikepalli • C.R. Middaugh • D.B. Volkin (✉)
Department of Pharmaceutical Chemistry, Macromolecule and Vaccine Stabilization Center, University of Kansas, 2030 Becker Drive, Lawrence, KS 66047, USA
e-mail: tingting.wang@ku.edu; joshi@ku.edu; okumru@ku.edu; srivalli@ku.edu; middaugh@ku.edu; volkin@ku.edu

L.O. Narhi (ed.), *Biophysics for Therapeutic Protein Development*, Biophysics for the Life Sciences 4, DOI 10.1007/978-1-4614-4316-2_9,

also increased. This situation leads to improved product quality with lower and more consistent impurity levels. Similarly, potentially more stable protein formulations may be developed in the future as our ability to monitor and control the formation of protein aggregates and particles improves with the introduction of more sensitive biophysical methods (Carpenter et al. 2010b; Mire-Sluis et al. 2011).

The case studies in this chapter are presented as a series of illustrative examples based on the size of the protein aggregates being examined. This approach permits better comparisons across laboratories as investigators study different types of aggregates from a wide range of proteins using a variety of different physicochemical methods (Narhi et al. 2011). Case studies examining the use of biophysical methods to better characterize soluble protein aggregates in the size range of 1–100 nm are first presented followed by examples utilizing different biophysical techniques to better characterize submicron-sized protein particles (0.1–1 μm). The last section then presents examples of monitoring and evaluating larger subvisible (1–100 μm) and visible (>100 μm) particles in protein formulations. For each section, a brief overview of the methods used in the case studies are presented followed by descriptions of recent literature examples that have employed a combination of biophysical methods to better characterize the nature and composition of protein aggregates and particles formed from different environmental stresses.

9.2 Case Studies Using Biophysical Methods to Count, Size, and Characterize Protein Soluble Aggregates

9.2.1 Overview of Analytical Methodologies Used to Evaluate Soluble Protein Aggregates Under Native-Like and Denaturing Conditions

A variety of analytical approaches have been used to examine the size, amount, and nature of soluble protein aggregates under both nondenaturing native and denaturing conditions. For example, size-exclusion HPLC (SE-HPLC), sedimentation velocity analytical ultracentrifugation (SV-AUC), dynamic light scattering (DLS), and field-flow fractionation (FFF) techniques are commonly performed using nondenaturing solution conditions to determine the extent and size of soluble protein aggregates. Under denaturing conditions, that is, in the presence of additives such as sodium dodecyl sulfate (SDS) or urea, protein aggregates are typically detected by electrophoresis (SDS-PAGE, capillary SDS) or chromatography (SE-HPLC with SDS or urea in the mobile phase, d-SEC). These electrophoretic and chromatographic approaches are generally used to better characterize the size, amount, and nature of soluble protein aggregates (e.g., presence or absence of native and nonnative covalent, disulfide cross-links). More recently, mass spectrometry has been employed to determine the size and nature of soluble protein aggregates. A brief overview of the key analytical techniques used in the specific case studies described in this chapter is provided below, followed by a few illustrative examples from the literature which have used combinations of these analytical techniques to evaluate the size, amount, and nature of different types of soluble protein aggregates in the size range of 1–100 nm.

9.2.1.1 Size-Exclusion High-Performance Liquid Chromatography

Due to its ease of use and high sample throughput, size-exclusion high-performance liquid chromatography (SE-HPLC) is the most commonly utilized technique for sizing and quantifying soluble protein aggregates in the molecular weight range of 5–1,000 kDa (Mahler et al. 2009; Philo 2009; Arakawa et al. 2010). Larger protein aggregates cannot be accurately evaluated for their size since they migrate in the void volume of the column or detected at all if they cannot pass through the column matrix. Typically, larger aggregates need to be removed by filtration or centrifugation prior to injection onto the SE-HPLC column. One disadvantage of SE-HPLC for use in sizing soluble protein aggregates is that the molecular weight (MW) calibration standards typically used with this method are globular in nature. In contrast, protein aggregates are not necessarily spherical resulting in inaccurate MW estimates. In addition, during sample preparation or during the chromatography process, protein aggregates may reversibly associate or dissociate and will therefore either not be detected or not represent the equilibrium state of interest (Carpenter et al. 2009; Mahler et al. 2009). Nonetheless, SEC protein stability studies are abundant and are a well-accepted approach (Oliva et al. 2001; Gabrielson et al. 2007; Kiese et al. 2008, 2010; Tyagi et al. 2009; Van Buren et al. 2009; Bond et al. 2010; Nayak et al. 2011). Since an often major limitation of SE-HPLC is the potential for aggregates to interact with the column matrix, orthogonal methods such as analytical ultracentrifugation, FFF, or DLS, which monitor the hydrodynamic properties of proteins free in solution, are now being more frequently used to verify SE-HPLC results (Carpenter et al. 2010a; Bond et al. 2010).

9.2.1.2 Sedimentation Velocity Analytical Ultracentrifugation

An analytical approach often used to confirm SE-HPLC results is SV-AUC (Liu and Shire 1999; Gabrielson and Arthur 2011). Briefly, an optical centrifugation cell is loaded in a rotor that contains two chambers, one sample and one reference, with detection of protein concentration in the cells by UV absorbance. The sample is centrifuged at high g-force forming a moving boundary that migrates to the outside of the rotor over time. Lower molecular weight species take longer to sediment than their higher molecular weight (HMW) counterparts and this behavior is monitored by absorbance at 280 nm over time. Software data analysis packages (e.g., SEDFIT) are then used to calculate the percentage of each protein species based on their relative sedimentation time (Dam and Schuck 2004; Schuck 2000, 2004). Compared to SE-HPLC, SV-AUC requires no dilution of the protein into a mobile phase and is performed in the absence of any separation matrices. The resolution of SV-AUC is usually much greater than SE-HPLC and thus can better separate closely related aggregated species in a protein sample. SE-HPLC is a higher-throughput technique with straightforward data analysis, while SV-AUC is a more time-consuming, lower-throughput technique with more care necessary in the analysis and interpretation of data.

9.2.1.3 Field-Flow Fractionation

An alternative approach to SV-AUC to detect and quantify soluble protein aggregates is FFF (also known as asymmetrical flow field-flow fractionation or FFF) (Cao et al. 2009). The FFF technique can also be useful for analyzing submicron-sized (0.1–1 μm) protein particulates based on their molecular mass and size (Mahler et al. 2009; Reschiglian et al. 2005). FFF methods separate molecules due to differences in their hydrodynamic radius (i.e., diffusion coefficients) when moving through different velocity regions of a parabolic flow in a channel (Reschiglian et al. 2005). Recent reviews by Roda et al. (2009) and Yohannes et al. (2011) describe novel applications and recent developments of the FFF technique with different biopolymers and bioparticles. Taylor dispersion analysis (TDA), a related technique based on monitoring dispersion of molecules when injected into a flow of solvent which is used to determine diffusion coefficients and hydrodynamic radii of molecules, has also recently been evaluated for its ability to monitor and size protein aggregates (Hawe et al. 2011). The coupling of electrospray mass spectrometer with differential mobility analyzers (ES-DMA) has been used to measure the size distribution of protein aggregates (3–250 nm) by monitoring their gas phase electrophoretic mobility (Pease et al. 2008). This technique separates molecules based on their charge-to-aerodynamic size ratio (vs. ES-MS which measures the charge-to-mass ratio).

9.2.1.4 Light Scattering

Various light scattering techniques have been employed to monitor protein aggregation in solution. DLS not only determines the size of protein aggregates in the 1 nm to 1 μm size range but also can provide information about the polydispersity of the sample (Kiese et al. 2008, 2010; Mahler et al. 2005). A DLS instrument records the time-dependent fluctuations in the amount of scattered light when particles are moving under Brownian motion. From the changes in these intensity fluctuations, it is possible through autocorrelation analysis to determine a diffusion coefficient of the particles and, through the Stokes–Einstein equation, their hydrodynamic radius (Lomakin et al. 1999). Sizes can be estimated by an averaging technique known as cumulant analysis or from various forms of deconvolution analysis that to a limited extent resolve individual populations that differ in size by a factor of two or more. As examples of the utility of DLS for monitoring protein aggregation, Mahler et al.

(2005) and Kiese et al. (2008) have monitored the aggregation behavior of IgG1 monoclonal antibodies during stirring and agitation under a variety of solution conditions. Static light scattering (SLS) can monitor protein aggregation by measuring changes in the average intensity of scattered light and if performed at different angles and concentrations can determine changes in a protein molecule's radius of gyration (Murphy 1997). The radius of gyration of species in protein solutions can also be determined by small angle neutron scattering (SANS), for example, in the detection of microaggregates in different insulin preparations (Heldt et al. 2011). There are now numerous examples using SE-HPLC in conjunction with UV absorbance, refractive index, and multiangle light scattering (MALS) detectors to obtain absolute molecular weight of soluble, aggregated protein species (Bond et al. 2010; Wen et al. 1996; Ye 2006). In addition, UV optical density spectroscopy is routinely used to monitor the presence of aggregation through increases in optical density at 320–350 nm (Mach and Middaugh 2011). This approach has been widely used to assess the ability of pharmaceutical excipients to minimize protein aggregate formation in a high-throughput format (Bhambhani et al. 2012).

9.2.1.5 SDS-PAGE, cSDS, and d-SEC

The size, extent, and nature of soluble protein aggregates can also be evaluated under denaturing conditions. Electrophoresis is a standard way to study protein aggregation, in the molecular weight range of 5–500 kDa (Mahler et al. 2009), using an electric field to separate macromolecules based on their charge and weight. The most common electrophoretic technique used to separate and quantify protein aggregates of different molecular weight is SDS polyacrylamide gel electrophoresis (SDS-PAGE). SDS-PAGE not only allows size estimation but can also be used to estimate purity levels and evaluate for the presence of covalent, reducible (e.g., disulfide-linked) aggregates (Mahler et al. 2009). Noncovalent protein aggregates, however, are generally not detected since they are usually disassociated during sample preparation due to the presence of SDS. Capillary electrophoresis (CE) offers an attractive alternative to conventional gel electrophoresis due to its robustness, speed, enhanced resolution, and reproducibility as well as its automation and quantitative features. CE techniques are now routinely used in the biopharmaceutical industry in the quality control environments for purity testing as well as assessing stability and integrity of protein drugs. Although a number of CE modes are available, the most common technique used is capillary electrophoresis SDS (CE-SDS). A detailed description of these techniques can be found elsewhere (Kraly et al. 2006; Rustandi et al. 2008; Kilar 2003). In addition to electrophoresis, SE-HPLC can be run with SDS, urea, or guanidine hydrochloride in the mobile phase (denaturing SE-HPLC or d-SEC) to separate disrupted protein aggregates under both reduced and nonreduced conditions (Michels et al. 2007). Often, dissociation of aggregates occurs prior to unfolding of the monomeric species as the concentration of the unfolding agent is increased permitting further analysis. CE-SDS is more commonly used in quality control environments to quantitatively determine protein aggregate size and purity as well as confirm the presence/absence of disulfide bonds. In contrast, d-SEC is often utilized during formulation development to rank order protein stability under different conditions. A number of studies have been published that demonstrate the utility and feasibility of using these three methods (SDS-PAGE, CE-SDS, d-SEC) for purity and stability assessment of protein drug candidates under denaturing conditions, especially with monoclonal antibodies (Rustandi et al. 2008; Salas-Solano et al. 2006; Han et al. 2006; Hunt et al. 1996; Qi et al. 2009; Costello et al. 1992; Lubiniecki et al. 2010).

9.2.1.6 Mass Spectrometry

High-resolution mass spectrometry techniques are commonly used to characterize the primary structure of proteins as well as posttranslational modifications (Grillberger et al. 2009). The ability of MS methods to detect and characterize protein aggregates is also rapidly progressing and provides high-resolution and site-specific information. Mass spectrometry (MS) is used to selectively detect, quantify, and determine the covalent structure of a given molecule by determining the mass-to-charge (*m*/*z*) ratio (Watson and Sparkman 2008). Analysis by MS requires the molecule to be ionized and present in the gas phase. A typical mass spectrometer includes four major components: an ionizer, a *m*/*z* analyzer, an ion detector, and a vacuum system. Until the development of electrospray ionization mass spectrometry (ESI-MS) and matrix-assisted laser desorption/ionization mass spectrometry (MALDI-MS), mass spectrometry had limited use in the analysis of biological molecules (Chait 2011). The advantages of ESI include (a) ability to ionize fragile molecules intact, (b) ease of coupling to liquid chromatography, and (c) efficient generation of multiple charged ions to measure high molecular weight biopolymers. On the other hand, a major limitation of ESI-MS is its intolerance of impurities and salts. The ability of ESI-MS to directly monitor soluble protein aggregate formulation during heat treatment of human antithrombin III was recently reported (Wang et al. 2012). While MALDI-MS possesses high speed and a relatively high tolerance to the presence of contaminants, it is also relatively nonquantitative. In addition to the selection of ionization techniques, there are many types of *m*/*z* analyzers, including time of flight, quadruple ion traps, and transmission quadrupole. The specific analytical application determines the optimal choice of setup and type of mass spectrometer.

Sample preparation for MS analysis of proteins can be divided into two approaches: bottom-up and top-down. In the bottom-up approach, proteins are first digested with an enzyme, and the resulting peptide fragments are separated and analyzed by MS and MS/MS to produce a unique "fingerprint" of an individual protein. This "peptide map" of a protein can be compared to the theoretical gene-derived amino acid sequence to establish or confirm a protein's identity as well as to determine if degradation or posttranslational modifications have occurred. The advantages of the bottom-up approach include its high accuracy in mass determination and ease of use (Chait 2011). In the top-down approach, intact proteins are directly analyzed in the mass spectrometer without prior enzymatic digestion. This approach gives a molecular mass of the intact protein, their major posttranslational modifications and can potentially detect the presence of protein aggregates.

Not only have both of these two approaches been used to examine protein aggregates, but they have also been coupled to the analysis of kinetic exchange rates of hydrogen for deuterium in the protein amide backbone (H/D exchange-MS), to gain insight into the role of protein flexibility and dynamics in protein aggregation pathways. The rate of H/D exchange in the amide backbone hydrogens in a protein differs in flexible or disordered regions (lacking stable hydrogen bonding) compared to more tightly folded regions (protected from H/D exchange) (Konermann et al. 2011). By quenching the H/D exchange reaction at different time points and enzymatically digesting the protein under quenched conditions, the rate of H/D exchange in different regions of the protein can be determined by LC–MS analysis of each peptide over time (Marcsisin and Engen 2010). Furthermore, regions protected by aggregation can be identified. Recent advances in chromatographic separations such as UPLC technology, combined with software advances in the analysis of H/D exchange rates in hundreds of peptides, has permitted the characterization of the conformational dynamics of IgG monoclonal antibodies by H/D exchange-MS (Houde et al. 2009).

9.2.1.7 Surface Plasmon Resonance

Surface plasmon resonance (SPR) is a method based on energy transfer from light (photons) to the electrons on a metal surface (Ahrer et al. 2003). As polarized light interacts with the surface of a biosensor chip coated with gold, an evanescent wave is created that is highly sensitive to the changes in the dielectric constant of the adjacent medium. This energy transfer can create an angle of minimum reflectance (the SPR angle) under certain conditions. The change in SPR angle is proportional to the mass of bound material at the chip surface (Thillaivinayagalingam et al. 2010). By analyzing real-time changes of the SPR angle (SPR sensorgrams) as fluid passes over the surface of the biosensor, both the extent of analyte binding to the surface as well as binding rate kinetics (k_{on}, k_{off}) and binding affinities (K_D) can be obtained. SPR has been extensively applied to the quantitative analysis of protein-ligand and protein-protein interactions and facilitated by the availability of automated equipment. One requirement for the application of SPR to biological molecules is the successful immobilization of sufficient amounts of ligand to the surface of the sensor chip. Chemical coupling methods (e.g., through amines, thiols) covalently bind the targeted ligand onto the biosensor surface. Since a lack of control over the orientation of protein ligands might affect subsequence analyte binding, indirect coupling methods use a second ligand (e.g., an antibody or avidin) in which the target ligand binds to the immobilized protein creating a more homogenously oriented bound ligand (Wear et al. 2005).

9.2.2 *Case Studies Characterizing Size and Amount of Soluble Protein Aggregates Under Native-Like and Denaturing Conditions*

This section will cover the three case studies illustrating a variety of analytical approaches to characterize different types of soluble protein aggregates, using methods under native-like (SE-HPLC and AUC) and denaturing conditions (SE-HPLC and CE with SDS). The first study uses a novel application of SE-HPLC, along with orthogonal testing by SV-AUC, to improve the sensitivity of aggregate detection for a variety of different types of aggregates of an IgG1 monoclonal antibody generated from different environmental stresses. The second study examines similar analytical approaches to characterize a high molecular weight aggregate of a recombinant therapeutic glycoprotein. The third utilizes analytical methods under denaturing conditions (SE-HPLC and capillary electrophoresis with SDS) to determine the level of reducible and nonreducible protein aggregates in stressed samples of a monoclonal antibody.

In the first case study, an IgG1 monoclonal antibody was subjected to different environmental conditions such as elevated temperature, decreased pH, and light exposure to generate different aggregates. Bond et al. were able to accurately, and with improved sensitivity compared to light scattering or MALS, quantify the amount of aggregate in force-degraded IgG1 monoclonal antibody solutions using a dual-wavelength approach (DW-SE-HPLC) with very good recovery (Bond et al. 2010). This optimized SE-HPLC technique monitored aggregate, monomer, and fragment peaks at two ultraviolet wavelengths (214 and 280 nm). It was determined that the limit of quantitation for this new method was 0.04%, which was much lower than the 0.2% found for the previously used single wavelength SE-HPLC method. In addition, the new approach extended the dynamic range ~five to ten fold. As shown in Fig. 9.1, different environmental stresses resulted in formation of different types of oligomeric species and soluble protein aggregates (dimer vs. multimer as well as covalent cross-links vs. noncovalent interactions). After assay optimization to minimize interactions with the column matrix, the DW-SE HPLC results correlated well with SV-AUC data in terms of the ability to quantify aggregate levels in solutions (Fig. 9.1). Interestingly, a high-concentration formulation of the IgG1 monoclonal antibody, when stressed at 40°C for 3 months, showed a low-level formation of a different type of oligomeric species, a noncovalently associated dimer (Bond et al. 2010).

Several investigations have now appeared using SV-AUC to confirm the results obtained with SE-HPLC, specifically to verify a lack of effect of SE-HPLC sample handling and mobile phase conditions on aggregate formation and dissolution (Berkowitz 2006; Hughes et al. 2009). In one interesting example with a complex protein, Hughes and colleagues quantitatively investigated the ability of SE-HPLC and SV-AUC to detect known amounts of protein aggregates of recombinant human acid alpha-glucosidase (rhGAA), an enzyme with seven N-lined glycosylation sites and an apparent molecular weight of 110 kDa. Overall, the authors found a

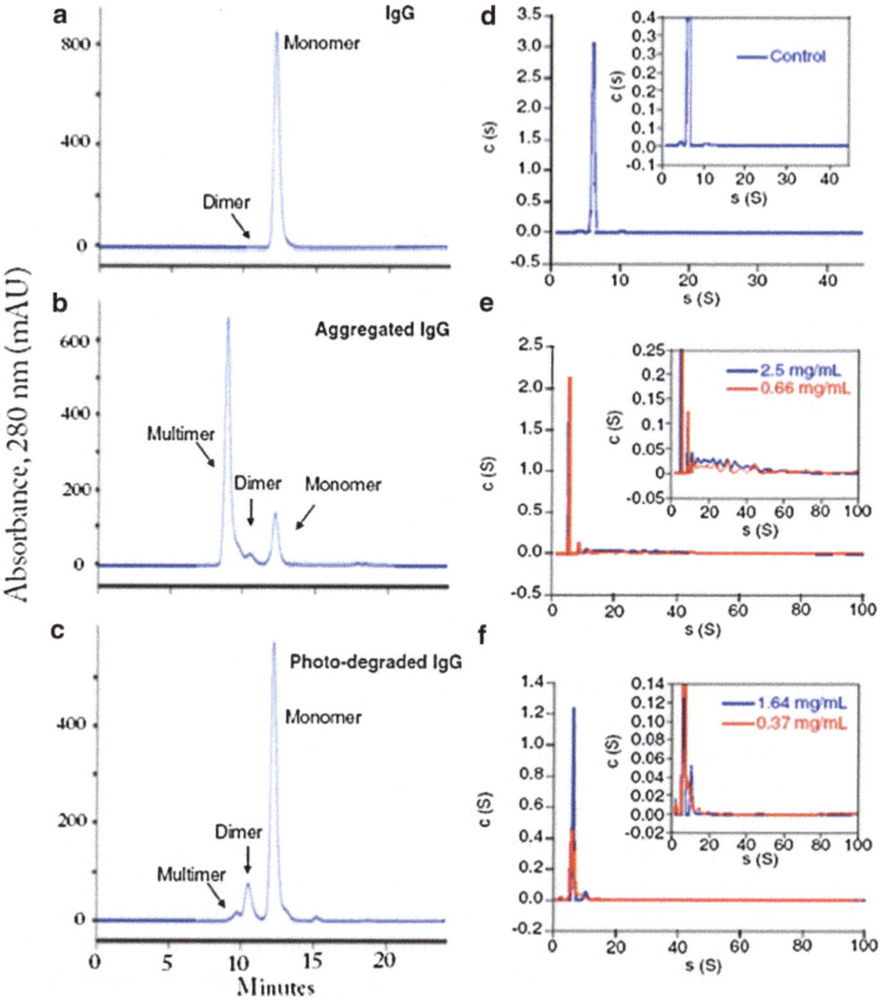

Fig. 9.1 Analytical characterization of IgG1 monoclonal antibody solutions containing soluble aggregates measured by orthogonal methods, SE-HPLC (**a, b, c**) and SV-AUC (**d, e, f**). Protein aggregates of varying size and composition were formed by exposure to different environmental stresses: (**a, d**) control, (**b, e**) heating under acidic pH conditions, and (**c, f**) light exposure. Adapted from Bond et al. (2010)

good correlation between the two methods, since the amount of the soluble aggregates measured was not statistically different (Hughes et al. 2009) (Table 9.1). It should be pointed out, however, that other studies in the literature have found significantly higher percentages of soluble protein aggregates when measured by SV-AUC compared to SE-HPLC. This suggests that the SE-HPLC approach may require additional method development to prevent aggregate adsorption to the column matrix with certain proteins. For example, in work by Gabrielson et al. with a

Table 9.1 Quantitative analysis of aggregate levels as determined by SV-AUC and SE-HPLC for forced degraded samples of recombinant human acid alpha-glucosidase (rhGAA)

	rhGAA forced aggregate (%)								
	0	0.25	0.5	1.25	2.5	4	5	10	20
% Aggregation (AUC), $N=6$[a]	0.4	0.9	0.6	1.3	2.5	3.1	4.2	8.5	18.1
SD	0.3	0.7	0.4	0.5	1.1	0.3	0.4	0.7	1.3
RSD (%)	67.0	84.0	69.4	35.7	44.0	8.8	10.4	8.1	7.0
% Aggregation (SEC), $N=9$	0.2	0.3	0.5	1.0	1.9	3.0[b]	4.1	8.4	17.5
SD	0.02	0.03	0.02	0.04	0.03	0.04	0.04	0.04	0.42
RSD (%)	9.2	10.8	3.7	3.7	1.6	1.2	1.0	0.5	2.4

Published in Hughes et al. (2009)
[a]Samples were analyzed in a single AUC run using six different cells
[b]These SEC data were obtained on six samples

recombinant humanized mAb (Gabrielson et al. 2007), SV-AUC detected one major peak and three minor peaks, while SE-HPLC only detected one major and one minor peak. In the same study, FFF and SV-AUC were compared as orthogonal methods to monitor the amount of soluble aggregates. It was shown that both SV-AUC and FFF not only detected much higher amounts of aggregates compared to SE-HPLC but also found additional HMW species. Nonetheless, it should be pointed out that although FFF has no stationary phase (Fraunhofer and Winter 2004; Liu et al. 2006), protein samples can undergo concentration and dilution stages in FFF in which protein aggregates may differentially interact with the medium and affect the amount of reversible aggregate in the sample (Mahler et al. 2009; Reschiglian et al. 2005; Liu et al. 2006).

A third example employs CE-SDS and d-SEC to characterize aggregate formation in a monoclonal antibody solution under denaturing conditions. One of the first studies to utilize both CE-SDS and SDS-PAGE for the quantitative analysis of a recombinant humanized monoclonal antibody (rhuMabHER2) was performed by Hunt et al. (1996). Both methods detected seven peaks in a nonreduced sample that correlated well with the seven bands observed for HER2 mAb on SDS-PAGE. The relative peak areas percent ascribed to high molecular mass aggregate observed on CE-SDS were consistent with that seen with SEC. The method, however, was found to be less sensitive than SDS-PAGE with silver staining in detecting minor species under reducing conditions. To improve sensitivity, precolumn derivatization with fluorogenic reagents coupled with laser-induced fluorescence (LIF) detection are now routinely used in CE-SDS analysis (Michels et al. 2007; Salas-Solano et al. 2006). Using this improved methodology, the ability of d-SEC and CE-SDS to determine the level of nonreducible, cross-linked protein in a degraded monoclonal antibody solution was evaluated (Fig. 9.2). SE-HPLC was initially used to establish the presence of monomer, dimer, and HMW aggregates. These species were then evaluated for the presence and absence of reducible and nonreducible cross-linked proteins under denaturing conditions by both d-SEC and CE-SDS (Fig. 9.2). Although both methods showed similar trends, the CE-SDS technique provided better resolution and an improved ability to quantify the levels of these aggregates (Michels et al. 2007).

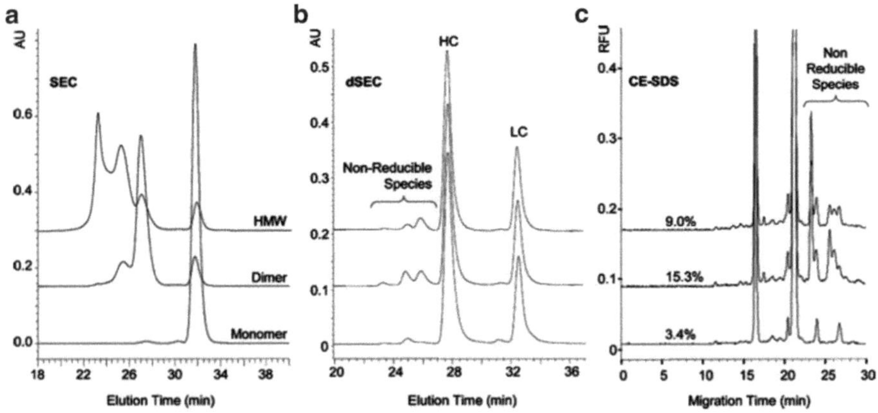

Fig. 9.2 Analytical characterization of monomers, dimers, and high molecular weight (HMW) aggregates of a monoclonal antibody as measured by SE-HPLC (native-like conditions) and d-SEC and CE-SDS (denaturing conditions). (**a**) SE-HPLC, (**b**) d-SEC, and (**c**) CE-SDS with indicated levels of percent nonreducible aggregates. Published in Michels et al. (2007)

9.2.3 Case Studies Characterizing Physical Composition and Biological Activity of Soluble Protein Aggregates

Despite recent advances in the use of biophysical methods to better characterize the higher-order structure and conformational stability of proteins, especially when combining multiple techniques with advanced data analysis approaches (Maddux et al. 2011), the determination of a protein's biological activity remains the "gold standard" in terms of monitoring subtle conformational changes. Two case studies are presented in this section, first demonstrating correlations of certain physical properties of protein aggregates with biological activity (Remmele et al. 2006) and second showing how biophysical binding assays using biosensors (SPR) can be employed to assess the Fc receptor binding activities of different aggregates and complexes of two different monoclonal antibodies (Luo et al. 2009).

In the case of an IgG1 monoclonal antibody (epratuzumab), samples were shown to contain both monomers (~150 kDa) and dimers (~300 kDa). The dimers contained ~70% covalent cross-links (as measured by CE-SDS) and consisted of three different assemblies including Fab-Fab, Fc-Fc, and Fab-Fc complexes. The biological activity of the dimers, in terms of relative potency in a cell-based bioassay, was shown to be twice that of the monomer (i.e., equal activity on a weight basis indicating that the dimer has twice the number of binding sites as the monomer) (Remmele et al. 2006). A more recent investigation examined the binding affinities of different mAb complexes and aggregates using an SPR biosensor (Luo et al. 2009). Similar to the previous study, monomeric and aggregated forms of the mAbs had similar antigen-binding properties in the Fab region. It was demonstrated, however, that the monomers and aggregates showed differences in their Fc region's ability to bind different Fc-gamma receptors. As shown in Fig. 9.3a, b, the monomeric form of

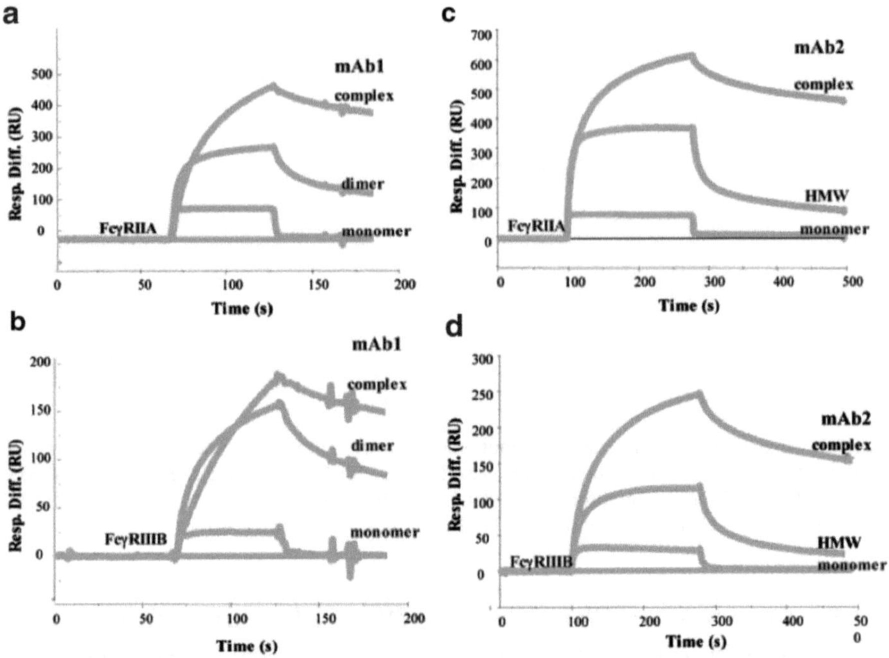

Fig. 9.3 The Fc receptor binding of the immune complexes and aggregates of mAb1 (**a** and **b**) and mAb2 (**c** and **d**) as measured by SPR binding assays. Two different Fc receptors were evaluated: FcγRIIA (**a** and **c**) and FcγRIIIB (**b** and **d**). The SPR sensorgrams for the binding of mAb monomer, dimer/HMW, and their immune complexes (containing 1:1 molar ratio of monomeric antibody and multivalent antigen) are overlaid. Published in Luo et al. (2009)

mAb1 manifested different binding properties to both the Fc-gamma RIIA and Fc-gamma RIIIB receptors compared to the dimeric aggregate and a multivalent immune complex formed between mAb1 and its antigen. Similar results are shown in Fig. 9.3c, d for monomeric mAb2, high molecular weight aggregates of mAb2, and a multivalent immune complex formed between the mAb2 and its antigen. The dimers and multimers of these mAbs showed higher in vitro binding affinities to Fc-gamma receptors as well (Luo et al. 2009).

9.2.4 Case Studies Characterizing Soluble Protein Aggregates by Mass Spectrometry

Although the use of mass spectrometry to better characterize soluble protein aggregates formed during storage has been more limited in its applications to date, MS has been pursued for many years to better characterize well-ordered amyloid fibril protein aggregates. For example, nanoflow ESI combined with Q-TOF mass spectrometer has been utilized to study the self-assembly, aggregation, and amyloid fibril formation of insulin under different conditions (Nettleton et al. 2000). Another example is the use of ESI-quadrupole MS to quantify β(beta)-amyloid monomers in the presence of potential therapeutic agents and to identify and rank order the compounds which inhibit the aggregation of β(beta)-amyloid monomers (Cheng and van Breemen 2005). The initial applications of MS technology to characterize protein aggregates under pharmaceutical conditions have combined chromatographic separations with subsequent MS analysis. For example, Van Buren et al. characterized dimers of an IgG after isolation and enzymatic digestions (Van Buren et al. 2009). A combination of SE-HPLC, sample dialysis, and ESI-TOF MS was recently used to identify and characterize different IgG aggregates (Kukrer et al. 2010). In addition to complex sample handling with limited throughput, these MS approaches have certain limitations including potential changes in the aggregate profile during sample dialysis and a bias of MS results toward lower MW species when monitoring a sample containing a mixture of oligomers (Kaltashov et al. 2012).

More sophisticated MS approaches are now being developed to examine and characterize protein aggregates without these experimental limitations as described in the following three case studies: (1) use of EDI-MS to directly monitor in real-time heat-induced protein deformation and aggregation (Wang et al. 2011), (2) use of ion mobility MS to characterize intact protein assemblies in the gas phase, and (3) the use of H/D exchange-MS to characterize different proteins (Kheterpal and Wetzel 2006; Zhang et al. 2011, 2012).

As temperature increases, loss of a protein's conformational integrity results in formation of high-charge-density ions and oligomers that can be easily identified by higher *m/z* ratios. A new design of a temperature-controlled ESI source not only permits direct monitoring of protein unfolding and aggregation but also ensures the efficiency of the heating process and prevents cooling of the protein solution during the sample introduction to the ESI interface (Wang et al. 2011). With this new approach, both reversible and irreversible protein unfolding events were captured by ESI-MS for cytochrome-C and glucocerebrosidase (GCase), respectively. The ability of ESI mass spectra to directly monitor the formation of dimers, trimers, tetramers, and pentamers during heat treatment of GCase is shown in Fig. 9.4.

Ion mobility spectrometry (IMS) is an analytical technique used to separate and identify ionized molecules in the gas phase based on their mobility in a carrier buffer gas. When coupled with mass spectrometry, ion mobility mass spectrometry (IM-MS) enables the separation and detection of isomers and conformers. More recently, the technology (also referred to as native MS) has been successfully used to characterize larger macromolecules, such as proteins and their biological complexes, in the gas phase (Uetrecht et al. 2010). Although Native MS has been employed to examine the composition and stability of numerous large biological complexes, confirmation of results with other methods (examining the same protein complex in the solution state) is typically performed (van Duijn 2010). Kaddis et al. used ESI-IMS to study the general size dimensions of proteins as protein molecules were physically transitioned from the solution to the gas phase (Kaddis et al. 2007). Traveling voltage-wave ion mobility mass spectrometry (TWIMS) is a recent

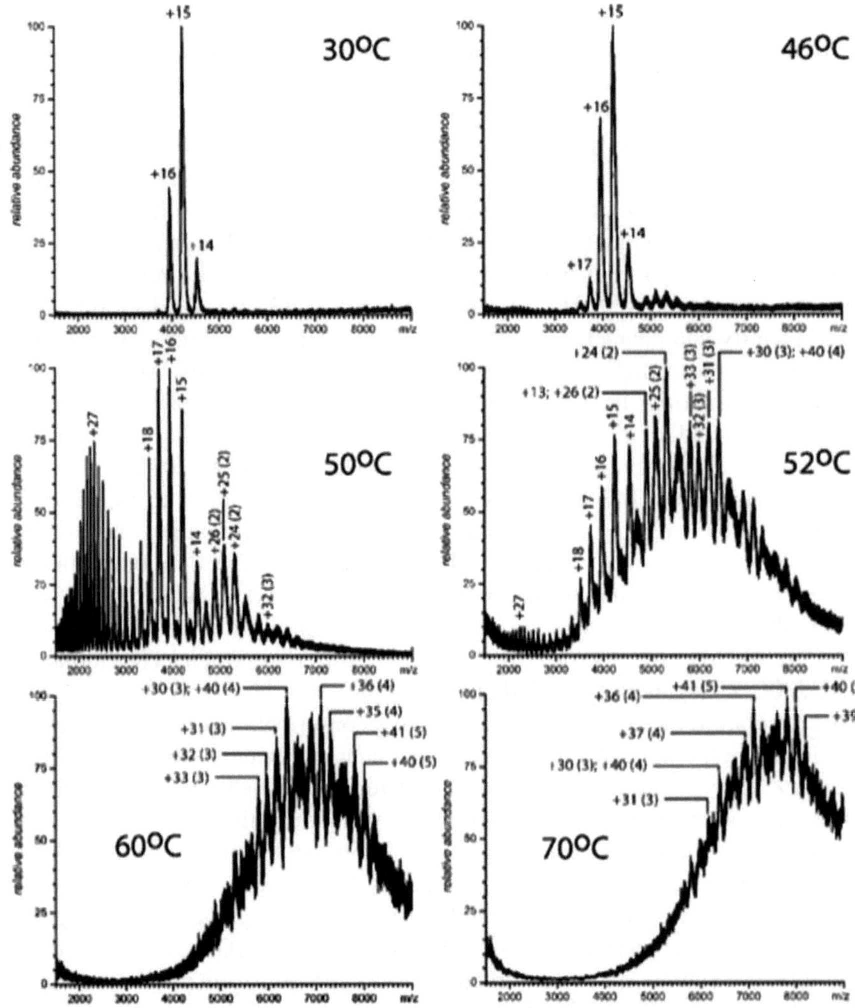

Fig. 9.4 Electrospray ionization (ESI) mass spectra of glucocebrosidase (GCase) in 20 mM ammonium acetate, pH 4.7 recorded at various solution temperatures. The *numbers without parentheses* indicate charge states of the GCase monomers; the *numbers in parentheses* indicate the size of the protein associations. Published in Wang et al. (2011)

advancement in the technology coupling an ion mobility unit to Q-TOF MS. The application of TWIMS to the characterization of hepatitis B viral capsids (MW of ~3,000 and ~4,000 kDa) as well as GroEL and its complexes with smaller proteins (MW of 801 and 857 kDa, respectively) has been recently reviewed (Uetrecht et al. 2010).

Hydrogen–deuterium exchange linked to MS (H/D exchange-MS) analysis has been used to probe the secondary structure of amyloid fibrils of insulin (Kheterpal and Wetzel 2006). This work combined an ESI-MS system with a bottom-up experimental approach (peptide mapping). First, protonated fibrils were collected by centrifugation, washed, resuspended, and incubated in deuterated buffer to promote exchange of labile protons with deuterium. Second, the exchange reaction was quenched and the fibrils were disaggregated. Finally, the monomers were subjected to proteolysis and analyzed by MS. The results showed that both the N- and C-terminal segments (1–19 and 35–40) are readily exchangeable (not involved in the core structure), whereas the fragment containing residues 20 34 was highly protected from exchange. H/D exchange-MS has also been used to the study of the effect of protein flexibility and sequences on protein aggregate formation with biopharmaceutical products including interferon-gamma (Tobler and Fernandez 2002) and, more recently, with the model enzyme lactate dehydrogenase (LDH) (Zhang et al. 2011) as well as a monoclonal antibody (Zhang et al. 2012). For smaller proteins such as interferon-gamma and LDH, the rate of H/D exchange can be monitored in the intact protein, as well as at the peptide level (after digestion with pepsin), to identify sites of increased and decreased flexibility. In contrast, due to their large size, mAbs are only analyzed after pepsin digestion and H/D exchange reactions must be carefully examined across hundreds of peptides. The H/D mass spectra of native and freeze-thaw (F/T)-induced aggregates of LDH are shown in Fig. 9.5. It can clearly be seen that F/T-induced aggregates consist primarily of unfolded protein. Peptide level analysis identified several short sequences (13–31, 109–117, and 133–143) with moderate protection from H/D exchange indicating that these sequences may be the regions of intermolecular interactions within the LDH aggregates induced by freeze-thaw (Zhang et al. 2011). In contrast, in the case of a monoclonal antibody, F/T aggregates had similar H/D protection maps compared to soluble protein, indicating a primarily native-like conformation (Zhang et al. 2012).

9.3 Case Studies Using Biophysical Methods to Count, Size, and Characterize Protein Submicron Particles

9.3.1 Overview of Analytical Methods Used in Selected Case Studies to Evaluate Submicron Protein Particulates

The development of improved analytical approaches to examine the size, amount, and nature of protein particles in the size range of 0.1–1 μm is an ongoing challenge (Carpenter et al. 2009). Some currently available approaches include DLS techniques, either as stand-alone instruments or connected to a microscope [nanoparticle tracking analysis (NTA)]. In addition, more sensitive microscopic techniques, most frequently used to characterize protein-based amyloid fibrils, are being evaluated for applicability to better characterize pharmaceutical protein aggregates. These include atomic force microscopy (AFM), transmission and scanning electron microscopy (TEM and SEM), as well as quartz crystal microbalances

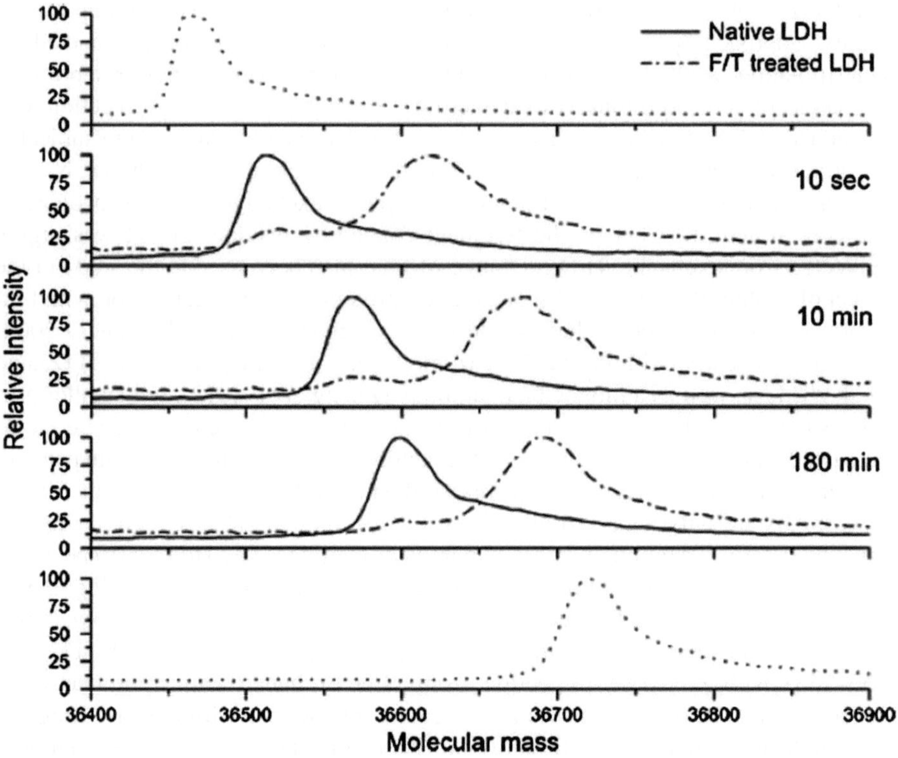

Fig. 9.5 Hydrogen–deuterium exchange mass spectrometry (H/D-MS) spectra of native lactate dehydrogenase (LDH) and freeze/thaw aggregated LDH obtained after 10 s, 10 and 180 min of deuterium labeling. The *dotted spectra* in the *top* and *bottom panels* are unlabeled and fully labeled controls, respectively. Published in Zhang et al. (2011)

(QCMs). A brief overview of the key analytical techniques used in the case studies presented in this chapter is provided below, followed by a few illustrative examples from the literature which have employed useful combinations of these techniques.

9.3.1.1 Nanoparticle Tracking Analysis

NTA is a laser-illuminated-based microscopy technique that detects and sizes submicron particles in the range of ~50–1,000 nm. NTA individually tracks and sizes particles moving under Brownian motion. With a knowledge of the viscosity and temperature of the sample, the instrument calculates the diffusion coefficient which can then be used via the Stokes–Einstein equation to calculate the hydrodynamic size. This new analytical technique is currently being assessed by several groups to detect protein aggregates and particulates that have not been accurately analyzed previously (Filipe et al. 2010, 2011; Engelsman et al. 2011).

9.3.1.2 Atomic Force Microscopy

An atomic force microscope (AFM) consists of a cantilever with an ultrasharp tip (probe), a sample stage, and an optical deflection system (Colton et al. 1997). During AFM measurements, the cantilever is brought into close (atomic) contact with the sample and scanned across the surface (contact mode). The interaction between probe and sample, which can include mechanical contact, van der Waals, capillary, electrostatic, and/or magnetic forces, causes the bending of the cantilever which is precisely recorded by an optical deflection system. The scanning is done under feedback control so that the bending of the cantilever remains constant to maintain a constant force. The up and down motion of the sample is a record of the sample topography. In the tapping mode, which senses the oscillation amplitude of cantilever instead of bending, AFM generally causes less damage to sample. For analysis of biological materials such as protein aggregates, tapping mode AFM is preferred. One key limitation of AFM is it can only analyze a small fraction of the total sample volume at any one time.

9.3.1.3 SEM and TEM

Scanning electron microscopy (SEM) and transmission electron microscopy (TEM) are standard tools used in biological research to examine a wide variety of samples due to their high resolution (from several nanometers to microns). Both SEM and TEM can also be used to characterize protein aggregates in the submicron size range. These techniques use an electron beam to bombard a specimen and generate various types of electron scattering (Joseph Goldstein et al. 2003; David Bernard Williams 2009). In the case of transmission electron microscope (TEM), electrons that go through a specimen (forward scattering) are collected and analyzed to provide image-based information on sample size and shape. The scanning electron microscope (SEM) collects the secondary and backscattered electrons to provide data not only on 2D sample size but also surface topology. Sample preparation remains a challenge for TEM and SEM (which requires fixation, staining, and drying) in terms of ensuring the size and morphology of the sample do not change during preparation and analysis. Fortunately, recent advances in cryotechniques (rapid freezing in liquid ethane near liquid nitrogen temperatures) can minimize such problems by analyzing samples in the frozen liquid state. TEM and SEM analysis is also limited by sampling frequency, requiring acquisition of numerous images to provide a representative analysis of the sample to determine particle number and size.

9.3.1.4 Quartz Crystal Microbalance

A new approach to monitor submicron protein particulates in solution is the QCM (Ferreira et al. 2009). A thin quartz disk is sandwiched between two metal electrodes. When an alternating electric field is applied, a shear deformation will be generated due to the piezoelectric nature of the quartz material. When the mechanical oscillations are close to the fundamental frequency of the crystal, a resonant oscillation is achieved. Changes in the resonant frequency are related to the mass accumulated on the crystal (Sauerbrey 1959). Commercial systems are designed to reliably measure mass changes up to ~100 μg (microgram) with a sensitivity of ~1 ng cm^{-2} (O'Sullivan and Guilbault 1999). Although QCM has been used successfully in other fields (e.g., vacuum deposition, thin film deposition control), application to biological samples has been limited due to its inherent nonspecificity. Recently, chemical modification of the electrode surface enabled the real-time study of bovine insulin aggregation to form amyloid fibrils (Knowles et al. 2011). Immersion of resonators into aqueous solutions, however, significantly reduces their resolution due to the damping effect of solution viscosity (Burg et al. 2007). A recently developed nanomechanical resonator, Archimedes, contains a solution channel inside a hollow resonator that is surrounded by vacuum to overcome this damping effect. As a result, Archimedes has improved mass resolution (orders of magnitude) over a commercial QCM for aqueous measurements (Burg et al. 2007).

9.3.2 Case Studies Involving Counting and Sizing Submicron Protein Particles

In a comparative study by Filipe et al. (2010), the ability of NTA and DLS to size submicron particulates using polystyrene bead standards of various sizes was initially examined. Although both techniques can accurately size monodisperse standards, NTA offers the advantage of also being able to count the number of particles in the sample. In addition, it was possible to visualize the particles in the samples using NTA. When particle standards of different sizes were mixed together, NTA can more easily distinguish their relative amounts compared to DLS. In this study, DLS detected the largest particle standard in a mixture, while NTA detected each of the standards and thereby did not skew the reported size distribution. The formation of submicron-sized protein aggregates was then monitored with heat treated IgG and metal-oxidized insulin. NTA was again better at analyzing sample polydispersity than DLS. As shown in Fig. 9.6, NTA can be used to monitor particle counts and size distribution in real time during the formation of IgG submicron-sized particulates during heating at 50°C. Recently, Filipe et al. used NTA in conjunction with fluorescence single-particle tracking analysis to obtain the size of fluorescently labeled monoclonal IgG and HSA when subjected to heat stress (Filipe et al. 2011).

Nanomechanical resonators, including QCMs, have achieved extraordinary sensitivity in the detection of mass accumulation on surfaces in a vacuum. The detection of most bimolecular samples, however, requires a solution environment. As described above, a recently developed nanomechanical resonator, Archimedes, includes a solution channel inside a hollow resonator that permits solution measurements (Burg et al. 2007). This device is able to measure the mass of objects that are not attached to the resonator surface, such as floating polystyrene beads, agglutinating microspheres, and bacteria (Burg et al. 2007; Rumi Chunara et al. 2007; Michel Godin et al. 2007). The instrument vendor's website (http://www.affinitybio.com/applications/protein_formulations.php) provides examples (from currently unpublished data) of the detection of submicron-sized IgG aggregates. Both size and number distributions of protein particles in solution can be displayed in the form of a histogram.

9.3.3 Case Studies Which Characterize the Morphology of Submicron Protein Particles

To better characterize the morphology and size of submicron protein particles, microscopic techniques such as AFM and TEM have been successfully used. Tapping AFM has been used for several years to study the in vitro formation of amyloid fibrils. For example, the process of in vitro generation of amyloid fibrils by seeded polymerization has been examined (Harper et al. 1997). In addition, a variety of biophysical techniques including AFM were utilized to study the aggregation and disaggregation of an amyloidogenic protein, human muscle acylphosphatase (Calamai et al. 2005). In this work, it was observed that globular aggregates from 100 to 200 nm and larger aggregates (>5 μm) were the dominant species as the aggregation time increased. Interestingly, upon diluting the protein solution, a variety of fibrillar structures was observed with AFM indicating disaggregation of the larger protein complexes.

In terms of pharmaceutical applications, tapping mode AFM has been recently used to better characterize the nature of the aggregation of monoclonal antibodies (Cao et al. 2010; Lee et al. 2011; Mach and Arvinte 2011). In one such study, most of the mAbs manifested similarly sized aggregation intermediates on the order of several monomers. The subsequent agglomeration of these intermediates into larger particulates, however, was dependent on solution conditions as evaluated by AFM images (Lee et al. 2011). For example, monomeric mAb1 had an average size of ~15 nm. Upon heat treatment, the presence of oligomeric structures was seen when the sample was diluted into water. In contrast, when the heat-treated mAb1 sample was diluted with saline, agglomerates of the oligomer structures began to appear. The morphology and size of the agglomerates appeared to differ between the different mAbs examined suggesting protein-specific and/or stability-dependent mechanisms. For example, as shown in Fig. 9.7, mAb3 formed submicron-sized particles with amorphous morphology containing agglomerates of smaller sized oligomers. These investigators point out some of the analytical limitations of studying protein particle formation with AFM including a very small fraction of total sample volume being examined as well as the sample preparation process potentially producing artifacts. As a result, at least one orthogonal method was recommended to confirm the results from AFM.

Although numerous groups have used TEM to study and characterize the structure of amyloid fibrils (Nettleton et al. 2000; Kheterpal and Wetzel 2006), TEM has not been widely used to characterize protein aggregates and particles from a pharmaceutical perspective. One group combined quantitative rheology measurements

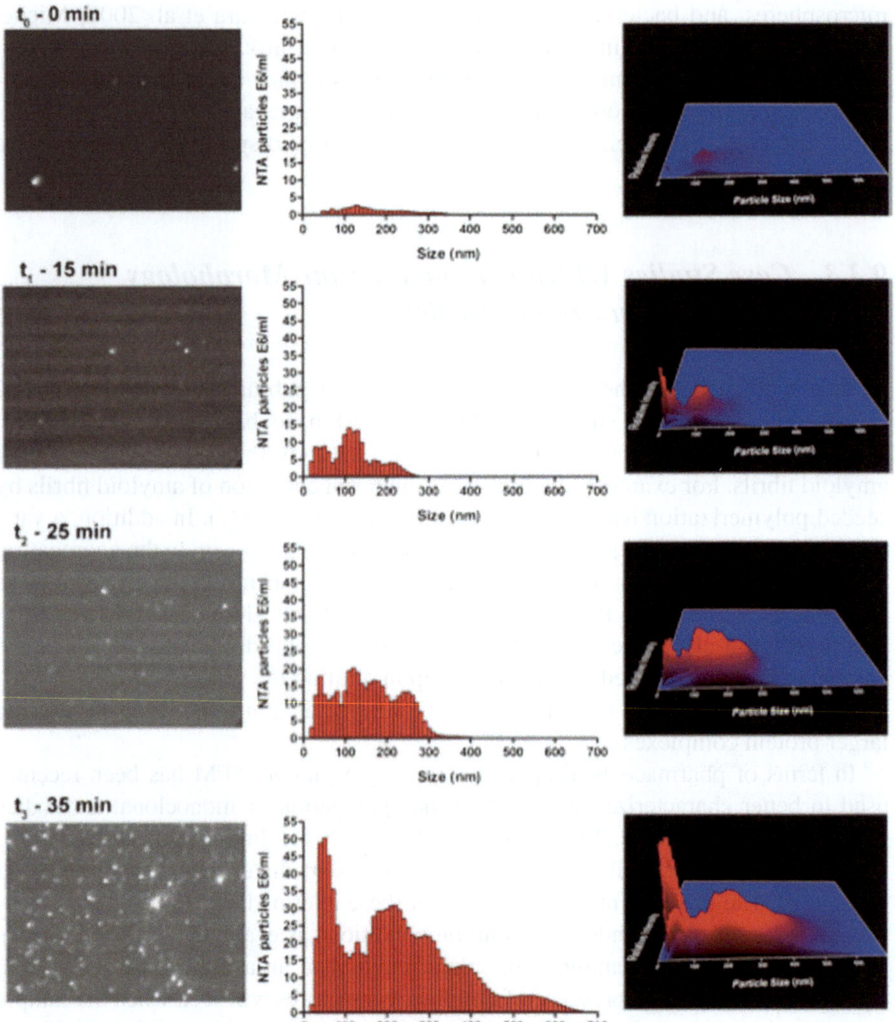

Fig. 9.6 Monitoring of IgG aggregation and submicron particle formation at 50°C in the NTA (NanoSight) instrument sample chamber. The size distribution (*middle panels*) with the corresponding NTA video frame (*left panels*) and 3D graph (size vs. intensity vs. concentration, *right panels*) are shown. Published in Filipe et al. (2010)

and negative stain TEM to study the assembly of monoclonal antibody into highly ordered structures in the presence of multivalent carboxylate ions (Esue et al. 2009). As shown in Fig. 9.8, TEM images demonstrate smaller bundles that were composed of straight, single filaments with an average diameter of 4 nm. In addition, the presence of larger bundles with diameters as large as 200 nm, containing midsized

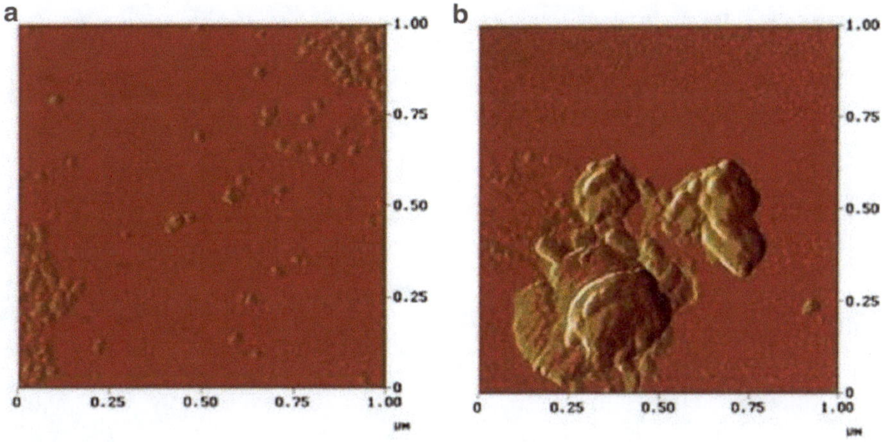

Fig. 9.7 Atomic force microscopy (AFM) visualization of submicron-size protein particles from a monoclonal antibody solution: (**a**) mAb aggregate (in water), (**b**) the same mAb aggregate in different viewing area. Published in Lee et al. (2011)

bundles from 29 to 55 nm in diameter, was observed by TEM. This type of detailed morphological information is currently only available from TEM studies. Sample preparation, however, involves the drying of the sample, again inducing possible sample preparation artifacts. The use of cryo-TEM to examine the morphology of protein particles in the liquid state will be of great interest in future studies.

9.4 Case Studies Using Biophysical Methods to Count, Size and Characterize Protein Subvisible and Visible Particles

9.4.1 Overview of Analytical Methodologies from Case Studies Used to Evaluate Subvisible and Visible Protein Particles

Current analytical methods utilized to count and size subvisible and visible particulates in the size range of 1–100 μm and over 100 μm, respectively, include coulter counter, light obscuration (LO), microflow digital imaging (MFI), light microscopy, and visual assessments. Techniques employed to better characterize the morphology and composition of subvisible and visible protein particulates include zeta potential measurement, flow cytometry, fluorescence, and FTIR microscopy as well as SEM-EDX. A brief discussion of each technique will be presented below followed by relevant case studies that have incorporated these techniques to size, count, and characterize subvisible and visible protein particles.

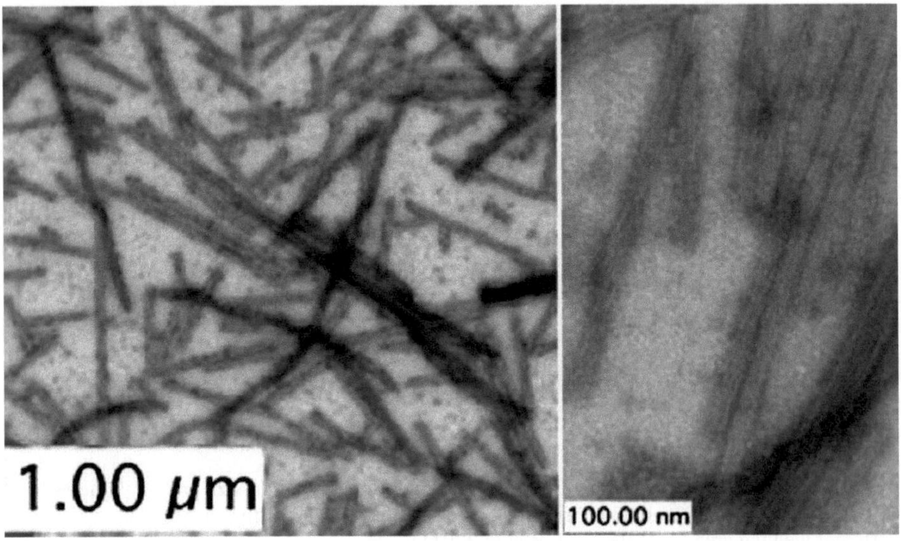

Fig. 9.8 Transmission electron microscopy (TEM) visualization of submicron and subvisible protein particles from a monoclonal antibody solution containing citrate. Analysis of micrographs showed large bundles containing smaller filament bundles. Individual straight filaments were of varying lengths and had an average diameter of 4 nm. Scale bars are 1 μm and 100 nm. Published in Esue et al. (2009)

9.4.1.1 Coulter Counter

The Coulter method counts subvisible particles in the size range of ~0.5–50 μm (micrometer). Unlike LO and MFI, the Coulter method does not rely on the optical characteristics of the particles, but rather on the perturbation of an electric field. Single particles flow through a pore and electrical pulses are generated, which produce a change in conductance. This change is directly proportional to the particle volume. To measure a change in conductance, the Coulter method requires a conductive buffer to permit an electric field to form. In some cases, this can be accomplished by adding a small amount of highly concentrated NaCl (or other electrolyte) to the buffer (Barnard et al. 2012). Since only single particles can transverse the pore, the sample must be diluted. Both dilution and the addition of salt during sample preparation are of concern when monitoring protein particulates since these steps may either disassociate aggregates or induce their formation. Until recently, Coulter counter measurements required large sample volumes (10 ml), but newer instruments have significantly decreased sample volumes to as little as ~0.1 ml (Barnard et al. 2012; Rhyner 2011; Roberts et al. 2012). These reduced sample volumes are far more practical for use with protein containing samples.

9.4.1.2 Light Obscuration

The principle of light obscuration (LO) is simple: the sample is passed through a chamber containing a light source. The amount of light that is blocked is directly proportional to the number and size of the particles. This technique is widely used as the "gold standard" for measuring subvisible particles in pharmaceutical applications since both US and EU pharmacopeia methods have been described. The concentration and size of particles ranging from 2 to 100 μm (micrometer) can be determined. Advantages of this technique include speed and the ability to cover the entire subvisible size range in a single measurement. Disadvantages of LO include a lack of morphological information and potential artifacts arising from air bubbles. Therefore, samples are typically degassed prior to analysis (Huang et al. 2009; Chrai et al. 1987). In addition, LO has been shown to undercount the number of proteinaceous subvisible particles as described in more detail below.

9.4.1.3 Microflow Digital Imaging

Microflow imaging (MFI) is a flow microscopy technique in which a sample is drawn through a flow cell by a peristaltic pump. As the sample slowly transverses the flow cell, bright-field images are illuminated by a 475 nm LED light source, magnified, captured by a digital camera, and then processed by the instrument software. Information about particle counts, size, and morphological characteristics can all be obtained from MFI measurements. Since MFI does not rely on the blockage of light, translucent protein particles are more readily detected and included by this analysis (Huang et al. 2009; Wuchner et al. 2010). Another advantage of MFI is that it can distinguish between silicone oil droplets, air bubbles, and protein aggregates based on morphological characteristics (Sharma et al. 2010a; Strehl et al. 2011; Liu et al. 2011).

9.4.1.4 Light Microscopy and Visual Assessments

During analysis by light microscopy, particles in solution are drawn through a grid-lined filter, dried, and either observed by a standard light microscope or subsequently stained with fluorescent or nonfluorescent dyes to enhance resolution (Li et al. 2007; Demeule et al. 2007a, 2009). Particles can then be counted manually or by automated methods. Advantages of staining protein aggregates include enhanced detection of smaller particles and the selective binding of certain dyes to proteinaceous particles (Li et al. 2007; Rosenberg et al. 2009). Artifacts resulting from sample preparation, handling, and/or filtering are again the major pitfalls in using microscopy. In addition, particles <10 μm can be difficult to visualize and care should be taken to ensure amorphous or fibril shaped particles do not pass through the filter.

Visual assessments of protein-containing solutions for visible particles suggest a straightforward process, but complications can arise due to a variety of factors, that is, the method setup (lighting, distance, background), training of the analyst (physical abilities of the individual performing the counting, time spent counting), and type of container and physical properties of the solution. Furthermore, particle morphology and refractive index can influence the ability to detect and count visible particles. Ideally, two independent, certified analysts need to count visible particles in defined illumination conditions using a black and white panel as background. Blanks and particle standards are also counted in conjunction with the unknowns to ensure accurate and unbiased counting (Wuchner et al. 2010; Madsen et al. 2009).

9.4.1.5 Zeta Potential

The zeta potential is a property of particles in suspension related to their surface charge. In an electric field, each particle and its most closely associated ions move through the solution as a unit, and the potential at the boundary between this unit and the bulk solution is known as the zeta potential. Zeta potentials can be experimentally determined by measuring the drift velocity of particles in an electrical field of known strength (Ross and Morrison 2002; Greenwood 2003) or by a method known as phase analysis light scattering. Since zeta potential reflects the effective charge on the particles and is thus related to the electrostatic repulsion between particles in suspension, it has proven to be extremely useful in predicting the stability of emulsions, colloids, as well as proteins in solution. A higher zeta potential value reflects a more stable suspension since the charged particles repel one another overcoming the natural tendency to aggregate. Conversely, lower values facilitate interactions between the particles. The zeta potential measurement is often a key to understanding dispersion and aggregation processes in various applications including protein formulation (Sejersen et al. 2007; Uskokovic et al. 2010; Schmitt et al. 2007). The zeta potential can be measured with particles ranging from a few nanometers to as large as 30 μm in diameter.

9.4.1.6 Fluorescence-Activated Cell Sorting

Fluorescence-activated cell sorting (FACS), often referred to as flow cytometry, is a technique used for quantitative measurements on single cells, cellular constituents, or particles. Flow cytometry has also been used to detect baculovirus particles and their aggregates (Jorio et al. 2006), monitor platelet aggregation events (Ginsberg et al. 1990; Harding et al. 2007), and characterize amyloid fibrils (Wall and Solomon 1999). Flow cytometry uses the principles of light scattering, light excitation, and the emission of fluorochrome molecules to generate specific multiparameter data from particles and cells in the size range of 0.5–100 μm (Shapiro 1995; Cram 2002). Using hydrodynamic focusing techniques, cells or particles are presented to a light source (usually a laser) and the light scattering and fluorescence emission of thousands of individual particles per second are measured using several detectors (forward scatter, side scatter and fluorescent detectors). Light scattering and/or fluorescence emission is unique to each cell/particle. Thus, a combination of the two can be used to distinguish different particles or cells in a heterogeneous sample. The capability of flow cytometer to characterize individual particles makes it an attractive technique for use in the study of subvisible particles/aggregates in protein formulations. This technique has been widely used in the field of cellular and molecular biology, but its potential for use in the study of subvisible particles in protein formulations has only recently been recognized (Ludwig et al. 2011; Mach et al. 2011).

9.4.1.7 Fluorescence Microscopy

This type of microscopy can identify cells, cellular constituents, and particulate matter with a high degree of specificity and is rapidly expanding in use for pharmaceutical applications. The technique is used to study samples that can fluoresce either by themselves or after treating with fluorescent probes/dyes (Lichtman and Conchello 2005). Fluorescence microscopy can be very useful for the detection of subtle changes in the aggregation state of the proteins (David Bernard Williams 2009; Demeule et al. 2007a). Some of the dyes used to selectively monitor intermolecular β(beta)-sheet interactions often present among proteins that form β(beta)-amyloid structures are Nile red, Congo red, and Thioflavin T (Hawe et al. 2008a; Khurana et al. 2001; Naiki et al. 1989; Kim et al. 2003). Covalent labeling of proteins with donor and acceptor fluorophores allows the use of total internal reflection fluorescence microscopy to image aggregated proteins (Hillger et al. 2007).

9.4.1.8 FTIR Microscopy

By combining Fourier transform infrared (FTIR) detection with optical microscopy, compositional analysis of particles/aggregates down to 10 μm (micrometer) in size is possible. FTIR spectroscopy is also a powerful tool to analyze protein secondary structures and conformational dynamics (Prati et al. 2010; Middaugh and Joshi 2011). When combined with microscopy, this technique provides added information to analytical results that cannot be obtained by spectroscopy or microscopy alone. Another advantage of FTIR microscopy over conventional FTIR spectroscopy is that it permits the collection of an IR spectrum from a small sample area. FTIR microscopy is both a sensitive and potentially comprehensive method to help understand the nature, composition, and morphology of the sample examined and has wide applications in the areas of forensic (Torraca and Wen 2005), biomedical (Erukhimovitch et al. 2002; Salman et al. 2002), and pharmaceutical sciences (Wuchner et al. 2010; Flores-Fernandez et al. 2009; Markovich et al. 1997). One new application of FTIR microscopy is chemical mapping and imaging. The instrumentation required to obtain reliable IR chemical images has only recently become available. Using this technology, multicomponent systems can now be mapped and the location, nature, shape, and size of each material present can be visualized through these chemical images. These images capture the spectral features unique to each material present and provide chemical information about the multicomponent system. Creating a map, however, is a time-consuming process especially if both high spatial resolution and high signal to noise ratio as required (Prati et al. 2010).

9.4.1.9 Energy-Dispersive X-Ray Spectrometer

Energy-dispersive X-ray spectrometer (EDX or EDS), which utilizes the X-ray generated when the electron beam bombards the sample to collect information on chemical composition of a sample, can be connected to both TEM and SEM (Joseph Goldstein et al. 2003; David Bernard Williams 2009).

9.4.2 Case Studies to Size and Count Subvisible and Visible Protein Particles

Although MFI is a relatively new technique, there are already many interesting case studies employing the technology to count and size subvisible particles in protein formulations. Examples include measuring subvisible-sized protein particles induced by contact with ceramic filling pumps or syringes (Nayak et al. 2011; Majumdar et al. 2011) as well as from exposure to different environmental stresses (Joubert et al. 2011; Luo et al. 2011; Fradkin et al. 2011; Barnard et al. 2010). Several studies also include comparisons of MFI or Coulter methods to light obscuration in terms of counting both protein particles and polystyrene bead standards (Huang et al. 2009; Wuchner et al. 2010; Sharma et al. 2010a, b). The major difference in MFI and LO results for counting subvisible protein particles is probably due to their translucent nature which results in undercounting by LO. A recent study by Demule et al. (2010) directly compared counting of protein particles obtained from LO, MFI, and Coulter instruments. One set of results is shown in Fig. 9.9. These data demonstrate that the smallest amount of protein subvisible particles detected is obtained with LO, followed by MFI and the Coulter method. The results were dependent on the dilution of the sample prior to analysis (Fig. 9.9) as well as the particle size cutoff used during the analysis (Demeule et al. 2010). The authors' explanation for obtaining higher subvisible counts with the Coulter method is its insensitivity to differences in the refractive index between the protein particles and the solvent. On the other hand, a study by Singh and colleagues showed that the MFI and the Coulter methods generated comparable data except with highly concentrated protein samples, although the authors acknowledge that sample handling or differences in the measurement methods can potentially affect the final results (Singh et al. 2010). Taken together, these studies again highlight the importance of using orthogonal techniques to count and size protein particulates in the subvisible range to determine which methods yield the most reproducible data for an individual protein formulation.

A detailed study by Wuchner and colleagues investigated the effect of long-term storage of a high-concentration mAb formulation on formation of subvisible and visible

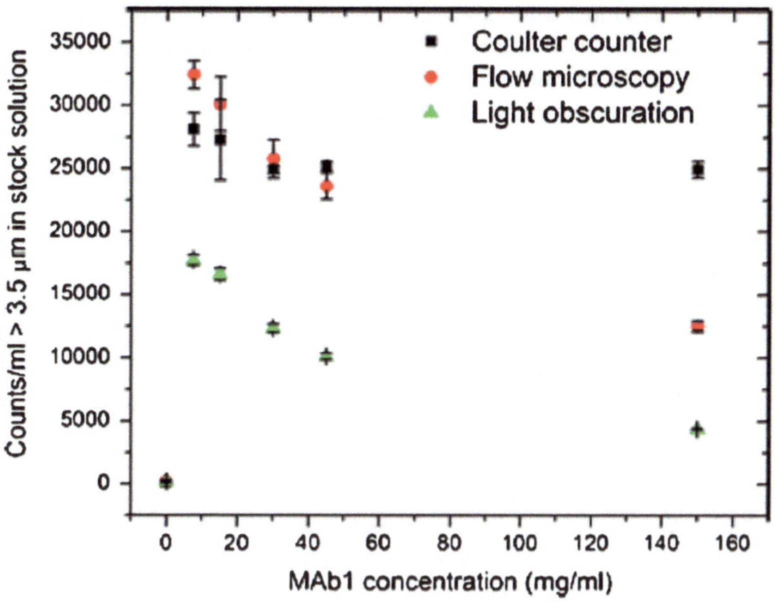

Fig. 9.9 Comparing the performance of the Coulter counter, flow microscopy (MFI), and light obscuration (LO) for measuring subvisible particles (>3.5 μm) in protein solutions. Results are reported as average ± standard deviation. Counting was performed at several protein concentrations with the reported particle numbers corrected for the dilution factor and represent the particle content of the stock (150 mg/ml) solution. The particle counts of the formulation buffer are included. Published in Demeule et al. (2010)

protein particles (Wuchner et al. 2010). Direct comparison of LO and MFI revealed five to tenfold higher subvisible protein particle counts in the MFI measurements, consistent with the results described above with other mAbs. This study proceeded to characterize the count and size of subvisible and visible particles during long-term storage as measured by MFI. The results found that a shift occurred in the particle size distribution over time, with larger subvisible particles accumulating during long-term storage (Fig. 9.10). This result was not observed by LO measurements. The accumulation of visible particles was also monitored by a combination of visible assessments, inverted microscopy, and MFI measurements (for protein particles larger than 100 μm; see Fig. 9.10). The three techniques showed good overall correlation in terms of the count of visible particle formed during long-term storage (Wuchner et al. 2010).

9.4.3 Case Studies Which Characterize the Morphology and Composition of Subvisible and Visible Protein Particles

Mach et al. (2011) demonstrated the use of a flow cytometer in a high-throughput format using a 96-well autosampler, to not only determine the number of subvisible particles in monoclonal antibody formulations but also differentiate protein vs. nonprotein particles. In this study, SYPRO Orange dye which is capable of binding to apolar surfaces of conformationally alerted proteins was employed to preferentially stain proteinaceous particles and measure their number. While low fluorescence signals were observed in the absence of the dye, the fluorescence increased almost 100-fold in the presence of SYPRO Orange due to binding of the dye to protein aggregates. A small number of particles did not stain presumably due to their nonprotein nature. The authors also demonstrated specificity of the dye for the protein particles using silicone oil droplets extracted from a commercial plastic syringe. It was convincingly demonstrated that the flow cytometry method can be reliably used to screen a large number of stability samples for the presence of subvisible particles with good precision and accuracy as well as moderate sample requirements. In a similar study (Ludwig et al. 2011), flow cytometry was used to detect and characterize subvisible particles in four different protein formulations contaminated with silicone oil droplets. The authors showed that this technique had the ability to distinguish homogenous protein particles/aggregates from heterogeneous particles consisting of silicone oil and protein.

Various extrinsic fluorescent dyes in combination with microscopic techniques have been used in the detection and characterization of protein aggregates and particulates. Examples include 1-anilinonaphthalene-8-sulfonate (ANS), a dimeric analog (bis-ANS), SYPRO Orange, Nile red, Congo red, and Thioflavin T. A review article by Hawe et al. (2008b) provides a detailed overview of solution properties and underlying mechanisms governing the fluorescent properties of these dyes. The use of such dyes in various fields of protein analysis has been discussed elsewhere (Filipe et al. 2011; He et al. 2010; Li et al. 2011). In general, they are extremely sensitive to polarity of their microenvironments and hence have become important tools in protein characterization studies. These apolar probes are typically essentially nonfluorescent in aqueous solution but become strongly fluorescent in less polar environments such as when a protein's hydrophobic regions became more exposed due to conformational alterations and/or formation of aggregates.

Fluorescence microscopy permits the evaluation of high-concentration protein formulations without dilution and with minimal impact on the local environment of the protein (Demeule et al. 2009). Therefore, it can be an extremely useful tool for the detection of protein aggregates/particulates. The use of Nile red for the early detection of small amounts of immunoglobulin aggregates using fluorescence microscopy has been shown by Demeule et al. (2007b). These studies were performed on an inverted microscope equipped with a mercury discharge lamp for fluorescence microscopy and a tungsten lamp for bright-field microscopy. Suspensions of a recombinant humanized monoclonal antibody were stained with Nile red and visualized by fluorescence microscopy. The same sample was examined by bright-field microscopy for comparison. A more efficient visualization of the protein aggregates was observed following addition of Nile red and subsequent observation by fluorescence microscopy compared to the samples examined with bright-field microscopy (see Fig. 9.11). The high specificity of Nile red for protein aggregates eliminated false positives due to the presence of impurities/dust. Multiple other

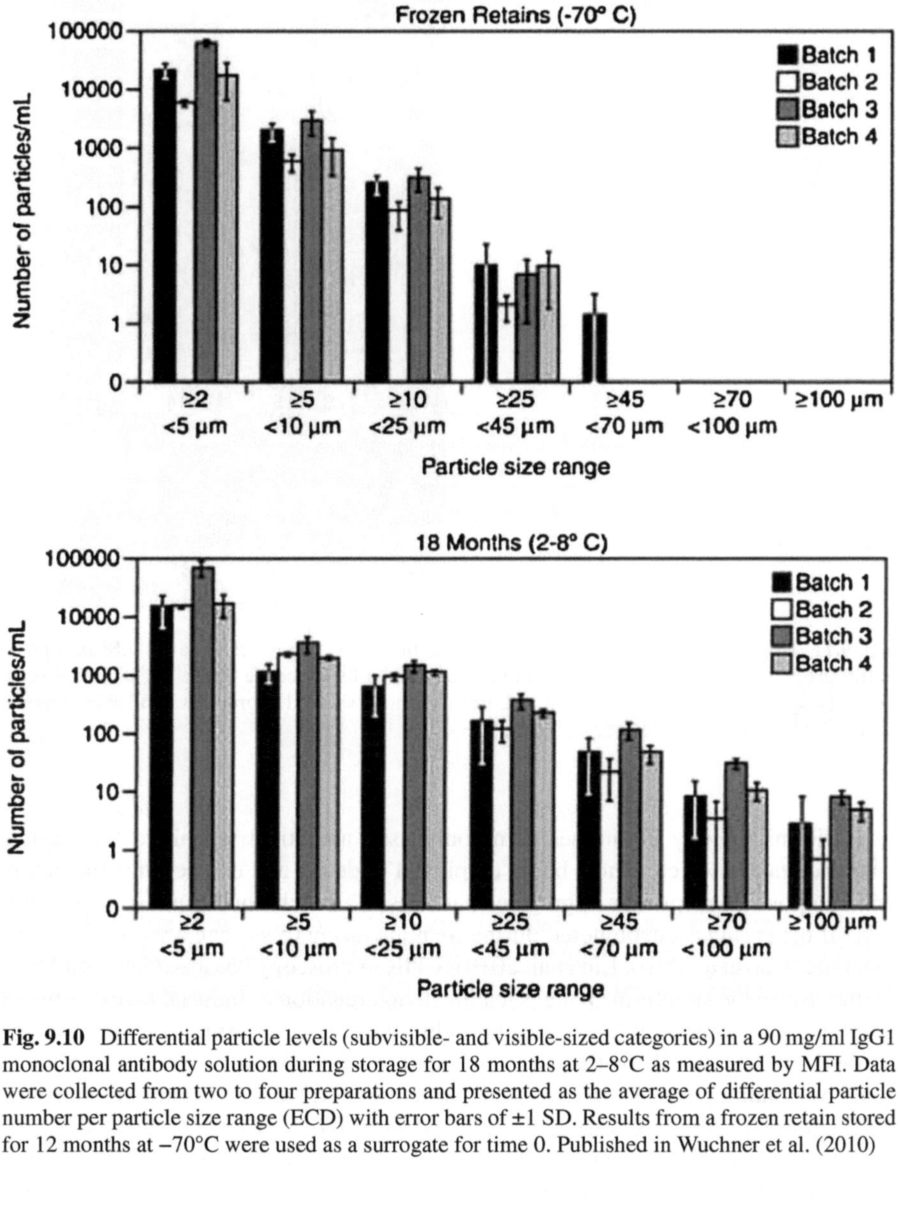

Fig. 9.10 Differential particle levels (subvisible- and visible-sized categories) in a 90 mg/ml IgG1 monoclonal antibody solution during storage for 18 months at 2–8°C as measured by MFI. Data were collected from two to four preparations and presented as the average of differential particle number per particle size range (ECD) with error bars of ±1 SD. Results from a frozen retain stored for 12 months at −70°C were used as a surrogate for time 0. Published in Wuchner et al. (2010)

publications have shown the use of Nile red staining and fluorescence microscopy for the detection and characterization of protein aggregates and particulates (Demeule et al. 2007a, 2009; Sutter et al. 2007). Caution, however, needs to be taken while using fluorescent dyes for protein characterization since these dyes may also induce formation of aggregates and particulates because of their tendency to perturb protein structure (Engelhard and Evans 1995).

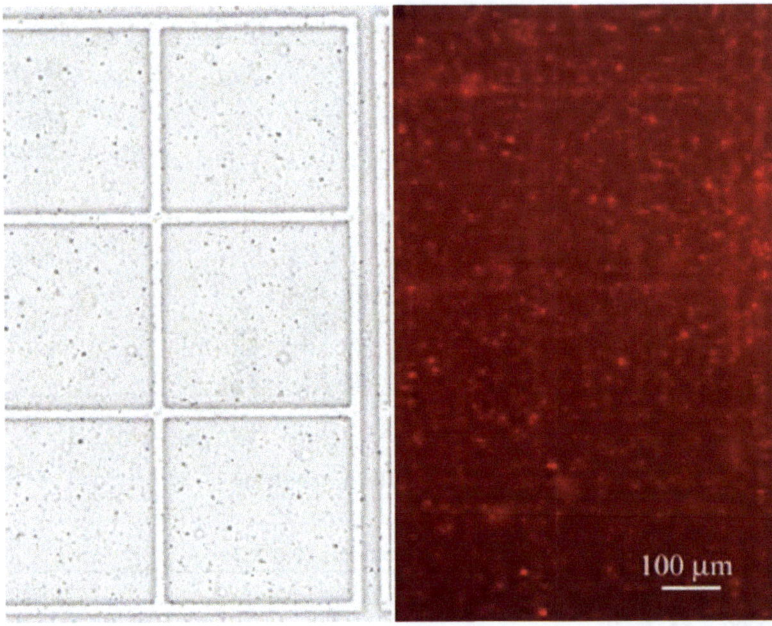

Fig. 9.11 Bright-field image (*left*) of unstained antibody A aggregates (in a 10 mM phosphate buffer pH 7 solution containing 0.8 mg/ml protein) and a fluorescence image (*right*) of the same antibody aggregates stained with *Nile red*. Both methods showed aggregates similar in size and shape. Published in Demeule et al. (2007a, b)

FTIR microscopy is another commonly used analytical technique for particle characterization which is now being employed to detect and characterize the nature and composition of protein aggregates/particles. This technique can also help determine if the particles encountered during manufacturing or storage are proteinaceous in nature (Cao et al. 2010; Liu et al. 2010). FTIR microscopy has also been employed to investigate the structural changes leading to aggregation as induced by exposure of lyophilized insulin to high levels of humidity (Giselle et al. 2009). In addition, this method has been used for the easy and rapid discrimination between bacteria and fungi by obtaining unique spectral peaks from these organisms (Erukhimovitch et al. 2002) as well as for detection of cells infected with Herpes virus (Salman et al. 2002).

Another experimental approach is capturing protein particles on gold-coated membrane filters. Using FTIR microscopy, aggregates can then be visualized and chemically mapped. Wuchner et al. (Wuchner et al. 2010) used a combination of FTIR microscopy and SEM-EDX to assess the composition of protein particulates formed during long-term storage of a high-concentration IgG1 monoclonal antibody formulation. Protein particles were isolated on a gold-coated membrane filter and the IR spectra were acquired by attenuated total reflectance (ATR) infrared spectroscopy employing a microscope coupled to an FTIR spectrometer. The authors were able to successfully show that the key amide bands (typical of proteins)

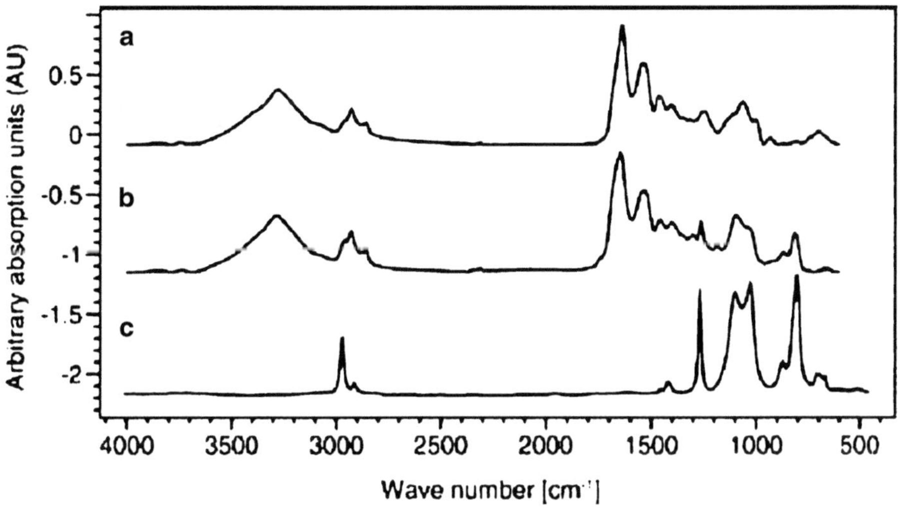

Fig. 9.12 Representative ATR FTIR spectrum of a protein particle (**b**), overlaid with IgG1 reference spectrum (**a**), and polydimethylsiloxane reference spectrum (**c**). Protein stability sample contained 90 mg/mL IgG1 stored for 16 months at 2–8°C. Published in Wuchner et al. (2010)

in the FTIR spectrum of protein particles at a 16-month stability time point were comparable to those obtained for a reference IgG1 preparation (Fig. 9.12). Additional bands characteristic of silicone were detected for most of the particles examined. The protein particulates were further characterized by SEM-EDX to study their composition (Wuchner et al. 2010) as shown in Table 9.2. In addition to silicon, some of the protein particles contained aluminum and fluorine. These compositional data from FTIR microscopy and SEM-EDX analysis indicated heterogeneity across the particles examined, suggesting protein particle formation is a heterogeneous process (Table 9.2).

9.5 Conclusion and Future Directions

This chapter has described recent case studies on the use of different analytical techniques to better characterize the nature of protein aggregates and particles. The focus was on recent work from different investigators who are exploring both new techniques as well as combinations of analytical approaches, to better understand the number, size range, and composition of protein aggregates and particles in protein formulations. It is well recognized that no one analytical measurement can detect all types and sizes of protein aggregates (Philo 2006). Some of the newer biophysical approaches used to monitor protein aggregate and particle formation have come from the field of protein aggregation diseases (Aguzzi and O'Connor 2010).

Table 9.2 Analytical characterization of protein particles isolated from long-term stability samples of a 90 mg/ml IgG1 monoclonal antibody solution formulation stored at 2–8°C

Sample	Age of material	Inverted microscopy	FTIR	SEM-EDX
Batch 1	11 mo	Some large and many small translucent particles of typical proteinaceous morphology	Protein (6/6), silicone (6/6)	C,N,O (4/4), Si (4/4)
Batch 2	14 mo	Some large and many small translucent particles of typical proteinaceous morphology	Protein (5/5), silicone (5/5)	C,N,O (3/3), Si (3/3), Al (1/3), F (1/3)[a]
Batch 3	11 mo	Some large and many small translucent particles of typical proteinaceous morphology	Protein (7/7), silicone (4/7)	C,N,O (6/6), Si (4/6), Al (4/6)
Batch 4	11 mo	Some large and many small translucent particles of typical proteinaceous morphology	Protein (10/10), silicone (10/10)	C,N,O (4/4), Si (3/4)

Published in Wuchner et al. (2010)

mo, months at 2–8°C

Inverted microscopy: large particles through microscope approximately >50 μm, small particles approximately <20 μm

FTIR microscopy: between 5 and 10 particles per sample were analyzed. The frequencies of main components (spectral matches) are indicated in parentheses (number of particles with component per total number or particles analyzed)

SEM-EDX: between 3 and 6 particles per sample were analyzed. The frequencies of the detected elements are indicated in parentheses (number of particles with element per total number of particles analyzed)

[a]Fluorine was detected at trace level

As new case studies are published demonstrating a further enhanced ability to detect protein aggregates and particles under varying conditions, our mechanistic understanding of the interrelationships of protein aggregation pathways (Philo and Arakawa 2009; Wang et al. 2010) as well as specific and nonspecific protein-protein interactions leading to protein self-association (Saluja and Kalonia 2008) will continue to improve. The interrelationships between chemical and physical stabilities are also emerging as an important new direction since some aggregates may contain an altered nature (e.g., mixed disulfide formation and oxidation) as well as the presence of nonprotein nucleating species such as metals (Carpenter et al. 2009; Narhi et al. 2011; Joubert et al. 2011; Luo et al. 2011). Similarly, although not the focus of this chapter, our mechanistic understanding of the interrelationships of protein aggregate composition and size to immunogenic potential is increasing (Carpenter et al. 2010b; Mire-Sluis et al. 2011; Rosenberg 2006; Maas et al. 2007). Based on this enhanced understanding, more rational approaches to the selection and optimization of stabilizing excipients to minimize protein aggregation and particle formation should result (Kamerzell et al. 2011). The combination of utilizing better analytical methods along with an enhanced understanding of protein aggregation pathways should ultimately lead to important improvements in the impurity levels and long-term stability of protein formulations.

References

Aguzzi A, O'Connor T (2010) Protein aggregation diseases: pathogenicity and therapeutic perspectives. Nat Rev Drug Discov 9(3):237–248

Ahrer K, Buchacher A, Iberer G, Josic D, Jungbauer A (2003) Analysis of aggregates of human immunoglobulin G using size-exclusion chromatography, static and dynamic light scattering. J Chromatogr A 1009(1–2):89–96

Arakawa T, Ejima D, Li T, Philo JS (2010) The critical role of mobile phase composition in size exclusion chromatography of protein pharmaceuticals. J Pharm Sci 99(4):1674–1692

Barnard JG, Singh S, Randolph TW, Carpenter JF (2010) Subvisible particle counting provides a sensitive method of detecting and quantifying aggregation of monoclonal antibody caused by freeze-thawing: insights into the roles of particles in the protein aggregation pathway. J Pharm Sci 100(2):492–503

Barnard JG, Rhyner MN, Carpenter JF (2012) Critical evaluation and guidance for using the coulter method for counting subvisible particles in protein solutions. J Pharm Sci 101(1):140–153

Berkowitz SA (2006) Role of analytical ultracentrifugation in assessing the aggregation of protein biopharmaceuticals. AAPS J 8(3):E590–E605

Bhambhani A, Kissmann JM, Joshi SB, Volkin DB, Kashi RS, Middaugh CR (2012) Formulation design and high-throughput excipient selection based on structural integrity and conformational stability of dilute and highly concentrated IgG1 monoclonal antibody solutions. J Pharm Sci 101(3):1120–1135

Bond MD, Panek ME, Zhang Z et al (2010) Evaluation of a dual-wavelength size exclusion HPLC method with improved sensitivity to detect protein aggregates and its use to better characterize degradation pathways of an IgG1 monoclonal antibody. J Pharm Sci 99(6):2582–2597

Burg TP, Godin M, Knudsen SM et al (2007) Weighing of biomolecules, single cells and single nanoparticles in fluid. Nature 446(7139):1066–1069

Calamai M, Canale C, Relini A, Stefani M, Chiti F, Dobson CM (2005) Reversal of protein aggregation provides evidence for multiple aggregated States. J Mol Biol 346(2):603–616

Cao S, Pollastrini J, Jiang Y (2009) Separation and characterization of protein aggregates and particles by field flow fractionation. Curr Pharm Biotechnol 10(4):382–390

Cao X, Masatani P, Torraca G, Wen ZQ (2010) Identification of a mixed microparticle by combined microspectroscopic techniques: a real forensic case study in the biopharmaceutical industry. Appl Spectrosc 64(8):895–900

Carpenter JF, Randolph TW, Jiskoot W et al (2009) Overlooking subvisible particles in therapeutic protein products: gaps that may compromise product quality. J Pharm Sci 98(4):1201–1205

Carpenter JF, Randolph TW, Jiskoot W, Crommelin DJ, Middaugh CR, Winter G (2010a) Potential inaccurate quantitation and sizing of protein aggregates by size exclusion chromatography: essential need to use orthogonal methods to assure the quality of therapeutic protein products. J Pharm Sci 99(5):2200–2208

Carpenter J, Cherney B, Lubinecki A et al (2010b) Meeting report on protein particles and immunogenicity of therapeutic proteins: filling in the gaps in risk evaluation and mitigation. Biologicals 38(5):602–611

Chait BT (2011) Mass spectrometry in the postgenomic era. Annu Rev Biochem 80:239–246

Cheng X, van Breemen RB (2005) Mass spectrometry-based screening for inhibitors of beta-amyloid protein aggregation. Anal Chem 77(21):7012–7015

Chrai S, Clayton R, Mestrandrea L et al (1987) Limitations in the use of HIAC for product particle counting. J Parenter Sci Technol 41(6):209–214

Colton RJ, Baselt DR, Dufrene YF, Green JB, Lee GU (1997) Scanning probe microscopy. Curr Opin Chem Biol 1(3):370–377

Costello MA, Woititz C, Defeo J et al (1992) Characterization of humanized anti-tac monoclonal-antibody by traditional separation techniques and capillary electrophoresis. J Liq Chromatogr 15(6–7):1081–1097

Cram LS (2002) Flow cytometry, an overview. Methods Cell Sci 24(1–3):1–9

Dam J, Schuck P (2004) Calculating sedimentation coefficient distributions by direct modeling of sedimentation velocity concentration profiles. Meth Enzymol 384:185–212

David Bernard Williams CBC (2009) The transmission electron microscope. Springer, New York

Demeule B, Gurny R, Arvinte T (2007a) Detection and characterization of protein aggregates by fluorescence microscopy. Int J Pharm 329(1–2):37–45

Demeule B, Lawrence MJ, Drake AF, Gurny R, Arvinte T (2007b) Characterization of protein aggregation: the case of a therapeutic immunoglobulin. Biochim Biophys Acta 1774(1):146–153

Demeule B, Palais C, Machaidze G, Gurny R, Arvinte T (2009) New methods allowing the detection of protein aggregates: a case study on trastuzumab. MAbs 1(2):142–150

Demeule B, Messick S, Shire SJ, Liu J (2010) Characterization of particles in protein solutions: reaching the limits of current technologies. AAPS J 12(4):708–715

Engelhard M, Evans PA (1995) Kinetics of interaction of partially folded proteins with a hydrophobic dye: evidence that molten globule character is maximal in early folding intermediates. Protein sci 4(8):1553–1562

Engelsman J, Garidel P, Smulders R et al (2011) Strategies for the assessment of protein aggregates in pharmaceutical biotech product development. Pharm Res 28(4):920–933

Erukhimovitch V, Talyshinsky M, Souprun Y, Huleihel M (2002) Spectroscopic characterization of human and mouse primary cells, cell lines and malignant cells. Photochem Photobiol 76(4):446–451

Esue O, Kanai S, Liu J, Patapoff TW, Shire SJ (2009) Carboxylate-dependent gelation of a monoclonal antibody. Pharm Res 26(11):2478–2485

Ferreira GN, da-Silva AC, Tome B (2009) Acoustic wave biosensors: physical models and biological applications of quartz crystal microbalance. Trends Biotechnol 27(12):689–697

Filipe V, Hawe A, Jiskoot W (2010) Critical evaluation of Nanoparticle Tracking Analysis (NTA) by NanoSight for the measurement of nanoparticles and protein aggregates. Pharm Res 27(5):796–810

Filipe V, Poole R, Kutscher M, Forier K, Braeckmans K, Jiskoot W (2011) Fluorescence single particle tracking for the characterization of submicron protein aggregates in biological fluids and complex formulations. Pharm Res 28(5):1112–1120

Flores-Fernandez GM, Sola RJ, Griebenow K (2009) The relation between moisture-induced aggregation and structural changes in lyophilized insulin. J Pharm Pharmacol 61(11):1555–1561

Fradkin AH, Carpenter JF, Randolph TW (2011) Glass particles as an adjuvant: a model for adverse immunogenicity of therapeutic proteins. J Pharm Sci 100(11):4953–4964

Fraunhofer W, Winter G (2004) The use of asymmetrical flow field-flow fractionation in pharmaceutics and biopharmaceutics. Eur J Pharm Biopharm 58(2):369–383

Gabrielson JP, Arthur KK (2011) Measuring low levels of protein aggregation by sedimentation velocity. Methods 54(1):83–91

Gabrielson JP, Brader ML, Pekar AH et al (2007) Quantitation of aggregate levels in a recombinant humanized monoclonal antibody formulation by size-exclusion chromatography, asymmetrical flow field flow fractionation, and sedimentation velocity. J Pharm Sci 96(2):268–279

Ginsberg MH, Frelinger AL, Lam SC et al (1990) Analysis of platelet aggregation disorders based on flow cytometric analysis of membrane glycoprotein IIb-IIIa with conformation-specific monoclonal antibodies. Blood 76(10):2017–2023

Giselle M, Flores-Fernandez G, Solá R, Griebenow K (2009) The relation between moisture-induced aggregation and structural changes in lyophilized insulin. J Pharm Pharmacol 61(11):1555–1561

Greenwood R (2003) Review of the measurement of zeta potentials in concentrated aqueous suspensions using electroacoustics. Adv Colloid Interface Sci 106:55–81

Grillberger L, Kreil TR, Nasr S, Reiter M (2009) Emerging trends in plasma-free manufacturing of recombinant protein therapeutics expressed in mammalian cells. Biotechnol J 4(2):186–201

Han M, Phan D, Nightlinger N et al (2006) Optimization of CE-SDS method for antibody separation based on multi-users experimental practices. Chromatographia 64:334–342

Harding SA, Din JN, Sarma J et al (2007) Flow cytometric analysis of circulating platelet-monocyte aggregates in whole blood: methodological considerations. Thromb Haemost 98(2):451–456

Harper JD, Lieber CM, Lansbury PT Jr (1997) Atomic force microscopic imaging of seeded fibril formation and fibril branching by the Alzheimer's disease amyloid-beta protein. Chem Biol 4(12):951–959

Hawe A, Friess W, Sutter M, Jiskoot W (2008a) Online fluorescent dye detection method for the characterization of immunoglobulin G aggregation by size exclusion chromatography and asymmetrical flow field flow fractionation. Anal Biochem 378(2):115–122

Hawe A, Sutter M, Jiskoot W (2008b) Extrinsic fluorescent dyes as tools for protein characterization. Pharm Res 25(7):1487–1499

Hawe A, Hulse WL, Jiskoot W, Forbes RT (2011) Taylor dispersion analysis compared to dynamic light scattering for the size analysis of therapeutic peptides and proteins and their aggregates. Pharm Res 28(9):2302–2310

He F, Phan DH, Hogan S et al (2010) Detection of IgG aggregation by a high throughput method based on extrinsic fluorescence. J Pharm Sci 99(6):2598–2608

Heldt CL, Sorci M, Posada D, Hirsa A, Belfort G (2011) Detection and reduction of microaggregates in insulin preparations. Biotechnol Bioeng 108(1):237–241

Hillger F, Nettels D, Dorsch S, Schuler B (2007) Detection and analysis of protein aggregation with confocal single molecule fluorescence spectroscopy. J Fluoresc 17(6):759–765

Houde D, Arndt J, Domeier W, Berkowitz S, Engen JR (2009) Characterization of IgG1 conformation and conformational dynamics by hydrogen/deuterium exchange mass spectrometry. Anal Chem 81(14):5966

Huang C-T, Sharma D, Oma P, Krishnamurthy R (2009) Quantitation of protein particles in parenteral solutions using micro-flow imaging. J Pharm Sci 98(9):3058–3071

Hughes H, Morgan C, Brunyak E et al (2009) A multi-tiered analytical approach for the analysis and quantitation of high-molecular-weight aggregates in a recombinant therapeutic glycoprotein. AAPS J 11(2):335–341

Hunt G, Moorhouse KG, Chen AB (1996) Capillary isoelectric focusing and sodium dodecyl sulfate-capillary gel electrophoresis of recombinant humanized monoclonal antibody HER2. J Chromatogr A 744(1–2):295–301

Jorio H, Tran R, Meghrous J, Bourget L, Kamen A (2006) Analysis of baculovirus aggregates using flow cytometry. J Virol Methods 134(1–2):8–14

Joseph Goldstein DN, Joy D, Lyman C, Echlin P, Lifshin E, Sawyer L, Michael J (2003) Scanning electron microscopy and X-ray microanalysis, 3rd edn. Plenum, New York

Joubert MK, Luo Q, Nashed-Samuel Y, Wypych J, Narhi LO (2011) Classification and characterization of therapeutic antibody aggregates. J Biol Chem 286(28):25118–25133

Kaddis CS, Lomeli SH, Yin S et al (2007) Sizing large proteins and protein complexes by electrospray ionization mass spectrometry and ion mobility. J Am Soc Mass Spectrom 18(7):1206–1216

Kaltashov IA, Bobst CE, Abzalimov RR, Wang G, Baykal B, Wang S (2012) Advances and challenges in analytical characterization of biotechnology products: mass spectrometry-based approaches to study properties and behavior of protein therapeutics. Biotechnol Adv 30(1):210–222

Kamerzell TJ, Esfandiary R, Joshi SB, Middaugh CR, Volkin DB (2011) Protein-excipient interactions: mechanisms and biophysical characterization applied to protein formulation development. Adv Drug Deliv Rev 63(13):1118–1159

Kheterpal I, Wetzel R (2006) Hydrogen/deuterium exchange mass spectrometry—a window into amyloid structure. Acc Chem Res 39(9):584–593

Khurana R, Uversky VN, Nielsen L, Fink AL (2001) Is Congo red an amyloid-specific dye? J Biol Chem 276(25):22715–22721

Kiese S, Papppenberger A, Friess W, Mahler HC (2008) Shaken, not stirred: mechanical stress testing of an IgG1 antibody. J Pharm Sci 97(10):4347–4366

Kiese S, Pappenberger A, Friess W, Mahler H-C (2010) Equilibrium studies of protein aggregates and homogeneous nucleation in protein formulation. J Pharm Sci 99(2):632–644

Kilar F (2003) Recent applications of capillary isoelectric focusing. Electrophoresis 24(22–23):3908–3916

Kim YS, Randolph TW, Manning MC, Stevens FJ, Carpenter JF (2003) Congo red populates partially unfolded states of an amyloidogenic protein to enhance aggregation and amyloid fibril formation. J Biol Chem 278(12):10842–10850

Knowles TPJ, Devlin GL, Dobson CM, Welland ME (2011) Probing protein aggregation with quartz crystal microbalances. In: Hill AF, Barnham KJ, Bottomley SP, Cappai R (eds) Protein folding, misfolding, and disease, vol 752, 1st edn. Humana Press, New York

Konermann L, Pan J, Liu YH (2011) Hydrogen exchange mass spectrometry for studying protein structure and dynamics. Chem Soc Rev 40(3):1224–1234

Kraly J, Fazal MA, Schoenherr RM et al (2006) Bioanalytical applications of capillary electrophoresis. Anal Chem 78(12):4097–4110

Kukrer B, Filipe V, van Duijn E et al (2010) Mass spectrometric analysis of intact human monoclonal antibody aggregates fractionated by size-exclusion chromatography. Pharm Res 27(10):2197–2204

Lee H, Kirchmeier M, Mach H (2011) Monoclonal antibody aggregation intermediates visualized by atomic force microscopy. J Pharm Sci 100(2):416–423

Li B, Flores J, Corvari V (2007) A simple method for the detection of insoluble aggregates in protein formulations. J Pharm Sci 96(7):1840–1843

Li Y, Mach H, Blue JT (2011) High throughput formulation screening for global aggregation behaviors of three monoclonal antibodies. J Pharm Sci 100(6):2120–2135

Lichtman JW, Conchello JA (2005) Fluorescence microscopy. Nat Methods 2(12):910–919

Liu J, Shire SJ (1999) Analytical ultracentrifugation in the pharmaceutical industry. J Pharm Sci 88(12):1237–1241

Liu J, Andya JD, Shire SJ (2006) A critical review of analytical ultracentrifugation and field flow fractionation methods for measuring protein aggregation. AAPS J 8(3):E580–E589

Liu W, Swift R, Torraca G et al (2010) Root cause analysis of tungsten-induced protein aggregation in pre-filled syringes. PDA J Pharm Sci Technol 64(1):11–19

Liu L, Ammar DA, Ross LA, Mandava N, Kahook MY, Carpenter JF (2011) Silicone oil microdroplets and protein aggregates in repackaged bevacizumab and ranibizumab: effects of long-term storage and product mishandling. Invest Ophthalmol Vis Sci 52(2):1023–1034

Lomakin A, Benedek GB, Teplow DB (1999) Monitoring protein assembly using quasielastic light scattering spectroscopy. Methods Enzymol 309:429–459

Lubiniecki A, Volkin DB, Federici M et al (2010) Comparability assessments of process and product changes made during development of two different monoclonal antibodies. Biologicals 39(1):9–22

Ludwig DB, Trotter JT, Gabrielson JP, Carpenter JF, Randolph TW (2011) Flow cytometry: a promising technique for the study of silicone oil-induced particulate formation in protein formulations. Anal Biochem 410(2):191–199

Luo Y, Lu Z, Raso SW, Entrican C, Tangarone B (2009) Dimers and multimers of monoclonal IgG1 exhibit higher in vitro binding affinities to Fcgamma receptors. MAbs 1(5):491–504

Luo Q, Joubert MK, Stevenson R, Ketchem RR, Narhi LO, Wypych J (2011) Chemical modifications in therapeutic protein aggregates generated under different stress conditions. J Biol Chem 286(28):25134–25144

Maas C, Hermeling S, Bouma B, Jiskoot W, Gebbink MF (2007) A role for protein misfolding in immunogenicity of biopharmaceuticals. J Biol Chem 282(4):2229–2236

Mach H, Arvinte T (2011) Addressing new analytical challenges in protein formulation development. Eur J Pharm Biopharm 78(2):196–207

Mach H, Middaugh CR (2011) Ultraviolet spectroscopy as a tool in therapeutic protein development. J Pharm Sci 100(4):1214–1227

Mach H, Bhambhani A, Meyer BK et al (2011) The use of flow cytometry for the detection of subvisible particles in therapeutic protein formulations. J Pharm Sci 100(5):1671–1678

Maddux NR, Joshi SB, Volkin DB, Ralston JP, Middaugh CR (2011) Multidimensional methods for the formulation of biopharmaceuticals and vaccines. J Pharm Sci 100(10):4171–4197

Madsen R, Cherris R, Shabushnig JG, Hunt DG (2009) Visible particulates in injections—a history and a proposal to revise USP general chapter injections (1). Pharmacop Forum 35:1383–1387
Mahler HC, Muller R, Friess W, Delille A, Matheus S (2005) Induction and analysis of aggregates in a liquid IgG1-antibody formulation. Eur J Pharm Biopharm 59(3):407–417
Mahler H-C, Friess W, Grauschopf U, Kiese S (2009) Protein aggregation: pathways, induction factors and analysis. J Pharm Sci 98(9):2909–2934
Majumdar S, Ford BM, Mar KD, Sullivan VJ, Ulrich RG, D'souza AJM (2011) Evaluation of the effect of syringe surfaces on protein formulations. J Pharm Sci 100(7):2563–2573
Marcsisin SR, Engen JR (2010) Hydrogen exchange mass spectrometry: what is it and what can it tell us? Anal Bioanal Chem 397(3):967–972
Markovich RJ, Taylor AK, Rosen J (1997) Drug migration from the adhesive matrix to the polymer film laminate facestock in a transdermal nitroglycerin system. J Pharm Biomed Anal 16(4):651–660
Michel Godin AKB, Burg TP, Babcock K, Manalis SR (2007) Measuring the mass, density, and size of particles and cells using a suspended microchannel resonator. Appl Phys Lett 91:3
Michels DA, Brady LJ, Guo A, Balland A (2007) Fluorescent derivatization method of proteins for characterization by capillary electrophoresis-sodium dodecyl sulfate with laser-induced fluorescence detection. Anal Chem 79(15):5963–5971
Middaugh CR, Joshi SB (2011) Spectroscopy of vaccines. In: Singh M, Srivastava IK (eds) Development of vaccines, from discovery to clinical testing. Wiley, Hoboken, pp 263–292
Mire-Sluis A, Cherney B, Madsen R, Polozova A, Rosenberg A, Smith H, Arora T, Narhi L (2011) Analysis and immunogenic potential of aggregates and particles. BioProcess Int 9(10):38–43
Murphy RM (1997) Static and dynamic light scattering of biological macromolecules: what can we learn? Curr Opin Biotechnol 8(1):25–30
Naiki H, Higuchi K, Hosokawa M, Takeda T (1989) Fluorometric determination of amyloid fibrils in vitro using the fluorescent dye, thioflavin T1. Anal Biochem 177(2):244–249
Narhi LO, Schmit J, Bechtold-Peters K, Sharma D (2011) Classification of protein aggregates. J Pharm Sci 101(2):493–498
Nayak A, Colandene J, Bradford V, Perkins M (2011) Characterization of subvisible particle formation during the filling pump operation of a monoclonal antibody solution. J Pharm Sci 100(10):4198–4204
Nettleton EJ, Tito P, Sunde M, Bouchard M, Dobson CM, Robinson CV (2000) Characterization of the oligomeric states of insulin in self-assembly and amyloid fibril formation by mass spectrometry. Biophys J 79(2):1053–1065
O'Sullivan CK, Guilbault GG (1999) Commercial quartz crystal microbalances – theory and applications. Biosens Bioelectron 14(8–9):663–670
Oliva A, Llabres M, Farina JB (2001) Comparative study of protein molecular weights by size-exclusion chromatography and laser-light scattering. J Pharm Biomed Anal 25:833–841
Pease LF 3rd, Elliott JT, Tsai DH, Zachariah MR, Tarlov MJ (2008) Determination of protein aggregation with differential mobility analysis: application to IgG antibody. Biotechnol Bioeng 101(6):1214–1222
Philo JS (2006) Is any measurement method optimal for all aggregate sizes and types? AAPS J 8(3):E564–E571
Philo JS (2009) A critical review of methods for size characterization of non-particulate protein aggregates. Curr Pharm Biotechnol 10(4):359–372
Philo JS, Arakawa T (2009) Mechanisms of protein aggregation. Curr Pharm Biotechnol 10(4):348–351
Prati S, Joseph E, Sciutto G, Mazzeo R (2010) New advances in the application of FTIR microscopy and spectroscopy for the characterization of artistic materials. Acc Chem Res 43(6):792–801
Qi P, Volkin DB, Zhao H et al (2009) Characterization of the photodegradation of a human IgG1 monoclonal antibody formulated as a high-concentration liquid dosage form. J Pharm Sci 98(9):3117–3130

Remmele RL Jr, Callahan WJ, Krishnan S et al (2006) Active dimer of Epratuzumab provides insight into the complex nature of an antibody aggregate. J Pharm Sci 95(1):126–145
Reschiglian P, Zattoni A, Roda B, Michelini E, Roda A (2005) Field-flow fractionation and biotechnology. Trends Biotechnol 23(9):475–483
Rhyner MN (2011) The Coulter principle for analysis of subvisible particles in protein formulations. AAPS J 13(1):54–58
Roberts GS, Yu S, Zeng Q et al (2012) Tunable pores for measuring concentrations of synthetic and biological nanoparticle dispersions. Biosens Bioelectron 31(1):17–25
Roda B, Zattoni A, Reschiglian P et al (2009) Field-flow fractionation in bioanalysis: a review of recent trends. Anal Chim Acta 635(2):132–143
Rosenberg AS (2006) Effects of protein aggregates: an immunologic perspective. AAPS J 8(3):E501–E507
Rosenberg E, Hepbildikler S, Kuhne W, Winter G (2009) Ultrafiltration concentration of monoclonal antibody solutions: development of an optimized method minimizing aggregation. J Membr Sci 342(1–2):50–59
Ross S, Morrison ID (2002) Colloidal dispersions. Wiley Interscience, New York
Rumi Chunara MG, Knudsen SM, Manalis SR (2007) Mass-based readout for agglutination assays. Appl Phys Lett 91:3
Rustandi RR, Washabaugh MW, Wang Y (2008) Applications of CE SDS gel in development of biopharmaceutical antibody-based products. Electrophoresis 29(17):3612–3620
Salas-Solano O, Tomlinson B, Du S, Parker M, Strahan A, Ma S (2006) Optimization and validation of a quantitative capillary electrophoresis sodium dodecyl sulfate method for quality control and stability monitoring of monoclonal antibodies. Anal Chem 78(18):6583–6594
Salman A, Erukhimovitch V, Talyshinsky M, Huleihil M, Huleihel M (2002) FTIR spectroscopic method for detection of cells infected with herpes viruses. Biopolymers 67(6):406–412
Saluja A, Kalonia DS (2008) Nature and consequences of protein-protein interactions in high protein concentration solutions. Int J Pharm 358(1–2):1–15
Sauerbrey G (1959) Use of quartz vibration for weighing thin films on a microbalance. J Phys 155:206–212
Schmitt C, Bovay C, Rouvet M, Shojaei-Rami S, Kolodziejczyk E (2007) Whey protein soluble aggregates from heating with NaCl: physicochemical, interfacial, and foaming properties. Langmuir 23(8):4155–4166
Schuck P (2000) Size-distribution analysis of macromolecules by sedimentation velocity ultracentrifugation and lamm equation modeling. Biophys J 78(3):1606–1619
Schuck P (2004) A model for sedimentation in inhomogeneous media. II. Compressibility of aqueous and organic solvents. Biophys Chem 108(1–3):201–214
Sejersen M, Salomonsena T, Ipsen R, Clark R, Rolin C, Engelsen S (2007) Zeta potential of pectin-stabilised casein aggregates in acidified milk drinks. Int Dairy J 17:302–307
Shapiro HM (1995) Practical flow cytometry. Wiley-Liss, New york
Sharma DK, Oma P, Pollo MJ, Sukumar M (2010a) Quantification and characterization of subvisible proteinaceous particles in opalescent mAb formulations using micro-flow imaging. J Pharm Sci 99(6):2628–2642
Sharma DK, King D, Oma P, Merchant C (2010b) Micro-flow imaging: flow microscopy applied to sub-visible particulate analysis in protein formulations. AAPS J 12(3):455–464
Singh SK, Afonina N, Awwad M et al (2010) An industry perspective on the monitoring of subvisible particles as a quality attribute for protein therapeutics. J Pharm Sci 99(8):3302–3321
Strehl R, Rombach-Riegraf V, Diez M et al (2011) Discrimination between silicone oil droplets and protein aggregates in biopharmaceuticals: a novel multiparametric image filter for sub-visible particles in microflow imaging analysis. Pharm Res 27:1–9
Sutter M, Oliveira S, Sanders NN et al (2007) Sensitive spectroscopic detection of large and denatured protein aggregates in solution by use of the fluorescent dye Nile red. J Fluoresc 17(2):181–192

Thillaivinayagalingam P, Gommeaux J, McLoughlin M, Collins D, Newcombe AR (2010) Biopharmaceutical production: Applications of surface plasmon resonance biosensors. J Chromatogr B Analyt Technol Biomed Life Sci 878(2):149–153

Tobler SA, Fernandez EJ (2002) Structural features of interferon-gamma aggregation revealed by hydrogen exchange. Protein sci 11(6):1340–1352

Torraca G, Wen Z (2005) Forensic investigation of biopharmaceutical manufacturing incidents by light microscopy, ftir microscopy, scanning electron microscopy and energy dispersive X-ray spectrometry. Microsc Microanal 11:1224–1225

Tyagi AK, Randolph TW, Dong A, Maloney KM, Hitscherich C Jr, Carpenter JF (2009) IgG particle formation during filling pump operation: a case study of heterogeneous nucleation on stainless steel nanoparticles. J Pharm Sci 98(1):94–104

Uetrecht C, Rose RJ, van Duijn E, Lorenzen K, Heck AJ (2010) Ion mobility mass spectrometry of proteins and protein assemblies. Chem Soc Rev 39(5):1633–1655

Uskokovic V, Castiglione Z, Cubas P, Zhu L, Li W, Habelitz S (2010) Zeta-potential and particle size analysis of human amelogenins. J Dent Res 89(2):149–153

Van Buren N, Rehder D, Gadgil H, Matsumura M, Jacob J (2009) Elucidation of two major aggregation pathways in an IgG2 antibody. J Pharm Sci 98(9):3013–3030

van Duijn E (2010) Current limitations in native mass spectrometry based structural biology. J Am Soc Mass Spectrom 21(6):971–978

Wall J, Solomon A (1999) Flow cytometric characterization of amyloid fibrils. Methods Enzymol 309:460–466

Wang W, Nema S, Teagarden D (2010) Protein aggregation–pathways and influencing factors. Int J Pharm 390(2):89–99

Wang G, Abzalimov RR, Kaltashov IA (2011) Direct monitoring of heat-stressed biopolymers with temperature-controlled electrospray ionization mass spectrometry. Anal Chem 83(8):2870–2876

Wang G, Johnson AJ, Kaltashov IA (2012) Evaluation of electrospray ionization mass spectrometry as a tool for characterization of small soluble protein aggregates. Anal Chem 84(3):1718–1724

Watson JT, Sparkman OD (2008) Introduction to mass spectrometry: instrumentation, applications, and strategies for data interpretation. Wiley, Chichester

Wear MA, Patterson A, Malone K, Dunsmore C, Turner NJ, Walkinshaw MD (2005) A surface plasmon resonance-based assay for small molecule inhibitors of human cyclophilin A. Anal Biochem 345(2):214–226

Wen J, Arakawa T, Philo JS (1996) Size-exclusion chromatography with on-line light-scattering, absorbance and refractive index detectors for studying proteins and their interactions. Anal Biochem 240:155–166

Wuchner K, Büchler J, Spycher R, Dalmonte P, Volkin DB (2010) Development of a microflow digital imaging assay to characterize protein particulates during storage of a high concentration IgG1 monoclonal antibody formulation. J Pharm Sci 99(8):3343–3361

Ye H (2006) Simultaneous determination of protein aggregation, degradation, and absolute molecular weight by size exclusion chromatography-multiangle laser light scattering. Anal Biochem 356(1):76–85

Yohannes G, Jussila M, Hartonen K, Riekkola ML (2011) Asymmetrical flow field-flow fractionation technique for separation and characterization of biopolymers and bioparticles. J Chromatogr A 1218:4104–4116

Zhang A, Qi W, Singh SK, Fernandez EJ (2011) A new approach to explore the impact of freeze-thaw cycling on protein structure: hydrogen/deuterium exchange mass spectrometry (HX-MS). Pharm Res 28(5):1179–1193

Zhang A, Singh SK, Shirts MR, Kumar S, Fernandez EJ (2012) Distinct aggregation mechanisms of monoclonal antibody under thermal and freeze-thaw stresses revealed by hydrogen exchange. Pharm Res 29(1):236–250

[illegible] (2006) [illegible] ICH guidelines Q [illegible] J Pharm Sci [illegible]

Chapter 10
Investigation of Nonconformance During Protein Therapeutic Manufacturing

Zai-Qing Wen, Gianni Torraca, Guiyang Li, Xiaolin Cao, and Chanel K. Yee

10.1 Introduction

Development and manufacturing of protein therapeutics involves complicated processes from molecular cloning, cell banking, cell culture fermentation, protein purification, drug product formulation, fill/finishing, packaging, and distribution (Nail and Akers 2002). Manufacturing nonconformances (NC) that involve particulate matter can happen during any step of the processes. However, the particulate matter in the upstream processes (fermentation and purification) may have relatively less product impact than those observed in the downstream processes (formulation and fill and finishing). In particular, manufacturing NC at the fill and finishing step can have direct impact on the final drug product quality. According to the USP compendia, product containers such as glass vials and syringes should be practically free of particles (United States Pharmacopeia 29-Formulary 24, 2012). If unexpected visible particulate matter is observed during the visual inspection, the container is rejected as part of quality control. The potential concern around microparticles in primary containers is patient safety. If microparticles in drug product were injected into patient, they could potentially cause inflammation depending on the physical size and chemical composition of the particles (Barber 2000). Another concern is the potential immunogenicity of microparticles, if they contain partially denatured protein (Rosenberg 2006).

Identification of the microparticles and material residues is the first step in these investigations. The analytical work should first determine the physical size and chemical identity of these particles, along with the classification of the particles as intrinsic or extrinsic. The intrinsic particles are normally generated by the product

Z.-Q. Wen (✉) • G. Torraca • G. Li • X. Cao • C.K. Yee
Department of Process and Product Development, Amgen Inc.,
One Amgen Center Drive, Thousand Oaks, CA 91320, USA
e-mail: zwen@amgen.com; gtorraca@amgen.com; gary.li@amgen.com; xiaolinc@amgen.com; chanely@amgen.com

L.O. Narhi (ed.), *Biophysics for Therapeutic Protein Development*, Biophysics for the Life Sciences 4, DOI 10.1007/978-1-4614-4316-2_10,

and the container per se. They include both visible (>100 μm) protein aggregates growing from single protein molecule either through self-association or partial denaturation and then aggregation (Narhi 2012) and/or particles generated from the materials of the primary container such as glass (Iacocca and Allgeier 2007), and silicone oil (Wen et al. 2009), which is used as a lubricant in prefilled syringes (PFSs). Extraneous particulate matter or foreign contaminants are material residues carried over into the primary container that were generated during the manufacturing process of the primary container by the vendor, as well as unexpected foreign particles from manufacturing environment occasionally fall into the container during filling process. For instance, tungsten oxide (Wen et al. 2007) and polymer residue left inside of PFSs (Liu et al. 2010) were from a tungsten pin and nylon pin used for PFS manufacturing. The identification of these microparticles plays a critical role in determining the root cause of the nonconformance, including the exact original source of the particle and the mechanism of particulate generation. This in turn enables process engineers and formulation scientists to take preventive and corrective action and to help quality and toxicological scientists to assess the product and safety impact, ultimately determining the disposition or rejection of that particular lot.

There are many analytical techniques that are employed to determine the physical size and chemical identities of the microparticles. The most frequently employed are optical microscopy to determine the morphology and physical size, vibrational microspectroscopies including Fourier transform infrared (FTIR) and Raman for the determination of the chemical identity (Humecki 1995; Li et al. 2009; Wen 2007; Wen et al. 2008), and scanning electron microscope (SEM) with energy dispersive spectroscopy (EDS) for elemental composition (Goldstein et al. 1992). In addition, FTIR and Raman microscopy can provide higher-order structural information of protein the microparticle may contain (Wen 2007; Wen et al. 2008). Moreover, information on whether the protein particles are denatured and/or aggregated can be assessed by the derivative vibrational spectra as well. Using SEM/EDS analysis, one can further provide a mapping of the distribution of individual elements throughout the microparticle to determine if the specific element is evenly distributed in the microparticle or merely heterogeneous among other components in the particle (Liu et al. 2010).

In this chapter, we will give a brief overview of microparticle analysis and therapeutic protein manufacturing NC investigation as currently performed in our laboratory and throughout the industry. The general approach to microparticle analysis and NC incident investigation will be outlined, and the principle of the analytical procedures will be presented. The goal of this chapter is to give insight into microparticle NC investigation, the challenges it presents, and how it fits into the overall application of biophysical and analytical methods to biopharmaceutical development and manufacturing. All of the cases presented in the chapter are from previous investigations that involve protein therapeutics contained in primary containers including prefilled glass syringes and glass vials. These incidents cover a broad range of materials that could be encountered during the fill and finishing processes during manufacturing of protein therapeutics.

10.2 Particle Analysis Methods

10.2.1 Microparticle Analytical Scheme

Figure 10.1 shows the flow chart of the scheme of microparticle analysis. The majority of the microparticle samples are obtained from manufacturing sites and quality control labs that conducted visual inspection of products contained in prefilled glass syringe or glass vials. The primary containers that have microparticles are normally sent to the forensic lab for analysis at a controlled temperature of 4–8°C. At the beginning of each investigation, a picture of this primary container is taken with a digital camera or a stereomicroscope to record the particles in the original state. It is recommended to record images of the entire container as well as

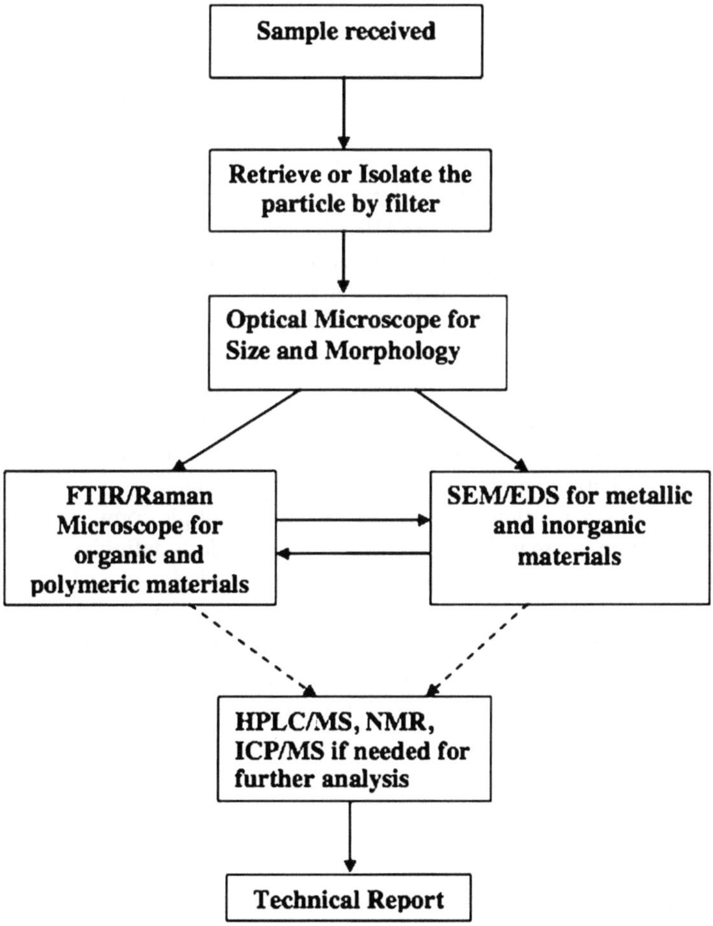

Fig. 10.1 Microparticle analysis procedure

the featured area that has the particles to make sure that their integrity was not compromised during transportation before the container is opened. In particular, for glass vials, a picture of the vial cap should be recorded before opening it. This is important to avoid potential controversy regarding if the vial had been punctured or not. This is particularly important for vials containing commercial drug product and returned as part of a customer complaint from patients or clinical centers. If the vial has been punctured by a needle, the source of the microparticles may not be caused by the manufacturing process.

The majority of the microparticles can be isolated through filtration using filter appropriate for the planned analysis. A gold-coated membrane filter (Rapid-ID, Berlin, Germany) facilitates FTIR-microscopic analysis of the particle directly on the filter without the need to pick up the particle and transfer it to a potassium bromide (KBr) disc. A polycarbonate membrane filter (Millipore Inc. Billerica) is better for SEM/EDS analysis, as it has no strong gold signal to interfere with other element analysis on the EDS spectrum. If the microparticle is larger than 100 μm, it may be retrieved directly with an appropriate probe (tungsten probe or plastic spatula) under a stereomicroscope in a laminar flow hood in a clean room (Class 100). The microparticles that are retained on the filters are then examined with a stereomicroscope to record their size and morphology, followed by FTIR-microscopic analysis. If they can be identified definitively by FTIR at this step (especially if they are organic in nature), further SEM/EDS analysis may not be required. However, for metallic and inorganic materials, SEM/EDS analysis is a must as these materials may not exhibit appropriately detailed FTIR or Raman spectra. Even for organic or polymeric materials, SEM/EDS analysis can provide additional information regarding the elemental composition of the materials to substantiate the vibrational spectroscopic analysis. In some cases the microparticles inside the glass container may be identified using in situ Raman microscopy (Cao et al. 2009, 2010). If the samples are amenable to in situ Raman analysis, this will be the ideal situation as Raman microscopy can provide identification without the need to open the container to isolate the microparticle and is a nondestructive technique.

For complex microparticles that are composed of multiple components, the above-mentioned techniques may not be sufficient to provide a definitive determination, and additional chromatographic and spectroscopic techniques such as inductively coupled plasma mass spectrometry (ICP-MS), gas chromatography (GC), high-performance liquid chromatography (HPLC/MS), and nuclear magnetic resonance (NMR) spectroscopy may be required. This may eventually lead to a full-scale comprehensive trace material analysis, which is beyond the scope of this chapter.

10.2.2 Optical Microscopy

Optical microscopy is the principal technique to observe and examine microparticles contained in a primary container. Optical microscopy is used to visualize the micron particles, in order to manipulate, pick up, and then transfer them for further FTIR and Raman analysis. Three major optical microscopes are commonly employed as the first step for microparticle analysis. A Carl Zeiss Stemi 2000C stereomicroscope can be employed to take pictures of microparticles that have been isolated onto filters. An Axioskop 2 MAT polarized microscope (Carl Zeiss Inc, Peabody) is employed to examine microparticles that may have birefringence such as liquid crystal polymer fiber. The polarized microscope can help to identify some materials that have a distinct birefringence such as cotton cellulose fibers, based on their optical properties. For in situ examination of microparticles contained inside the prefilled glass syringe and large-size glass vials, a long working distance and high-resolution optical microscope is required. These requirements are fulfilled with a Keyence high-magnification optical microscope VHX-600 or similar instrument (Keyence, Itasca), which can take high-magnification-quality images of very tiny particles in situ. The three microscopes are selectively employed according to the specific needs of investigations and the type of containers and samples being analyzed.

10.2.3 Vibrational Microspectroscopy

FTIR and Raman microscopy are two vibrational microspectroscopic techniques that provide complementary information on the molecular structure of materials (Humecki 1995; Li et al. 2009; Wen 2007; Wen et al. 2008). They are the result of a marriage between vibrational spectroscopy and optical microscopy. The combination of a microscope and a vibrational spectrometer offers the power to focus easily on a specific microparticle using the microscopic unit and then to record the vibrational spectrum of this selected particle with the spectrometer. They are the primary techniques for determining the identities of microparticles composed of polymeric or organic compounds at the molecular level. Some inorganic materials can also be examined by FTIR and Raman microscopy if they exhibit strong vibrational spectra.

The principles of FTIR and Raman spectroscopy are detailed in Chap. 3. FTIR-microscopy is generally preferred as the analytical tool for microparticle analysis for particles that have been isolated on a filter as the FTIR-microscope has been developed into a very sophisticated tool that is relatively easy to operate. Moreover, a large FTIR database of reference spectra with more than a quarter of a million compounds is now available from Bio-Rad Know-it-all (Shermann and Brodbelt 2002) that can be used to aid in particle identification. For microparticles composed of a single component, the FTIR spectrum of the microparticle is compared with the reference FTIR spectral library; if a perfect match is found to a reference spectrum from the library, the identity of the microparticle can be established. In cases where there is no perfect match, the chemical and structural information from a set of best matching reference spectra could still provide information on the category or structural family of the material, and to narrow down the possibilities and the direction the investigation should take. Further analysis relies on the experienced vibrational spectroscopist to interpret the FTIR spectra based on the structural chemistry using the empirical group frequency analysis approach, to determine the most likely composition based on the combination of functional groups and any knowledge relevant to the particles (Li et al. 2009; Smith 1999). In many cases, an identification of the type or category of a material can be very helpful in finding the root cause of the microparticles. This type of information can also be used to assess the potential chemical toxicity of the particles and for making decisions on product quality impact.

Raman microscopy is employed as a complementary technique to FTIR-microscopy. However, it has the advantage over FTIR-microscopy in that it can perform in situ analysis without the particle isolation procedure (Goldstein et al. 1992; Cao et al. 2009) for microparticles contained in glass vials and prefilled glass syringes. This is because the visible laser beam used in Raman can penetrate through the container without any interference from the glass or the aqueous solution. The glass (or plastic) containers strongly absorb infrared radiation and thus prevent FTIR measurements in situ. Another unique feature of Raman microscopy is that molecules that have rich electron density, such as conjugated aromatic compounds, show strong Raman bands (Chaps. 3, 13, and 16). Inorganic crystal material can exhibit strong Raman scattering in the low-frequency region as well due to the vibrational modes of the crystal lattice. The weakness of Raman microscopy is relatively low instrument sensitivity compared to FTIR-microscopy, resulting in longer acquisition time of a spectrum. Samples with a strong fluorescent background can also interfere with the ability to obtain a Raman spectrum. Currently the reference database for Raman spectra is also much smaller (15,000 Raman spectra) than the FTIR spectral database (a quarter of a million). Fortunately, the Raman database of organic and polymeric materials has been expanding rapidly in recent years with the continued improvement of the instrumentation, the increased ease of use, and the accumulation of Raman spectra of various materials in labs around the world. Raman microscopy will play an even greater role for microparticle analysis in the pharmaceutical industry in the near future.

10.2.4 Scanning Electron Microscope/Energy Dispersive Spectroscopy

SEM coupled with the energy dispersive X-ray spectroscopy (EDS) is a chemical microanalysis technique that utilizes X-rays emitted from the sample upon bombardment with an electron beam to characterize the elemental composition of the materials (Goldstein et al. 1992). The SEM function provides a high-resolution micrograph of the microparticle and the EDS provides a spectrum of the elementary composition. When the microparticle is irradiated by the electron beam, electrons are ejected from the atoms of the sample. The resulting electron vacancy is filled by electrons from higher orbit shells, and the characteristic X-rays of the element are emitted. The EDS spectrum consists of the characteristic X-rays of the elements, which is used for qualitative determinations of the elements in the sample (Goldstein et al. 1992). SEM/EDS is particularly useful for identification of microparticles from metallic and inorganic materials that do not produce molecular vibrational spectra. It is also a powerful tool for identifying composite materials of multiple elements. Another excellent feature of SEM/EDS instruments is the Cameo mapping function of an individual element in a complex microparticle composed of multiple elements. This feature can generate a distribution map of a selected element in the microparticle, showing whether the specific element is evenly distributed throughout the entire microparticle or concentrated in certain local areas. The combined use of optical microscopy, FTIR, and Raman microscopy as well as SEM/EDS offers a comprehensive toolbox that can be used for the analysis of the physical–chemical properties of the microparticles, in particular, for multicomponent materials.

10.3 Cases Studies

In the following sections, we will describe a few case studies of manufacturing NC investigation where visible microparticles were observed in primary containers of protein therapeutics.

10.3.1 Tungsten/Protein Particulates

Microparticles were initially observed in a protein drug product in prefilled glass syringes during a stability quality inspection, and the PFSs containing the particulates were then sent to the forensic lab for analysis. The top panel of Fig. 10.2 shows a picture of the microparticles retained on a membrane filter after isolation. The largest microparticle is about 100 μm and others are smaller. The bottom panel of Fig. 10.2 shows the FTIR spectrum of one representative particle, which is typical of protein FTIR spectrum. However, the FTIR spectrum exhibited a few extra distinctive bands at 954, 906, and 808 cm^{-1}, respectively; the normal protein FTIR spectrum does not have sharp bands in this region. This suggested that there might be other components in the microparticle in addition to the protein. Subsequent analysis of the microparticle by SEM/EDS revealed that there was tungsten evenly distributed throughout the particle. Figure 10.3 displays the SEM micrograph and the EDS spectrum of a typical microparticle observed in the sample, showing the characteristic tungsten peaks at 1.775 and 8.396 keV, respectively. The identification of tungsten in the particle helped to explain the presence of the three small peaks at 954, 906, and 808 cm^{-1} in the FTIR spectrum; they can be assigned to tungsten oxide (Iacocca and Allgeier 2007). However, it was surprising to observe tungsten in protein particles in a PFS which is made of glass with no tungsten as its material components. The immediate question was, where did the tungsten come from and what was the role of tungsten in the formation of these protein particulates?

Consulting with the prefilled glass syringe manufacturer revealed that tungsten pins were employed during the manufacturing of the prefilled glass syringe barrel at

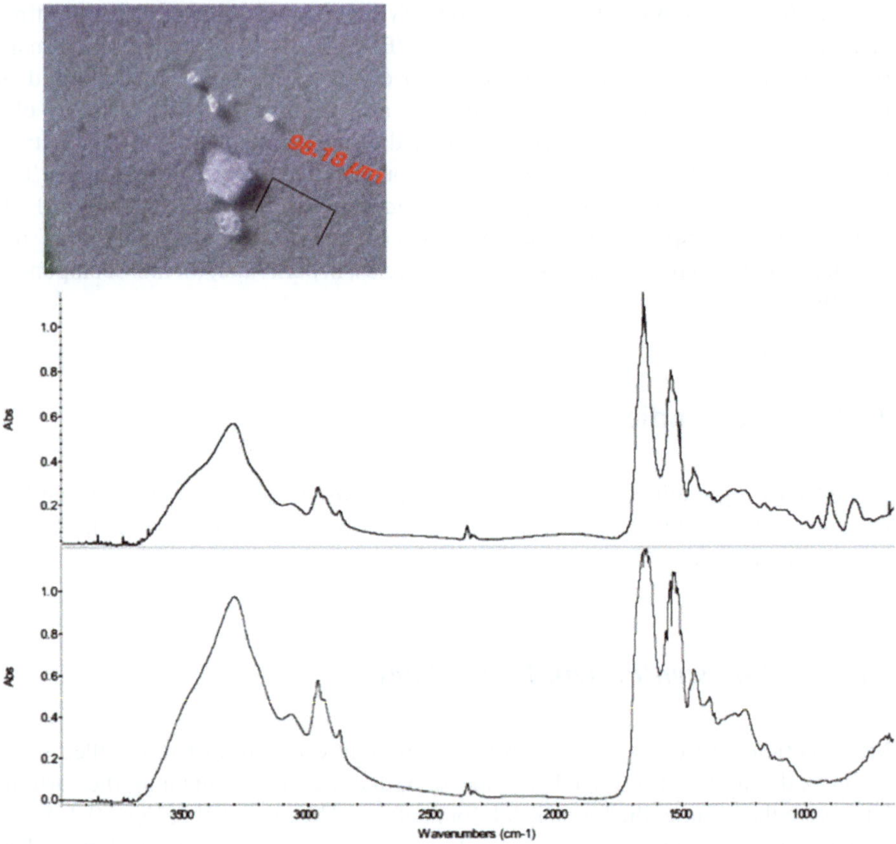

Fig. 10.2 The micrograph of microparticles (*top panel*) and the FTIR spectrum (*up trace*) of a microparticle in a pre-filled syringe and the reference FTIR spectrum of a protein (*lower trace*)

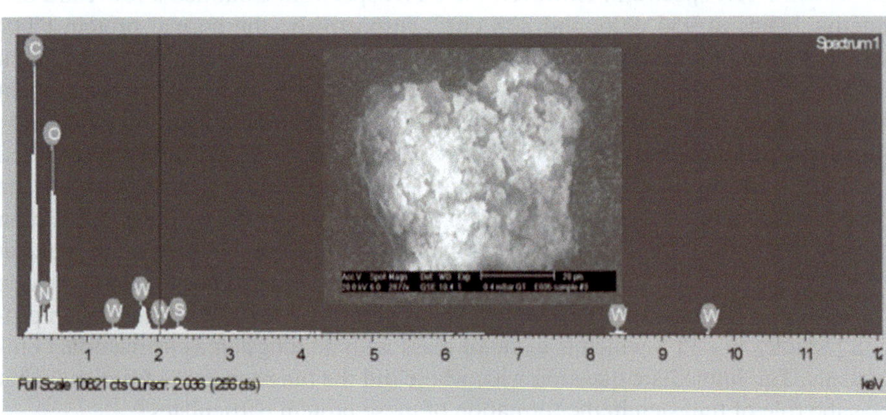

Fig. 10.3 The SEM micrographs and the EDS spectra of a microparticle and a tungsten chip found in a pre-filled syringe

high temperatures in the range of 1000°C. The tungsten pin was used to create the hole of the glass barrel where the needle would later be inserted as the glass was heated at temperatures approaching the glass melting point. Oxidization of the tungsten pin surface can generate tungsten oxide, since the glass syringe barrel was made in an open air environment at this high temperature. These tungsten-rich compounds can deposit (most likely via sublimation) on the interior of the glass syringe, primarily near the tip of the barrel (Liu et al. 2010). During the manufacturing process of the PFSs, the barrels were rinsed with jets of 80°C water for injection prior to final packaging and sterilization; however, the tungsten deposits may not be sufficiently soluble to be completely removed by the washing process (Liu et al. 2010). The possibility of the presence of tungsten residues in the syringe was confirmed by the analysis of white deposits seen in the funnel area of empty syringe barrels. The root cause of tungsten element in the microparticles was thus attributed to the tungsten pins used at high temperature during the manufacturing process of the glass barrel (Wen et al. 2007). The residual tungsten species remained in the barrel and were carried over to the product during the fill and finishing process, at which point they came in contact with the protein therapeutic. This resulted in particle formation on stability which was then found by visual inspection.

To understand the role of tungsten compounds in protein particulation, subsequent experiments spiking tungsten oxide compounds and used tungsten pin extracts into a protein solution were conducted (Jiang et al. 2009). These studies demonstrated that some proteins under specific solution conditions indeed precipitated into microparticles with the addition of tungsten compounds at a level above 1 ppm of used tungsten pin extractables. The induced microparticles contained both proteins and tungsten as confirmed by SEM/EDS (Jiang et al. 2009). Moreover, the tungsten was also distributed evenly throughout these microparticles as revealed by the EDS Cameo mapping of tungsten (see Fig. 10.4) (Liu et al. 2010). Biophysical characterization of the protein–tungsten aggregates in solution using Raman, FTIR, and DLS confirmed that the protein particles induced by tungsten in this case were reversible. They were associated with the formation of tungsten oxide colloids, which require specific solution conditions and a stoichiometry ratio (1:1) between protein and tungsten (Jiang et al. 2009). To prevent microparticle formation in protein drug products induced by tungsten residues, a specification of below 0.5 ppm tungsten was established for the lots of PFSs. The manufacturer of the PFSs also changed their manufacturing processes to minimize the tungsten residues in PFSs.

10.3.2 Glass Gel-Like Particulates

In this case, visible particles were initially observed in a few frozen placebo vials of the protein therapeutic. The particles in the placebo vials were visible to the naked eye, appeared translucent to white, and amorphous. However, no similar particles were seen in the protein therapeutic vials which had the exact same formulation as the placebo. The particles were first isolated on a membrane filter and examined with optical microscopy followed by FTIR-microscopic analysis.

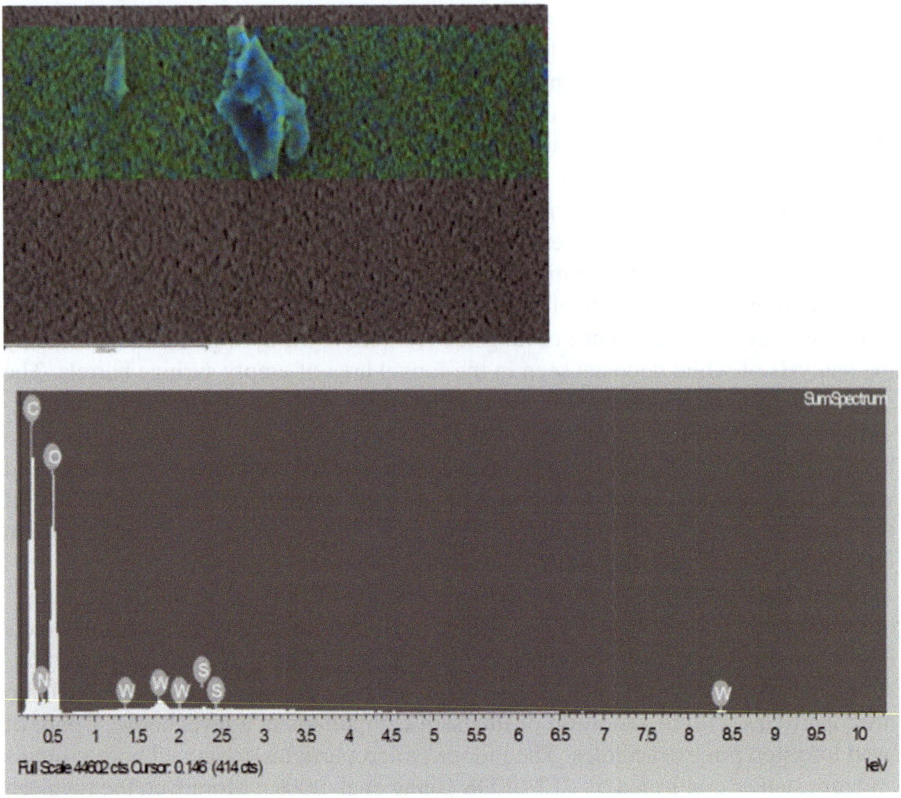

Fig. 10.4 The cameo mapping of Tungsten element in a microparticle and its EDS spectrum, which confirms that the tungsten element is distributed evenly on the entire microparticle, which is a mixture of protein and tungsten element

Figure 10.5 shows the results of the optical microscopic and FTIR-microscopic analyses. The top panel of Fig. 10.5 shows the micrograph of the microparticles in the size range of 20–250 μm. The bottom panel of Fig. 10.5 is the FTIR spectra of the particles together with a reference FTIR spectrum of polysorbate-20 and a reference FTIR spectrum of silica gel from the Bio-Rad FTIR spectral library (Shermann and Brodbelt 2002). The FTIR spectrum of the particle showed bands at 1,736 cm^{-1} (C=O stretching), 2,926 and 2,853 cm^{-1} (CH_2 stretching), 1,460 cm^{-1} (CH_2 bending), and 1,070 cm^{-1} (CO stretching), respectively. These FTIR features of the particles are very similar to the FTIR reference spectrum of polysorbate-20 except that the strongest band is centered at 1,070 cm^{-1} in the particle spectrum, while polysorbate-20 shows two strong bands at 1,100 and 2,800–2,950 cm^{-1} region. Since polysorbate-20 was part of the placebo formulation, it was initially suspected to be the major component and the source of the particles. However, careful examination of the FTIR spectrum of the particle found that there are a few bands of medium intensity at 958 and 802 cm^{-1} that could not be attributed to polysorbate-20. Moreover, the CH_2 stretching bands at 2,925 and 2,853 cm^{-1} had much weaker signals than those of polysorbate-20, which has many CH_2 groups from its polyoxyethylene

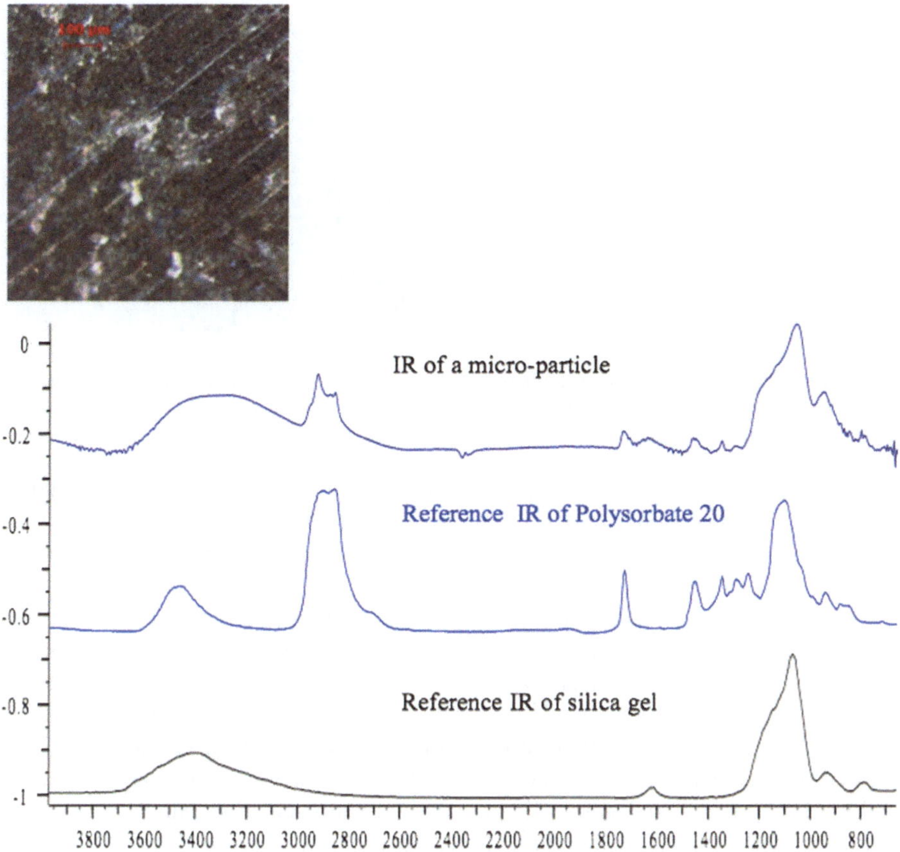

Fig. 10.5 The optical micrograph of the microparticles (*top*) and FTIR spectra of a microparticle, the reference FTIR spectra of Polysorbate-20 and of a silica gel

chain and the long fatty acid ester moiety. This suggested that the particulates may have contained polysorbate-20 but may have other components which contribute to the FTIR bands that cannot be assigned to surfactant (Fig. 10.6).

Subsequent SEM/EDS analysis of the particles revealed an extra element Si in addition to the two major expected organic elements of C and O. This indicated that the particles included silicon-containing material. Trace aluminum was also found in some of the particles (data not shown here) by SEM/EDS. Moreover, a Cameo mapping of the particle with Si element showed that the silicon was distributed evenly throughout the particle (see Fig. 10.5). The vial was made of borosilicate glass, and silicon can be dissolved from the glass vial in aqueous solution (Perera and Doremus 1991). Quantitative comparative analysis of the silicon and other glass elements (boron, aluminum, sodium) by ICP-MS revealed that the placebo glass vial containing visible particles had 30 times silicon in solution than a control placebo plastic vial (below the detection limit of 150 ppb) (Ricci et al. 2011). These analyses strongly suggested that the visible particles were composed of silica

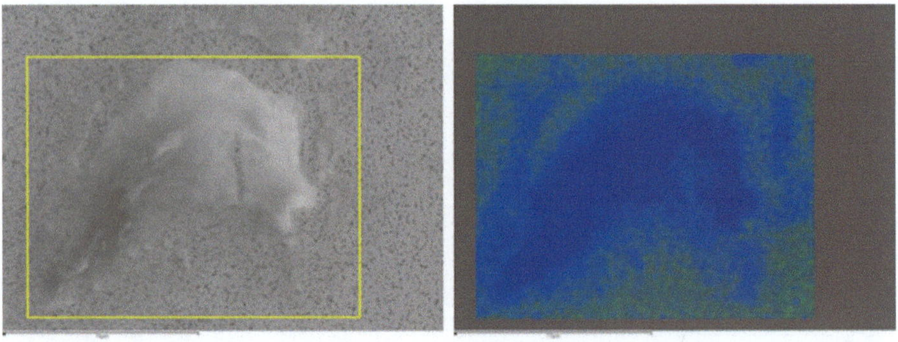

Fig. 10.6 Secondary electron image of a representative particle on gold coated filter (*Left*) the silicon element map displays Si in blue (*Right*). Note that Si is evenly distributed throughout the particle and it matches the particle shape well shown on the left

and polysorbate-20 and that the glass container itself was the source of the silica particles. The root cause analysis then focused on why the placebo vials had the visible particles, but the product vials did not, and what caused the placebo vial to generate visible particulates that contain silicon.

Reviewing the manufacturing history of the protein product and placebo vials revealed some subtle differences in the post-fill handling and storage temperatures between the product and placebo vials. To maximize the stability of drug product, it was frozen at −30°C immediately after fill. In contrast, the placebo was considered to be robust and therefore their storage temperature was not deemed critical. For practical and logistical considerations, the placebo vials were stored at 2–8°C after fill until they were needed for clinical labeling and packaging. Since the demand varied over the course of the clinical studies, the time of storage at 2–8°C varied across the placebo lots by months. On the other hand, the product vials were frozen and stored at −30°C both after fill and packaging and at the clinical trial sites.

Moreover, it was realized that the product and placebo were formulated at neutral pH 7, a pH that can accelerate glass dissolution (Perera and Doremus 1991). Reports in the literature indicated that silica, the dominant component of the glass vials, is less stable in neutral pH than under relatively acidic conditions (Perera and Doremus 1991). The combination of neutral pH and storage of the placebo vials at 2–8°C for a period apparently increased the dissolution of glass into solution in the form of silicic acids, which then polymerized into silica gel (Hsu-Chiang et al. 2005). The polysorbate-20 molecules were merely associated to the silica gel particulates, but not the cause. Scanning of the inner surface of the placebo glass vials by SEM revealed that there were defects in the glass surface of the liquid contact area, but not on the nonliquid contact area above the fill line (data not shown here). This further reinforces the hypothesis that the silica gel particles were indeed formed from the glass vial through glass dissolution. Since the protein drug product was frozen at −30°C immediately after fill and was stored at −30°C, the silica dissolution was greatly minimized at lower temperature for the product container (Ricci et al. 2011). Therefore, the storage condition of the placebo was changed to match that of the protein therapeutic product, reducing the storage time of the placebo at 2–8°C to less than 48 h.

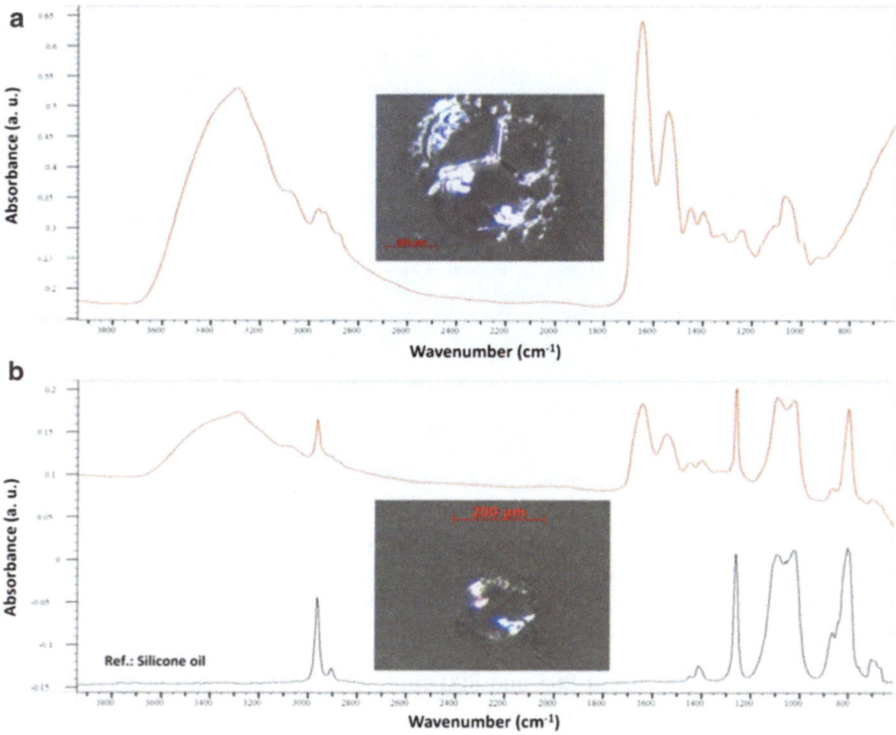

Fig. 10.7 Optical micrograph, FTIR spectra of the bubble-like particle (top trace FTIR is the bubble like microparticle and the bottom trace FTIR is the reference spectrum of silicone oil

10.3.3 Silicone Oil/Protein Particulates

During a product comparability study, bubble-like visible particles were observed in a product that was contained in PFSs. FTIR spectra, along with the corresponding optical images (insets), of the isolated particles are shown in Fig. 10.7. The particles, varying from microns to millimeters, appeared to be "bubble-like" with most of the particles dried on the edge. The infrared spectra of the particles revealed two different spectral characteristics. The particle shown in Fig. 10.7a exhibits a typical protein IR spectrum with characteristic bands of N–H stretching (3,292 cm^{-1}), the CH_2 asymmetric stretching (2,937 cm^{-1}), the CH_3 symmetric (2,876 cm^{-1}) and asymmetric (2,966 cm^{-1}) stretching modes, and the amide I (1,646 cm^{-1}) and amide II (1,539 cm^{-1}) bands. The FTIR spectrum of the particle shown in Fig. 10.7b, on the other hand, revealed the spectral characteristics of a mixture containing protein and another component. In addition to the major protein bands mentioned above, the IR spectrum showed typical silicone oil spectral signatures, indicated by the CH_3 asymmetric stretching (2,965 cm^{-1}) band, the Si-CH_3 deformation band at 1,261 cm^{-1}, and the twin peaks at 1,094 and 1,026 cm^{-1} associated with the Si–O–Si and the Si–O–C stretching vibrations, respectively (Smith 1999). Also present were

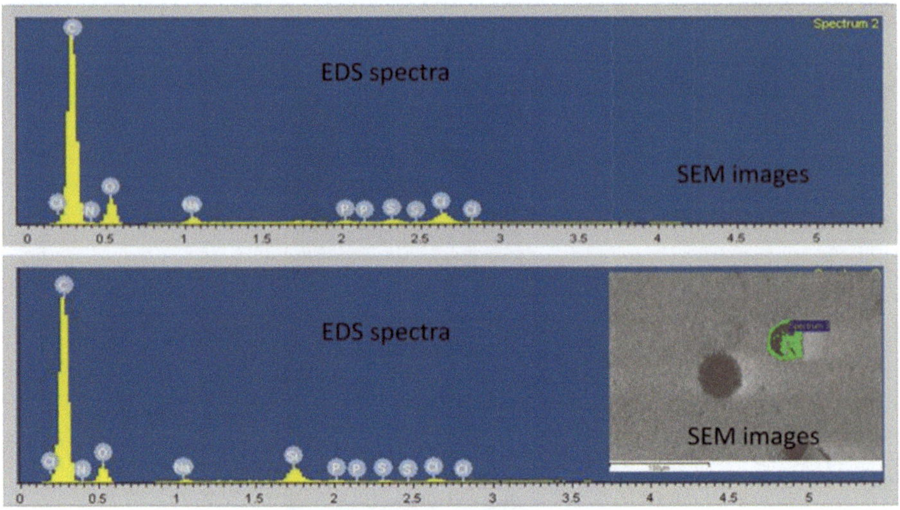

Fig. 10.8 The two SEM/EDS spectra are obtained from the bubble like particles. The *top panel* of the EDS spectrum was obtained on the protein component of the bubble-like particle. The *bottom panel* of the EDS spectrum was obtained on the bubble-like micro-particle that containing silicone oil and protein

the Si–C stretching bands at 865 and 800 cm^{-1}. The FTIR spectra were consistent with a protein–silicone oil mixture.

The SEM/EDS analyses, shown in Fig. 10.8, supported the FTIR results as the EDS spectra of the two particles differed only in the presence of Si, which is attributed to the presence of silicone oil in the particle shown in Fig. 10.7b. Other than Si, both particles were found to contain C, N, O, and S. This is consistent with the FTIR results as a proteinaceous material containing the basic elements of sulfur, carbon, oxygen, and nitrogen, although some C and O signals could also be contributed by the underlying polycarbonate membrane filter. The EDS spectra also showed P, Na, and Cl, which can be attributed to the formulation buffer. The SEM micrographs demonstrated morphological similarities of both "dried bubble" particles. The source of the silicone oil was the PFS, as silicone oil is coated in a very thin layer of a few hundred nanometers on the glass barrel surface as a lubricant (Wen et al. 2009). The generation of these bubble-like particles was attributed to the silicone oil/protein/buffer interaction during the transportation (vibration) of the products via dewetting processes as silicone oil is hydrophobic, but is in contact with aqueous solution. Tiny silicone oil droplets on the glass surface can migrate together to form relatively large visible bubble-like particles associated with protein molecules.

10.4 Conclusions

NC investigations and root cause analyses are critical to protein therapeutic development and commercialization. Physical particulate contaminants are one of the types of NC that can occur during the fill and finishing process of biopharmaceutical products. When this occurs, the physical–chemical nature of the particulate matter must be identified as one of the first steps in determining the origin and the root cause to allow remediation. This requires the use of the appropriate tools from the biophysical toolbox as described above.

It is worthwhile to point out that many physical–chemical contaminants found in the final biopharmaceutical products were originally generated during the processing of the drug product container and were associated with the container manufacturing equipment, being carried over into the final commercial drug product. Other times, it is the glass container per se that interacted with the formulation components of the protein products to generate particulate issues such as glass delamination. To quickly resolve NC investigations, detailed analysis of materials present, and determination of their root cause and source are mandatory. Moreover, close collaboration among the raw material vendor, primary container manufacturer, and the protein pharmaceutical manufacturer is required for resolution of the NCs and to take appropriate preventive measures.

References

Barber TA (2000) Control of particulate matter contamination in healthcare manufacturing. Interpharm, Denver

Cao X, Wen ZQ, Vance A, Torraca G (2009) Raman microscopic applications in the biopharmaceutical industry: in situ identification of foreign particulates inside glass containers with aqueous formulated solutions. Appl Spectrosc 63:830–834

Cao X, Masatani P, Torraca G, Wen ZQ (2010) Identification of a mixed microparticle by combined microspectroscopic techniques: a real forensic case study in the biopharmaceutical industry. Appl Spectrosc 64:895–900

Goldstein JI, Newbury DE, Echlin P, Joy DC, Romig AD Jr, Lyman CE, Fiori C, Lifshin E (1992) Scanning electron microscopy and X-ray microanalysis, 2nd edn. Plenum, New York

Humecki HJ (1995) Practical guide to infrared microspectroscopy. CRC, New York

Iacocca RG, Allgeier M (2007) Corrosive attack of glass by a pharmaceutical compound. J Mater Sci 42:801–811

Jiang Y, Nashed-Samuel Y, Li C, Liu W, Pollastrini J, Mallard D, Wen ZQ, Fujimori K, Pallitto M, Donahue L, Chu G, Torraca G, Vance A, Mire-Sluis T, Freund E, Davis J, Narhi LO (2009) Tungsten-induced protein aggregation: solution behavior. J Pham Sci 98:4695–4710

Kuan H-C, Su H-Y, Ma C-C (2005) Synthesis and characterization of polysilicic acid nanoparticles/waterborne polyurethane nanocomposite. J Mater Sci 40(23):6063–6070

Li G, Torraca G, Jing W, Wen Z-Q (2009) Applications of FTIR in identification of foreign materials for biopharmaceutical clinical manufacturing. Vib Spectrosc 50:152–159

Liu W, Swift R, Torraca G, Nashed-Samuel Y, Wen ZQ, Jiang Y, Vance A, Mire-Sluis A, Freund E, Davis J, Narhi LO (2010) Root cause analysis of tungsten-induced protein aggregation in pre-filled syringes. PDA Pharm J Sci Technol 64:11–19

Nail SL, Akers MJ (2002) Development and manufacture of protein pharmaceuticals. Springer, New York

Narhi LO, Schmit J, Bechtold-Peters K, Sharma D (2012) Classification of protein aggregates. J Pharm Sci 101:493–498

Perera G, Doremus RH (1991) Dissolution rates of silica glasses in water at pH 7. J Am Ceram Soc 74(6):1269–1274

Ricci MS, Hair A, Thirumangalath R, Dharmavaram V, Piedmonte D, Latypov R, Lau H, Bjorklund J, Callahan W, Purtell J, Abel J, Li G, Fujimori K, Torraca G, Jing W, Wen Z-Q, Nashed-Samuel Y, Ratnaswamy G, Swift R (2011) Glass dissolution and delamination in formulation development: implications for neutral pH product development. In: American Chemical Society Division of Biochemical Technology 241st ACS national meeting, Anaheim, 27–31 March 2011

Rosenberg AA (2006) Effects of protein aggregates: an immunologic perspective. AAPS J 8(3):E501–E507

Shermann C, Brodbelt J (2002) Knowitall analytical system. J Am Chem Soc 124:7

Smith B (1999) Infrared spectral interpretation: a systematic approach. CRC, New York

United States Pharmacopeia 29-Formulary 24 (2012), p 2722 (Chap. 788) USP 35-NF30

Wen ZQ (2007) Raman spectroscopy of protein pharmaceuticals. J Pharm Sci 96:2861–2878

Wen Z-Q, Torraca G, Yee C, Li G (2007) Investigation of contaminants in protein pharmaceuticals in pre-filled syringes by multiple microspectroscopies. Am Pharm Rev 10:101–109

Wen ZQ, Cao X, Vance A (2008) Conformation and side chains environments of recombinant human interleukin-1 receptor antagonist (rh-IL-1ra) probed by Raman, Raman optical activity, and UV-resonance Raman spectroscopy. J Pharm Sci 97:2228–2241

Wen ZQ, Vance A, Vega F, Cao X, Eu B, Schulthesis R (2009) Distribution of silicone oil in pre-filled glass syringes probed with optical and spectroscopic methods. PDA Pharm J Sci Technol 63:149–158

Chapter 11
Higher-Order Structure and Protein Aggregate Characterization of Protein Therapeutics: Perspectives from Good Manufacturing Practices and Regulatory Guidance

Evi B. Struble, John F. Cipollo, Chava Kimchi-Sarfaty, Zuben E. Sauna, Jack A. Ragheb, and Ewa Marszal

11.1 Introduction

Proteins are complex molecules that affect gene expression, cellular structure, metabolism, reproduction, inter-/intracellular signaling, immune responses, and virtually every aspect of life. Function of proteins depends on the orientation of the functional groups of their building blocks. For example, the presence and orientation of the three critical amino acids (Ser, His, and Asp) in the catalytic triad of serine proteases, such as coagulation factors, determines their ability to perform their biologic role, i.e., cleave a peptide bond and, in the case of the coagulation cascade, activate the substrate. Furthermore, nearby regions (secondary and tertiary structures) are

Disclaimer The findings and conclusions in this chapter have not been formally disseminated by the Food and Drug Administration and should not be construed to represent any agency determination or policy.

E.B. Struble • C. Kimchi-Sarfaty • Z.E. Sauna • E. Marszal (✉)
Division of Hematology, Office of Blood Research and Review,
Center for Biologics Evaluation and Research (CBER), Food and Drug Administration (FDA), 29 Lincoln Drive, Bethesda, MD 20892, USA
e-mail: evi.struble@fda.hhs.gov; chava.kimchi-sarfaty@fda.hhs.gov; zuben.sauna@fda.hhs.gov; ewa.marszal@fda.hhs.gov

J.F. Cipollo
Division of Bacterial, Parasitic and Allergenic Products,
Office of Vaccine Research and Review, CBER, FDA, 29 Lincoln Drive, Bethesda, MD 20892, USA
e-mail: john.cipollo@fda.hhs.gov

J.A. Ragheb
Division of Therapeutic Proteins, Office of Biotechnology Products, Center for Drug Evaluation and Research, FDA, 29B Lincoln Drive, Bethesda, MD 20892, USA
e-mail: jack.ragheb@fda.hhs.gov

L.O. Narhi (ed.), *Biophysics for Therapeutic Protein Development*, Biophysics for the Life Sciences 4, DOI 10.1007/978-1-4614-4316-2_11,

essential for substrate specificity and, in the case of coagulation factors, enhancement of the biologic activity by cofactors and attenuation of the effect by inhibitors.

Unlike the bonds that determine the structures of small molecules, the interactions driving the complex structure of proteins—hydrogen bonds, electrostatic interactions, and van der Waals forces—are noncovalent in nature. As such, the energy that maintains the stability of a protein's three-dimensional structure is inherently low, in many cases not much greater than the thermal energy at room temperature (Pfeil and Privalov 1976). Thus, increases in temperature result in protein unfolding (Griko et al. 1988). Other environmental conditions such as pH (Anderson et al. 1990), solvents (Grothe et al. 2009), salts, small molecule additives, impurities (Bolen and Baskakov 2001; Maity et al. 2009; Rosgen 2007; England and Haran 2011), and even other proteins (Hartl and Hayer-Hartl 2009) can influence the three-dimensional structure and result in complete or partial unfolding of the protein.

Destabilized proteins often form aggregates, which may vary from dimers with a size in the low nanometer range to visible agglomerates of thousands of protein molecules reaching a size of several millimeters. Aggregated proteins lose physiologic activity (Meager et al. 2011; Carpenter et al. 2010a) and may acquire new, toxic attributes. Examples include diseases associated with protein aggregation such as systemic amyloidoses (Obici et al. 2005), neurological disorders (Aguzzi and O'Connor 2010), and, as recently shown, cancer (Xu et al. 2011). Interestingly, it has been postulated that in some diseases toxicity is associated with oligomeric intermediates rather than the large aggregates (Haass and Selkoe 2007; Klein et al. 2001).

The formation of protein aggregates is a major concern in the manufacture and clinical use of protein therapeutics. Protein aggregates have been associated with enhanced immune responses both in animal studies and clinical trials (Dresser 1962; Braun et al. 1997; Moore and Leppert 1980). Aggregates may stimulate the innate immune system to promote an adaptive immune response by enhancing antigen presentation and the release of cytokines, leading to immunogenicity. Cytokine release may also directly induce more immediate infusion-associated reactions.

The effects of the resulting antidrug antibodies (ADA) can range from a small alteration in the pharmacokinetic properties of the therapeutic protein to a decrease or complete loss of efficacy due to immune-mediated clearance or neutralization of the biologic activity of the drug. When both the ADA and the therapeutic protein are present at sufficiently high levels, such as may occur with therapeutic monoclonal antibodies, toxicity can occur due to deposition of circulating immune complexes. In the most severe cases, the immune reaction to the offending therapeutic protein (e.g., erythropoietin) can cause a breakdown of immune tolerance, resulting in neutralization of the patient's endogenous counterpart to the therapeutic.

Given the consequences to both activity and toxicity, conformational changes and protein aggregation affect product quality and can affect product safety. Product safety and efficacy are determined during clinical trials and must not decline over the shelf life of a given lot or the market life of a product. It is, thus, essential that the structural and functional features of the product be characterized to an acceptable extent during product development and that each batch of the product has similar characteristics at release and throughout its shelf life. In part, this is ensured by product release testing, but because this testing may not be capable of detecting subtle changes that can occur in the product due to variations in manufacturing processes, product consistency over time must be ensured by tight control of the manufacturing process. In addition, any major manufacturing change should be supported by detailed product characterization and comparability studies, including a comparison of product stability profiles.

Factors that affect protein structure, aggregation, and stability are numerous, and, while we acknowledge that not all of them can be controlled during the manufacture of biotherapeutic products, the principles outlined in this chapter are meant to be science-based good manufacturing considerations to be applied when and if appropriate and feasible for a given specific product.

It cannot be overemphasized that assuring the quality of biologic products is a complex process that can be achieved only through tight control of process parameters and in-depth characterization beginning with product development and continuing through the lifecycle of the product, including routine testing of every batch and long-term monitoring of product stability. Recent technical developments allow for the manufacture of biosynthetic protein therapeutics under tightly defined conditions. This has increased our potential to limit impurities and control biosynthetic protein characteristics to a greater extent than is possible for naturally derived products. When coupled with extensive characterization of protein structure, these advancements better ensure protein comparability after a manufacturing change.

11.2 Intrinsic Factors Affecting Protein Higher-Order Structure and Stability

11.2.1 Protein Sequence and DNA Sequence

Since Anfinsen's Nobel Prize winning work (Anfinsen 1973), it is well established that the primary sequence of the polypeptide chain determines a protein's three-dimensional structure. Since then, the discovery of the protein basis of "inborn errors of metabolism" (Garrod 1908; McCarthy 2011; Benoit et al. 2010; Elia and Albanese 2010; Percy and Rumi 2009; Bentham and Bhattacharya 2008; Scriver 2007; Pasinelli and Brown 2006; Harrison et al. 2008) and the experience acquired in the development and clinical use of engineered proteins (Alexander et al. 2007; Meeker et al. 1996; Stepanenko et al. 2008) have underscored that even seemingly small changes in protein primary sequence can affect structure and stability. In the case of therapeutic proteins made from recombinant sources, this understanding serves to reinforce the importance of the amino acid sequence, its implications for aggregation propensity, and, as a result, product efficacy and safety.

However, the amino acid sequence is not the only determinant of the protein three-dimensional structure. Indeed, folding kinetics and the structure of proteins can also be affected by synonymous mutations resulting in phenotypic changes such as an increased resistance to anticancer drugs (Kimchi-Sarfaty et al. 2007). One of the possible mechanisms for this effect is via codon usage and cognate tRNAs frequency (Sharp et al. 1988; Ikemura 1985). Organism and cell-specific distribution of tRNAs, together with biased codon usage, could provide pauses during translation at critical sites, such as at the end of individual domains. This may allow a protein to fold correctly by curtailing disruptive interactions between unfolded regions of the protein. Thus, in order to achieve the desired fold in a recombinant protein, the conditions for its synthesis should be optimized with respect to codon usage and tRNA abundance. Other parameters of its synthesis, which are addressed in more detail in Sect. 3.1 of this chapter, should also be stringently controlled, especially since minor structural changes that may result from the suboptimal rate of translation and the composition of the cellular milieu may be difficult to identify.

11.2.2 Protein Glycosylation

Glycosylation is another determinant in the formation of protein higher-order structure and can help to improve protein stabilization. The added stability that glycosylation may provide can serve as a counterbalance to many of the degrading forces that might be encountered during therapeutic protein manufacturing and clinical application, such as differences in pH, temperature, or chemical conditions. The stabilizing effects include reduced short- and long-range mobility (Sola et al. 2007; Liu et al. 2006), increased intramolecular electrostatic interactions, formation of additional hydrogen bonds and van der Waals interactions, and increased solubility (Mitra et al. 2006; Rademacher et al. 1988). Increased solvent contacts also decrease intermolecular electrostatic interactions, thus further diminishing the tendency toward aggregation. Glycosylation can also act as a shield against some chemical degradative processes such as oxidation and cross-linking. Examples of these include free radical scavenging in erythropoietin (Uchida et al. 1997) and suppression of transamidation in glyco-engineered insulin (Baudys et al. 1995). In general, glycosylation diminishes the immunogenicity of therapeutic proteins.

Pharmacokinetics can be positively or negatively affected by changes in glycosylation. The addition of glycosylation sites or greater glycosylation site occupancy has been reported to add considerably to serum half-life and clearance times (Fares et al. 1992; LaPolt et al. 1992). However, loss of glycan or inappropriate glycan composition can negatively impact clearance (Appa et al. 2010; Wright and Morrison 1998). As we will discuss in the next section, manufacturing conditions can dramatically affect glycan composition and therapeutic efficiency. Therefore, both glycan characterization and rigorous control over manufacturing conditions that influence this parameter are necessary for stable therapeutic protein pharmacokinetic profiles.

11.3 Extrinsic Factors Affecting Protein Higher-Order Structure and Stability

11.3.1 Recombinant Protein Production Process

Protein folding and its glycosylation can be affected by conditions encountered during the upstream manufacturing process, including both the intra- and extracellular milieu. Chaperones such as the Hsp70s play an essential role in correct folding and secretion of proteins (Kramer et al. 2009). During protein synthesis and posttranslational modifications, binding to chaperones protects the polypeptide chain from collapse and incorrect folding that could occur in the absence of the yet to be translated, downstream sections or from aggregation in the highly crowded cell environment. Although chaperones are found across all organisms, there are mechanistic differences as well as differences in expression levels amongst commonly used recombinant protein production systems and cell lines (Hartl and Hayer-Hartl 2009). To add to this complexity, the availability of chaperones depends on other factors such as temperature, nutrient state (Neuhofer et al. 1999), and the expression level of the recombinant protein. Chaperone availability in turn can affect folding, aggregation, and yield of the product. These factors should be carefully analyzed not only when the process is being developed but also when changes in media formulation and other process parameters are being considered or implemented.

Many cell culture process variables can affect glycosylation. Manufacturing scale, fermentation type, and fermentation conditions all must be considered. Bioreactor pH, manganese ions, dissolved oxygen, ammonia concentration, and temperature have all been shown to affect glycosylation patterns (Hossler et al. 2009). The three major modes of production—batch, fed-batch, and perfusion—all have different waste product accumulation and nutrient depletion profiles, which can have profound effects on glycosylation profiles. Accumulation of ammonia, principally through glutamine and asparagine metabolism, raises media pH, which can diminish the activity of the late Golgi glycosyltransferases (Thorens and Vassalli 1986). Manganese is required for the activity of some glycosyltransferases, including those involved in late carbohydrate modifications such as the addition of *N*-acetylneuraminic acid and galactose. Nucleotide sugar precursors and dolichol, which are substrates for glycosylation biosynthesis, can become depleted, thus affecting glycosylation efficiency (Castro et al. 1995; Jenkins et al. 1994; Kochanowski et al. 2008). The mode of production dictates the necessary level of control over these factors, and therefore, its choice must be weighed with care. Shear stress has also been shown to affect late Golgi processing (Senger and Karim 2003) and should be taken into account, especially when the scale of production is changed (Hossler et al. 2009). All of these factors should be considered early in the development of the manufacturing process as differences in cellular stresses due to culture scale and production mode can have profound effects on glycosylation profiles and thus the conformation, efficacy, and safety of the protein product.

11.3.2 Purification Process

Protein therapeutic products are derived from highly complex starting materials—human plasma or blood, transgenic animal milk, bacterial, plant, insect, yeast, and mammalian cells or tissues. Multistep purification protocols are often used that may include precipitation and filtration steps, chromatography steps, and freeze/thawing. In addition, viral inactivation operations involving filtration, elevated temperatures, low pH exposure, and/or the use of solvent/detergents are frequently employed.

Virtually every step of the manufacturing process may have an effect on protein structure (Vazquez-Rey and Lang 2011; Cromwell et al. 2006). Therapeutic proteins are often purified from a mixture containing cell and/or tissue components based on different surface characteristics that affect solubility and interactions with solid media. Conditions selected for precipitation operations and chromatography steps involve pH changes, addition of a nonpolar solvent, or a change in ionic strength. These changes may alter the distribution of charges on the protein surface and the properties of the solvent, affecting the interactions of the protein with both itself and the solvent and thus potentially impact the integrity of the protein (Zimmerman 2006; Lewis and Nail 1997; Shukla et al. 2007). Temperature also has an effect on the strength of these interactions, with a rise in temperature strengthening hydrophobic and reducing ionic interactions. However, ultimately, an increase in temperature results in higher kinetic energy and will eventually lead to partial unfolding. Unfolded protein species are highly susceptible to formation of aggregates which are more stable in common formulation media than the native form of the protein. Additionally, low temperatures may induce cold denaturation, and interaction with ice during lyophilization, an operation that involves freezing and dehydration, may also destabilize protein structure (Rathore and Rajan 2008; Tang and Pikal 2004). Mechanical stress, which affects therapeutics when the protein solutions are pumped, mixed, filtered, or filled, may lead to structural distortions and aggregation of the protein product (Rathore and Rajan 2008). In addition, chemical modifications (e.g., oxidation and deamidation) occurring during the purification process or upon storage may lead to conformational changes, activity reduction, and aggregation (Luo et al. 2011; Patel et al. 2011). Extractables and leachables are other potential sources of chemical modifications that may impact protein stability (Huang et al. 2011). Lastly, lyophilization, used in the manufacture of many biologics, can result in local concentration and pH changes that may induce conformational changes in proteins leading to aggregation.

11.3.3 Storage and Transportation

Protein stability upon storage and transportation greatly depends on the protein's physical state, i.e., on whether the protein is dissolved or lyophilized and on the composition of the product formulation. The formulation should be carefully optimized not only to ensure long shelf life but also to enable successful completion of the last steps of manufacturing (Jorgensen et al. 2009; Shire 2009). Other important factors include temperature (and potential temperature excursions), presence of interfaces such as a liquid–air interface, and interactions with the container-closure system, including leachables, extraneous particles such as glass, metal, and silicone oil; the latter of which is commonly used as a lubricant in syringes and on stoppers (Bee et al. 2011). Product agitation during transport and handling may also induce aggregation of the protein product (Kiese et al. 2008; Thirumangalathu et al. 2009). Thus, in addition to thorough characterization of the final container product, careful analysis of the effect of transportation parameters on the protein conformation and aggregation state of the product is important.

11.3.4 Aggregation During Reconstitution and Administration

Lyophilized biotherapeutics are reconstituted before administration. Reconstitution may involve the use of transfer devices that help to transfer sterile water for injection to vials containing the lyophilized product. If siliconized, the elements of infusion sets, such as transfer devices, needles, and syringes, may contribute to formation of protein aggregates in the product. Different devices may carry different amounts of silicone. Therefore, similar devices from different suppliers may not be equivalent or interchangeable.

Patient exposure to protein aggregates may be limited by the use of in-line filters, which are recommended for many protein therapeutics. Nonetheless, to ensure high quality of the product, manufacturers should minimize the level of aggregates in the final container.

11.4 Product Control and Process Control: Elements of Product Quality Assurance

Protein therapeutics, in contrast to small molecule drugs, can never be fully characterized. Their structure is complex, small structural changes are difficult to recognize, and contaminating factors that co-purify with the protein product and which exist in minute amounts may not be well defined. The quality of biological materials is verified by final lot release testing of the product and by ensuring the consistency of manufacture. The latter is achieved by establishing in-process parameters and product limits that have been demonstrated to deliver a defined quality product. Extended product characterization is performed for licensure and at the time of any major manufacturing change. As product quality must be maintained throughout shelf life, quality attributes are evaluated during stability studies.

11.4.1 Analytical Methods for Characterization of the Primary, Secondary, and Higher-Order Structure

In addition to functional characterization, to help ensure process control, comparability, and manufacturing consistency, recombinant therapeutic proteins should be characterized to an acceptable extent by assessing the protein's primary, secondary, and higher-order structure. Analytical techniques used to assess protein structure should be scientifically sound, reflect the current state of scientific understanding, and have high sensitivity and precision. Considering the complexity of protein structure and the inherent limits of any given analytical tool, at least two orthogonal methods should be used for structural characterization. For example, the primary sequence can be determined by using mass spectrometric analysis, which is based on fragmentation of the polypeptide chain and the rate of fragment migration in an electromagnetic field, and by N-terminal sequencing following the Edman degradation reaction or by C-terminal sequencing following chemical degradation with thiocyanate.

The secondary structure can be characterized using circular dichroism (CD), which is based on the differences in absorption of polarized light in proteins, or by analyzing the spectra of frequencies of vibration for a molecule using infrared and Raman spectroscopy. The secondary and tertiary structure can also be assessed by NMR spectroscopy, which measures chemical shifts of the magnetically active nuclei (^{1}H, ^{13}C, ^{15}N, etc.) when the protein solution is placed in a magnetic field. Tertiary and quaternary structure can be probed with methods such as intrinsic and dye fluorescence spectroscopy which respectively measure fluorescence of aromatic amino acids such as tryptophan and fluorescent reporter molecules bound to the protein, thus providing information on the local environment of the fluorescing moiety. X-ray crystallography, based on calculating the electron density of atoms from the diffraction pattern of the crystallized protein, is a powerful technique for determining the secondary, tertiary, and quaternary structure of a protein. Hydrogen/deuterium (H/D) exchange mass spectrometry enables detection of conformational differences based on solvent accessibility to the protein amino acid residues. Chromatography techniques, measuring the mobility in a chromatographic column, can also provide information on changes in protein structure and surface characteristics.

Protein aggregate content can be detected by means of visual inspection, microscopy (optical, fluorescence, electron, atomic force), light obscuration, flow imaging, conductivity-based counting (Coulter method), static and dynamic light scattering, nanoparticle tracking analysis, turbidity, nephelometry, X-ray scattering, and mass spectrometry. Separation methods such as size exclusion chromatography, field-flow fractionation, polyacrylamide gel electrophoresis, capillary electrophoresis, and analytical ultracentrifugation can also be used. Recently, fluorescence single-particle tracking was used to size submicron matter in undiluted whole blood (Braeckmans et al. 2010), and the technique has been adopted to evaluate its potential to detect and determine the size of submicron protein aggregates in serum, plasma, and pharmaceutical formulations containing albumin (Filipe et al. 2011).

The characterization, utility, and limitations of these and other techniques have been discussed in numerous reports (see, e.g., Philo 2009; Huang et al. 2009; Fraunhofer and Winter 2004; Filipe et al. 2010 and Chap. 3 and 4 of this volume). Although the techniques have evolved in sophistication and sensitivity, analyzing therapeutic proteins or their aggregates in biological media remains a major challenge.

There is an interest in industry to use high throughput screens to assess higher-order structure. Among the established methods of measuring protein stability, thermal and chemical denaturation monitored by intrinsic fluorescence were the earliest to be automated (Stites et al. 1995) and are constantly being improved upon (Edgell et al. 2003; Aucamp et al. 2005). This technology has been reduced to the nanoliter scale (Gaudet et al. 2010).

A technique that does not depend on the intrinsic fluorescence of a protein, which can vary widely depending on the local environment and number of fluoroactive residues, is differential scanning fluorimetry (DSF). This method, based on monitoring fluorescence of a specific dye that has increased affinity for the hydrophobic regions of protein surfaces, has been found to be applicable to a broad range of proteins (Ericsson et al. 2006, Chap. 2 of this volume), and good correlations have been found between melting temperature values determined from CD thermal denaturation and DSF (Lavinder et al. 2009).

One goal of adopting high throughput assays for determining protein stability is to sort out the stable variants in a library. To reduce the search space, considerable effort has been expended in developing first-principles computational methods to predict the stability of proteins. These programs have relied, to a large extent, on knowledge-based potentials based on distances between residue pairs or backbone torsion angles (Gilis and Rooman 2000), statistical potentials (Parthiban et al. 2006), empirical potentials that describe the physical interactions that contribute to protein stability (Yin et al. 2007), and different types of machine learning tools (Capriotti et al. 2005; Cheng et al. 2006; Masso and Vaisman 2008). While most programs require knowledge of the structure of the target protein, some methods can predict stability changes in proteins of unknown structure with reasonable accuracy (Capriotti et al. 2005). While these algorithms are unlikely to produce a comprehensive predictive model in the foreseeable future, they can be used in conjunction with experimental methods.

The above method enumeration is by no means exhaustive. A recent industry effort to review and evaluate analytical methods available for protein aggregates detection and characterization was reported upon by den Engelsman et al. (den Engelsman et al. 2011). Furthermore, there are a large number of publications on biophysical methods that can be used as a guide to select an appropriate set of methods. There are a few methods which, although at present are not used widely by industry, show great potential for evaluating therapeutic biologics and should be explored further. For example, H/D exchange in conjunction with MALDI mass spectrometry, another robust and well-characterized method, has recently been adapted to measure protein stabilities in living cells (Ghaemmaghami and Oas 2001). It was also recently employed to detect conformational changes associated with posttranslational modifications of a monoclonal antibody preparation.

Interestingly, the detected changes correlated with changes in receptor binding (Houde et al. 2010).

Due to inherent differences in method sensitivity, complexity of the measurement, and other considerations such as inter-operator precision, different methods to determine higher-order structure may give vastly different results. For example, slight changes in the amino acid composition of proteins can greatly affect NMR spectra but do not affect the overall structure as determined by X-ray crystallography (Struble et al. 2008). While some differences in results are to be expected, method validation and the use of appropriate controls and standards should reduce variability. Furthermore, for many of these methods, data fitting by using mathematical models and interpretation still poses a challenge and is an area where an integrated effort is needed to overcome existing limitations. This need will intensify in the coming years with increased demand for optimized bioengineered products that are safe and efficacious, as well as convenient for patients. There has been considerable progress in the last decade, but there continues to be unmet needs with respect to this important issue for the biotechnology and pharmaceutical industries.

11.5 Industry Standards and Regulatory Considerations

Several guidances promulgated by the FDA[1] or the International Conference on Harmonization of Technical Requirements for Registration of Pharmaceuticals for Human Use (ICH),[2] to which the FDA is a signing party, address issues central to ensuring the structural and functional integrity (in short the "quality") of a biological therapeutic product. The long-standing emphasis on understanding the manufacturing process and the biological product is critical for successful and controlled manufacture of biologics, including protein therapeutics. The recent Quality by Design paradigm provides a structured, systematic approach to understanding the process and product, which can provide substantial knowledge and understanding that is useful throughout the product lifecycle (ICH Q8; ICH Q9; ICH Q10). Although the control of the process is an important and inseparable component of the product quality assurance, it is beyond the scope of this chapter. Below, we focus solely on the product characterization and quality consistency assessment.

---

[1] FDA guidance documents contain recommendations that reflect FDA's current thinking on given issues and are intended to assist the industry in carrying out its obligations under statutes and regulations. They do not create or confer any rights for or on any person and do not operate to bind FDA or the public. They are publically available on the FDA website. http://www.fda.gov/Drugs/GuidanceComplianceRegulatoryInformation/Guidances/default.htm.

[2] ICH is an organization that brings together regulators and pharmaceutical industry representatives from Europe, Japan, and the USA. The ICH mission is to facilitate development of safe, effective, and high-quality drugs in the most resource-efficient manner by harmonization of guidelines and requirements for product registration.

11.5.1 Structural Characterization, Purity, Specifications, Stability, and Product Consistency Throughout Lifecycle

For approval of a new pharmaceutical product, the ICH M4Q(R1) guidance for industry ICH M4Q(R1) recommends the submission of details on primary, secondary, and higher-order structure for "the desired product and product-related substances." In addition to information on biological activity of the product, the guidance recommends the submission of information on the molecular mass, details of "posttranslational forms (e.g., glycoforms)," and a schematic amino acid sequence showing glycosylation sites or other posttranslational modifications.

Part B of ICH Q5 guidance discusses the need for characterization of the nucleic acid coding sequence or the transcription products, as appropriate. This guidance, published in 1995, although it does not specifically address the effect of the sequence on protein folding, points out that "the genetic sequence of recombinant proteins produced in living cells can undergo mutations that could alter the properties of the protein with potential adverse consequences to patients." As we know today, even an alteration in nucleotide sequence that has no effect on the amino acid sequence may affect the rate of translation and folding of a protein.

ICH Q6B guidance focuses on product physicochemical characterization, which generally includes determination of the composition, physical properties, and primary structure of the product. The guidance provides examples of product attributes that can be considered for structural characterization, such as amino acid composition, terminal sequencing, peptide mapping, determination of the primary sequence, location of disulfide bonds, and characterization of the carbohydrate content and structure. Physicochemical characterization includes determination of the molecular weight, isoform pattern, extinction coefficient, electrophoretic pattern, liquid chromatography patterns, and spectroscopic profiles. The guidance emphasizes that optimal physicochemical characterization varies from product to product and that new technologies are still being developed and should be applied when appropriate.

Demonstration of product purity presents a challenge, and the results may depend on the method used (ICH Q6B). The relative purity of the product is often determined in terms of specific activity expressed in units of biological activity present in a mass unit of the product. Specific activity determined in such a way is also highly method dependent. Thus, the purity is assessed by using a combination of analytical techniques. However, small conformational changes in the protein structure may be difficult to identify, and guidance on how to efficiently select a method(s) that identifies such changes in an individual protein is lacking.

ICH Q6B also provides guidance on the types of impurities that may be found in a product. Impurities may be either process related (i.e., derived from the manufacturing process) or product related, e.g., degradation products that may be inactive and affect the safety profile of the product. Protein aggregates are included in the latter category. Impurities can have a known structure(s), may be partially characterized, or remain unidentified. When adequate quantities of impurities are present or can be generated by stressing the protein, the guidance advises that they be characterized to the greatest possible extent and, where feasible, their biological activity evaluated.

Product characterization sets a basis for establishing specifications, which represent a set of test methods and acceptance criteria that are used in routine product testing and are chosen not to characterize the product but rather to confirm its quality. In contrast to methods used for product characterization that need to be "scientifically sound and provide results that are reliable" (ICH Q5E) (i.e., qualified), methods used in determining specifications and stability should be validated (ICH Q6B; ICH Q5C; ICH Q2 (R1)). This limits the selection of the methods that can be considered for routine product testing because some of the biophysical methods are difficult to validate.

The structural and functional properties of a protein product can change with time. Thus, it is essential to establish criteria for an acceptable product stability pattern and to determine the product shelf life. The stability of a biological product is evaluated following ICH Q1A and ICH Q5C guidances. It is recommended that attributes at risk for change during storage and likely to influence quality, safety, and/or efficacy be monitored. The manufacturer should develop specifications that provide assurance that changes in the purity and potency of the product are detected. ICH Q1A states that "the testing should cover, as appropriate, the physical, chemical, biological, and microbiological attributes" and that "validated stability-indicating analytical procedures should be applied."

The ICH Q6B guidance notes that the inherent degree of structural heterogeneity in proteins due to the biosynthetic processes used by living organisms or resulting from manufacture and/or storage necessitates that the pattern of heterogeneity be defined and lot-to-lot consistency and comparability between lots used in preclinical and clinical studies be demonstrated. The guidance states that "if a consistent pattern of product heterogeneity is demonstrated, an evaluation of the activity, efficacy, and safety (including immunogenicity) of individual forms may not be necessary." When process changes and degradation products result in heterogeneity patterns which differ from those observed in the material used during preclinical and clinical development, the guidance recommends evaluation of the significance of such alterations. The ICH Q6B guidance also notes that under certain circumstances, physicochemical tests may replace a biological assay to measure the biological activity. However, such instances are limited to cases where sufficient physicochemical information about the drug exists, including when the higher-order structure can be thoroughly established by qualified methods, a relevant physicochemical correlation with biologic activity is demonstrated, and a well-established manufacturing history exists.

Changes to the manufacturing process may have an effect on product composition and its structural properties. A determination of comparability of the product pre- and post-manufacturing change can be based on a combination of analytical testing and, in some cases, nonclinical and clinical data when physicochemical or biological assays are not considered adequate to confirm that there has been no adverse effect on the product. ICH Q5E guidance states that "generally, quality data on the pre- and post-change product are generated, and a comparison is performed that integrates and evaluates all data collected, e.g., routine batch analyses, in-process control, process validation/evaluation data, characterisation and stability, if appropriate. The comparison of the results to the predefined criteria should allow an objective assessment of whether or not the pre- and post-change product are comparable."

Since manufacturing process changes may have an impact on protein structure, ICH Q5E advises manufacturers to "attempt to determine that higher-order structure (secondary, tertiary, and quaternary structure) is maintained." If the appropriate higher-order structural information cannot be obtained, the guidance indicates that a relevant biological activity assay with appropriate precision and accuracy could indicate that changes in conformational structure have not occurred. However, it should be noted that the mass of protein aggregates in the product can be so small that their presence has no meaningful impact on the functional test results. The small mass of protein aggregates, especially in the subvisible size range, and their heterogeneity present a significant analytical challenge.

ICH Q5E advises further that to address the full spectrum of physicochemical properties, more than one analytical technique may be needed to evaluate the same quality attribute, e.g., the secondary and tertiary structures and presence of impurities. In such cases, each method should be based on different physicochemical or biological principles to maximize the chance that differences in the product caused by a change in the manufacturing process are detected.

Proteins may be sensitive to changes in buffer composition, processing and holding conditions, and the use of organic solvents. Even slight modifications to manufacturing, storage, and handling may affect the stability of the product. For example, the presence of trace amounts of metal ions might activate proteases or catalyze chemical modifications of the product that may lead to protein aggregation. Therefore, ICH Q5E recommends initiation of real-time/real temperature stability studies on the product potentially affected by the change. Stability studies performed under accelerated conditions, such as increased temperature, light intensity, agitation, and freeze–thaw cycles, may be particularly helpful in identifying subtle changes that are not detectable by routine product characterization, such as may occur following a manufacturing change.

11.5.2 Characterization of Protein Aggregates

The inherent instability of proteins makes them susceptible to conformational changes and aggregation during manufacture, storage, and any changes in the manufacturing conditions. Protein aggregates in biologics present a highly heterogeneous group of protein agglomerates which differ in solubility, size, and structure, all of which makes comprehensive characterization difficult if not impossible given the current state of the available analytical resources. The size of protein aggregates ranges from nanometers to micrometers. The smallest aggregates, with sizes below 100 nm, are often quantified by using SE-HPLC with UV detection. While such an assay is often used for product release, it is not free from limitations (Carpenter et al. 2010b). These limitations include sample dilution, which has the potential to shift the equilibrium for soluble aggregate formation, and exclusion/adsorption of aggregates on the SE-HPLC matrix. Also, extinction coefficients may be conformation dependent, and using the same extinction coefficient for different protein species may lead to erroneous results. Solutions proposed include the use of orthogonal methods, determination of mass recovery, confirmation of method suitability by using stressed protein samples, and development of new media. For HPLC and other methods used for protein aggregates characterization, development of protein standards would be a useful and important advance toward standardizing various characterization methods.

Until recently, protein aggregates with sizes between 100 nm and 10 μm were neither quantified nor characterized (Carpenter et al. 2009), in part due to the lack of relevant analytical techniques. Recently, there is increasing interest in characterization of protein aggregates in this size range. It has been postulated that aggregates of this size may be of particular importance due to risk of an immune response stimulated by their presence. Fortunately, recent advances in the analytical field, including the development of new methods (e.g., resonant mass measurement-based techniques, nanoparticle tracking analysis) and expanding the dynamic range of existing methods (flow microscopy), may eventually allow this gap in our understanding of the impact of subvisible particles on product quality and safety to be bridged.

Standards for the amount of particulate matter in the final container of injectable therapeutics are recommended by the pharmacopeias (ICH Q6B). United States Pharmacopeial Convention (USP)[3] requirements for particulate matter in parenteral products are described in general chapters "Injections" (USP <1>) and "Particulate Matter in Injections" (USP <788>). According to the USP, therapeutic protein solutions need to be "essentially free from visible particles." Each container of the product should be visually inspected for the presence of observable particulate and other foreign matter ("visible particulates"). Every container showing evidence of visible particulates should be rejected. However, this standard was set for "particulate matter" defined as "extraneous mobile undissolved particles, other than gas bubbles, unintentionally present in the solutions." As such, it cannot be directly applied to all biological products because protein aggregates are often unavoidable in protein therapeutics, some of which may contain protein aggregates even in the visible size range.

The meaning of the term "essentially free" has been under discussion for an extended period of time, and attempts to implement some numerical standards have been made (Madsen et al. 2009). Recently, USP has proposed a new general chapter "Visible particulates in injections" (USP <790>), which includes a specification for determining whether a product meets the criterion of being "essentially free" of particles. The chapter also states that "some products, such as those derived from proteins, may contain intrinsic particles of agglomerates" and in such cases recommends meeting the requirements of individual monographs.

[3]USP is a standard setting organization that can aid in ensuring the quality and safety of drugs, dietary supplements, and food ingredients.

Particles with sizes ≥10 μm have been quantified by using the USP chapter <788> methods, light obscuration, and/or microscopic test. The methods and the acceptance criteria were developed for extraneous foreign particles and should not be applied to biologics without critical analysis. The microscopy method, which is applicable to non-proteinaceous products, should not be used for characterizing protein aggregates because they are fragile and translucent, may not be visible on the membrane used for filtration, or may break down and be filtered out.

The limitations of the light obscuration method include the need for large sample volume, which given biological product final container volumes may render the method very costly. The details of sample preparation are also problematic for biological materials; protein aggregates may dissociate or form under the sample degassing conditions (sonication) recommended in chapter <788>. Also, the method cannot distinguish between protein aggregates and extraneous, non-proteinaceous matter (e.g., silicone oil). More objective and data-driven ways to determine the acceptance criteria for particulates in biologic products are being developed based on process performance together with the product safety profile.

The exceptions that are not subject to the requirements of chapter <788> include parenteral products for which the labeling specifies the use of a final filter prior to administration, provided scientific data exist to justify the exemption, such as in-use qualification of the filter. However, we note that the presence of protein aggregates represents not only a potential safety problem but is also a product quality issue. In-line filters may become clogged by protein aggregates, exposing patients to the inconvenience and potential safety risk of having filters changed during infusion of a large volume biologics; thus, the quantity of the aggregates should be controlled even if the product is filtered.

Some of the above limitations were considered during the development of a new USP general chapter "Subvisible particulate matter in therapeutic protein injections" (USP <787>), which addresses the presence of subvisible particles in biotherapeutics and has been published in the Pharmacopeial Forum for comments. Also of note are two recent improvements in methodology. Due to technical advancements, the light obscuration method can be validated to monitor particle size down to 2 μm. In addition, new and promising flow imaging methods may help characterize particulate matter content in the size range that earlier was mainly assessed by light obscuration.

Quantification and characterization of protein aggregates is not an easy task. The complexity is related to differences in protein propensity to aggregate, heterogeneity of aggregation, and unique aggregate-associated potential risk. As discussed above, even with the existing gaps, many analytical methods are available for use, although identifying the relevant test(s) for a specific product is not trivial given the complexities of the protein therapeutic products. It is extremely encouraging that members of the biologics community, which includes industry, academic, and government scientists and regulators, are working together to develop a better understanding of protein aggregates in protein therapeutics and to devise effective methods to identify and mitigate the associated risks (Marszal and Fowler 2012).

11.6 Conclusion

Protein therapeutic products are characterized by a complex three-dimensional structure. This structure both defines and determines a highly specific biologic activity while itself being defined and dependent on the parameters and conditions used for protein production, purification, handling, and storage. Structural changes that may occur during the production and distribution processes will affect product quality and may affect product safety, necessitating the need for a tight control of these processes.

The safety and efficacy of marketed products are established in clinical trials. When changes to the manufacturing process are being implemented, detailed structural characterization, determination of product stability, and demonstration of product structural comparability are necessary to ensure that product quality, and thus safety and efficacy, are comparable to that material used in the clinical trials.

Structural characterization is complex; it requires the use of numerous methods that look at different structural features and are based on different physical principles. Understanding the advantages and limitations of different methods is essential for obtaining meaningful structural information. New methods and improvements to existing methods are being developed to fill the gaps resulting from the limitations of current methods.

Various publications discussed in this chapter, including those promulgated or used by regulatory agencies, provide an overview of the scientific and good manufacturing principles that can be used to facilitate the selection of product parameters to be characterized. However, given the complexity and dynamic nature of proteins, structural characterization can never be complete. It is knowledge of the manufacturing process design space and tight control over process parameters that, in addition to testing and characterization, ensure product quality.

References

Aguzzi A, O'Connor T (2010) Protein aggregation diseases: pathogenicity and therapeutic perspectives. Nat Rev Drug Discov 9:237–248

Alexander PA, He Y, Chen Y, Orban J, Bryan PN (2007) The design and characterization of two proteins with 88 % sequence identity but different structure and function. Proc Natl Acad Sci USA 104:11963–11968

Anderson DE, Becktel WJ, Dahlquist FW (1990) pH-induced denaturation of proteins: a single salt bridge contributes 3–5 Kcal/Mol to the free energy of folding of T4 lysozyme. Biochemistry 29:2403–2408

Anfinsen CB (1973) Principles that govern the folding of protein chains. Science 181:223–230

Appa RS, Theill C, Hansen L, Moss J, Behrens C, Nicolaisen EM, Klausen NK, Christensen MS (2010) Investigating clearance mechanisms for recombinant activated factor VII in a perfused liver model. Thromb Haemost 104:243–51

Aucamp JP, Cosme AM, Lye GJ, Dalby PA (2005) High-throughput measurement of protein stability in microtiter plates. Biotechnol Bioeng 89:599–607

Baudys M, Uchio T, Mix D, Wilson D, Kim SW (1995) Physical stabilization of insulin by glycosylation. J Pharm Sci 84:28–33

Bee JS, Randolph TW, Carpenter JF, Bishop SM, Dimitrova MN (2011) Effects of surfaces and leachables on the stability of biopharmaceuticals. J Pharm Sci 100:4158–4170

Benoit G, Machuca E, Heidet L, Antignac C (2010) Hereditary kidney diseases: highlighting the importance of classical Mendelian phenotypes. Ann N Y Acad Sci 1214:83–98

Bentham J, Bhattacharya S (2008) Genetic mechanisms controlling cardiovascular development. Ann N Y Acad Sci 1123:10–19

Bolen DW, Baskakov IV (2001) The osmophobic effect: natural selection of a thermodynamic force in protein folding. J Mol Biol 310:955–963

Braeckmans K, Buyens K, Bouquet W, Vervaet C, Joye P, De VF, Plawinski L, Doeuvre L, Angles-Cano E, Sanders NN et al (2010) Sizing nanomatter in biological fluids by fluorescence single particle tracking. Nano Lett 10:4435–4442

Braun A, Kwee L, Labow MA, Alsenz J (1997) Protein aggregates seem to play a key role among the parameters influencing the antigenicity of interferon alpha (IFN-Alpha) in normal and transgenic mice. Pharm Res 14:1472–1478

Capriotti E, Fariselli P, Casadio R (2005) I-Mutant2.0: predicting stability changes upon mutation from the protein sequence or structure. Nucleic Acids Res 33:W306–W310

Carpenter JF, Randolph TW, Jiskoot W, Crommelin DJ, Middaugh CR, Winter G, Fan YX, Kirshner S, Verthelyi D, Kozlowski S et al (2009) Overlooking subvisible particles in therapeutic protein products: gaps that may compromise product quality. J Pharm Sci 98:1201–1205

Carpenter J, Cherney B, Lubinecki A, Ma S, Marszal E, Mire-Sluis A, Nikolai T, Novak J, Ragheb J, Simak J (2010a) Meeting report on protein particles and immunogenicity of therapeutic proteins: filling in the gaps in risk evaluation and mitigation. Biologicals 38:602–611

Carpenter JF, Randolph TW, Jiskoot W, Crommelin DJ, Middaugh CR, Winter G (2010b) Potential inaccurate quantitation and sizing of protein aggregates by size exclusion chromatography: essential need to use orthogonal methods to assure the quality of therapeutic protein products. J Pharm Sci 99:2200–2208

Castro PM, Ison AP, Hayter PM, Bull AT (1995) The macroheterogeneity of recombinant human interferon-gamma produced by Chinese-hamster ovary cells is affected by the protein and lipid content of the culture medium. Biotechnol Appl Biochem 21:87–100

Cheng J, Randall A, Baldi P (2006) Prediction of protein stability changes for single-site mutations using support vector machines. Proteins 62:1125–1132

Cromwell ME, Hilario E, Jacobson F (2006) Protein aggregation and bioprocessing. AAPS J 8:E572–E579

den Engelsman J, Garidel P, Smulders R, Koll H, Smith B, Bassarab S, Seidl A, Hainzl O, Jiskoot W (2011) Strategies for the assessment of protein aggregates in pharmaceutical biotech product development. Pharm Res 28:920–933

Dresser DW (1962) Specific inhibition of antibody production. II. Paralysis induced in adult mice by small quantities of protein antigen. Immunology 5:378–388

Edgell MH, Sims DA, Pielak GJ, Yi F (2003) High-precision, high-throughput stability determinations facilitated by robotics and a semiautomated titrating fluorometer. Biochemistry 42: 7587–7593

Elia AE, Albanese A (2010) Emerging parkinsonian phenotypes. Rev Neurol 166:834–840

England JL, Haran G (2011) Role of solvation effects in protein denaturation: from thermodynamics to single molecules and back. Annu Rev Phys Chem 62:257–277

Ericsson UB, Hallberg BM, Detitta GT, Dekker N, Nordlund P (2006) Thermofluor-based high-throughput stability optimization of proteins for structural studies. Anal Biochem 357:289–298

Fares FA, Suganuma N, Nishimori K, LaPolt PS, Hsueh AJ, Boime I (1992) Design of a long-acting follitropin agonist by fusing the C-terminal sequence of the chorionic gonadotropin beta subunit to the follitropin beta subunit. Proc Natl Acad Sci USA 89:4304–8

Filipe V, Hawe A, Jiskoot W (2010) Critical evaluation of nanoparticle tracking analysis (NTA) by nanosight for the measurement of nanoparticles and protein aggregates. Pharm Res 27:796–810

Filipe V, Poole R, Kutscher M, Forier K, Braeckmans K, Jiskoot W (2011) Fluorescence single particle tracking for the characterization of submicron protein aggregates in biological fluids and complex formulations. Pharm Res 28:1112–1120
Fraunhofer W, Winter G (2004) The use of asymmetrical flow field-flow fractionation in pharmaceutics and biopharmaceutics. Eur J Pharm Biopharm 58:369–383
Garrod A (1908) The Croonian lectures on inborn errors of metabolism, lecture II: alkaptonuria. Lancet 2:73–79
Gaudet M, Remtulla N, Jackson SE, Main ER, Bracewell DG, Aeppli G, Dalby PA (2010) Protein denaturation and protein: drugs interactions from intrinsic protein fluorescence measurements at the nanolitre scale. Protein Sci 19:1544–1554
General Chapter <1>Injections (2012) USP 35-NF 30, pp 33–37
General Chapter <787>Subvisible particulate matter in therapeutic protein injections [New] (USP36-NF31 1S) (2012) Pharmacopeial Forum 38(3)
General Chapter <788>Particulate Matter in Injections (2012) USP 35-NF 30, pp 339–342
General Chapter <790>Visible particulates in injections [New] (USP36-NF31 1S) (2012) Pharmacopeial Forum 38(2)
Ghaemmaghami S, Oas TG (2001) Quantitative protein stability measurement in vivo. Nat Struct Biol 8:879–882
Gilis D, Rooman M (2000) PoPMuSiC, an algorithm for predicting protein mutant stability changes: application to prion proteins. Protein Eng 13:849–856
Griko YV, Privalov PL, Venyaminov SY, Kutyshenko VP (1988) Thermodynamic study of the apomyoglobin structure. J Mol Biol 202:127–138
Grothe HL, Little MR, Cho AS, Huang AJ, Yuan C (2009) Denaturation and solvent effect on the conformation and fibril formation of TGFBIp. Mol Vis 15:2617–2626
Haass C, Selkoe DJ (2007) Soluble protein oligomers in neurodegeneration: lessons from the Alzheimer's amyloid beta-peptide. Nat Rev Mol Cell Biol 8:101–112
Harrison M, Maresso K, Broeckel U (2008) Genetic determinants of hypertension: an update. Curr Hypertens Rep 10:488–495
Hartl FU, Hayer-Hartl M (2009) Converging concepts of protein folding in vitro and in vivo. Nat Struct Mol Biol 16:574–581
Hossler P, Khattak SF, Li ZJ (2009) Optimal and consistent protein glycosylation in mammalian cell culture. Glycobiology 19:936–49
Houde D, Peng Y, Berkowitz SA, Engen JR (2010) Post-translational modifications differentially affect IgG1 conformation and receptor binding. Mol Cell Proteomics 9:1716–1728
Huang CT, Sharma D, Oma P, Krishnamurthy R (2009) Quantitation of protein particles in parenteral solutions using micro-flow imaging. J Pharm Sci 98:3058–3071
Huang M, Horwitz TS, Zweiben C, Singh SK (2011) Impact of extractables/leachables from filters on stability of protein formulations. J Pharm Sci 100:4617–4630
ICH Harmonized Tripartite Guideline M4Q(R1) (2002) The common technical document for the registration of pharmaceuticals for human use: quality
ICH Harmonized Tripartite Guideline Q10 (2008) Pharmaceutical quality system
ICH Harmonized Tripartite Guideline Q1A(R2) (2003) Stability testing of new drug substances and products
ICH Harmonized Tripartite Guideline Q2(R1) (2005) Validation of analytical procedures: text and methodology
ICH Harmonized Tripartite Guideline Q5B (1995) Quality of biotechnological products: analysis of the expression construct in cells used for production of r-DNA derived protein products
ICH Harmonized Tripartite Guideline Q5C (1995) Quality of biotechnological products: stability testing of biotechnological/biological products
ICH Harmonized Tripartite Guideline Q5E (2004) Comparability of biotechnological/biological products subject to changes in their manufacturing process
ICH Harmonized Tripartite Guideline Q6B (1999) Specifications: test procedures and acceptance criteria for biotechnological/biological products
ICH Harmonized Tripartite Guideline Q8(R2) (2009) Pharmaceutical development

ICH Harmonized Tripartite Guideline Q9 (2005) Quality risk management

Ikemura T (1985) Codon usage and tRNA content in unicellular and multicellular organisms. Mol Biol Evol 2:13–34

Jenkins N, Castro P, Menon S, Ison A, Bull A (1994) Effect of lipid supplements on the production and glycosylation of recombinant interferon-gamma expressed in CHO cells. Cytotechnology 15:209–15

Jorgensen L, Hostrup S, Moeller EH, Grohganz H (2009) Recent trends in stabilising peptides and proteins in pharmaceutical formulation—considerations in the choice of excipients. Expert Opin Drug Deliv 6:1219–1230

Kiese S, Papppenberger A, Friess W, Mahler HC (2008) Shaken, not stirred: mechanical stress testing of an IgG1 antibody. J Pharm Sci 97:4347–4366

Kimchi-Sarfaty C, Oh JM, Kim IW, Sauna ZE, Calcagno AM, Ambudkar SV, Gottesman MM (2007) A "silent" polymorphism in the MDR1 gene changes substrate specificity. Science 315:525–528

Klein WL, Krafft GA, Finch CE (2001) Targeting small abeta oligomers: the solution to an Alzheimer's disease conundrum? Trends Neurosci 24:219–224

Kochanowski N, Blanchard F, Cacan R, Chirat F, Guedon E, Marc A, Goergen JL (2008) Influence of intracellular nucleotide and nucleotide sugar contents on recombinant interferon-gamma glycosylation during batch and fed-batch cultures of CHO cells. Biotechnol Bioeng 100: 721–33

Kramer G, Boehringer D, Ban N, Bukau B (2009) The ribosome as a platform for co-translational processing, folding and targeting of newly synthesized proteins. Nat Struct Mol Biol 16:589–597

LaPolt PS, Nishimori K, Fares FA, Perlas E, Boime I, Hsueh AJ (1992) Enhanced stimulation of follicle maturation and ovulatory potential by long acting follicle-stimulating hormone agonists with extended carboxyl-terminal peptides. Endocrinology 131:2514–2520

Lavinder JJ, Hari SB, Sullivan BJ, Magliery TJ (2009) High-throughput thermal scanning: a general, rapid dye-binding thermal shift screen for protein engineering. J Am Chem Soc 131:3794–3795

Lewis JD, Nail SL (1997) The influence of ion exchange chromatography conditions on aggregation of equine IgG. Process Biochem 32:279–283

Liu H, Bulseco GG, Sun J (2006) Effect of posttranslational modifications on the thermal stability of a recombinant monoclonal antibody. Immunol Lett 106:144–153

Luo Q, Joubert MK, Stevenson R, Ketchem RR, Narhi LO, Wypych J (2011) Chemical modifications in therapeutic protein aggregates generated under different stress conditions. J Biol Chem 286:25134–25144

Madsen RE, Cherris RT, Habushnig JG, Hunt DG (2009) Visible particulates in injections-a history and a proposal to revise USP general chapter injections <1> Pharmacopeial Forum 35:1383–1387

Maity H, O'Dell C, Srivastava A, Goldstein J (2009) Effects of arginine on photostability and thermal stability of IgG1 monoclonal antibodies. Curr Pharm Biotechnol 10:761–766

Marszal E, Fowler E (2012) Overview: workshop on predictive science of the immunogenicity aspects of particles in biopharmaceutical products. J Pharm Sci 101:3555–9

Masso M, Vaisman II (2008) Accurate prediction of stability changes in protein mutants by combining machine learning with structure based computational mutagenesis. Bioinformatics 24:2002–2009

McCarthy EF (2011) Genetic diseases of bones and joints. Semin Diagn Pathol 28:26–36

Meager A, Dolman C, Dilger P, Bird C, Giovannoni G, Schellekens H, Thorpe R, Wadhwa M (2011) An assessment of biological potency and molecular characteristics of different innovator and noninnovator interferon-beta products. J Interferon Cytokine Res 31:383–392

Meeker AK, Garcia-Moreno B, Shortle D (1996) Contributions of the ionizable amino acids to the stability of staphylococcal nuclease. Biochemistry 35:6443–6449

Mitra N, Sinha S, Ramya TN, Surolia A (2006) N-linked oligosaccharides as outfitters for glycoprotein folding, form and function. Trends Biochem Sci 31:156–163

Moore WV, Leppert P (1980) Role of aggregated human growth hormone (hGH) in development of antibodies to hGH. J Clin Endocrinol Metab 51:691–697

Neuhofer W, Muller E, Grunbein R, Thurau K, Beck FX (1999) Influence of NaCl, urea, potassium and pH on HSP72 expression in MDCK cells. Pflugers Arch 439:195–200

Obici L, Perfetti V, Palladini G, Moratti R, Merlini G (2005) Clinical aspects of systemic amyloid diseases. Biochim Biophys Acta 1753:11–22

Parthiban V, Gromiha MM, Schomburg D (2006) CUPSAT: prediction of protein stability upon point mutations. Nucleic Acids Res 34:W239–W242

Pasinelli P, Brown RH (2006) Molecular biology of amyotrophic lateral sclerosis: insights from genetics. Nat Rev Neurosci 7:710–723

Patel J, Kothari R, Tunga R, Ritter NM, Tunga BS (2011) Stability considerations for biopharmaceuticals, Part 1. Overview of protein and peptide degradation pathways. BioProcess Int 9:20–31

Percy MJ, Rumi E (2009) Genetic origins and clinical phenotype of familial and acquired erythrocytosis and thrombocytosis. Am J Hematol 84:46–54

Pfeil W, Privalov PL (1976) Thermodynamic investigations of proteins. III. Thermodynamic description of lysozyme. Biophys Chem 4:41–50

Philo JS (2009) A critical review of methods for size characterization of non-particulate protein aggregates. Curr Pharm Biotechnol 10:359–372

Rademacher TW, Parekh RB, Dwek RA (1988) Glycobiology. Annu Rev Biochem 57:785–838

Rathore N, Rajan RS (2008) Current perspectives on stability of protein drug products during formulation, fill and finish operations. Biotechnol Prog 24:504–514

Rosgen J (2007) Molecular basis of osmolyte effects on protein and metabolites. Methods Enzymol 428:459–486

Scriver CR (2007) The PAH gene, phenylketonuria, and a paradigm shift. Hum Mutat 28:831–845

Senger RS, Karim MN (2003) Effect of shear stress on intrinsic CHO culture state and glycosylation of recombinant tissue-type plasminogen activator protein. Biotechnol Prog 19:1199–209

Sharp PM, Cowe E, Higgins DG, Shields DC, Wolfe KH, Wright F (1988) Codon usage patterns in Escherichia coli, Bacillus subtilis, Saccharomyces cerevisiae, Schizosaccharomyces pombe, Drosophila melanogaster and Homo sapiens; a review of the considerable within-species diversity. Nucleic Acids Res 16:8207–8211

Shire SJ (2009) Formulation and manufacturability of biologics. Curr Opin Biotechnol 20:708–714

Shukla AA, Gupta P, Han X (2007) Protein aggregation kinetics during Protein A chromatography: case study for an Fc fusion protein. J Chromatogr A 1171:22–28

Sola RJ, Rodriguez-Martinez JA, Griebenow K (2007) Modulation of protein biophysical properties by chemical glycosylation: biochemical insights and biomedical implications. Cell Mol Life Sci 64:2133–2152

Stepanenko OV, Verkhusha VV, Shavlovsky MM, Kuznetsova IM, Uversky VN, Turoverov KK (2008) Understanding the role of Arg96 in structure and stability of green fluorescent protein. Proteins 73:539–551

Stites WE, Byrne MP, Aviv J, Kaplan M, Curtis PM (1995) Instrumentation for automated determination of protein stability. Anal Biochem 227:112–122

Struble EB, Ladner JE, Brabazon DM, Marino JP (2008) New crystal structures of ColE1 Rom and variants resulting from mutation of a surface exposed residue: implications for RNA-recognition. Proteins 72:761–768

Tang X, Pikal MJ (2004) Design of freeze-drying processes for pharmaceuticals: practical advice. Pharm Res 21:191–200

Thirumangalathu R, Krishnan S, Ricci MS, Brems DN, Randolph TW, Carpenter JF (2009) Silicone oil- and agitation-induced aggregation of a monoclonal antibody in aqueous solution. J Pharm Sci 98:3167–3181

Thorens B, Vassalli P (1986) Chloroquine and ammonium chloride prevent terminal glycosylation of immunoglobulins in plasma cells without affecting secretion. Nature 321:618–620

Uchida E, Morimoto K, Kawasaki N, Izaki Y, Abdu Said A, Hayakawa T (1997) Effect of active oxygen radicals on protein and carbohydrate moieties of recombinant human erythropoietin. Free Radic Res 27:311–323

Vazquez-Rey M, Lang DA (2011) Aggregates in monoclonal antibody manufacturing processes. Biotechnol Bioeng 108:1494–1508

Wright A, Morrison SL (1998) Effect of C2-associated carbohydrate structure on Ig effector function: studies with chimeric mouse-human IgG1 antibodies in glycosylation mutants of Chinese hamster ovary cells. J Immunol 160:3393–3402

Xu J, Reumers J, Couceiro JR, De SF, Gallardo R, Rudyak S, Cornelis A, Rozenski J, Zwolinska A, Marine JC et al (2011) Gain of function of mutant P53 by coaggregation with multiple tumor suppressors. Nat Chem Biol 7:285–295

Yin S, Ding F, Dokholyan NV (2007) Modeling backbone flexibility improves protein stability estimation. Structure 15:1567–1576

Zimmerman TP (2006) Yield improvement for manufacture of α_1-proteinase inhibitor. Vox Sang 91:309–315 [illegible] M, [illegible] (20[illegible]) [illegible]

[illegible] M, [illegible] A (201[illegible]) [illegible]

[illegible]

[illegible] Nature [illegible] (201[illegible]) [illegible]

[illegible] (19[illegible]) [illegible] 576

[illegible]

Index

L.O. Narhi (ed.), *Biophysics for Therapeutic Protein Development*, Biophysics for the Life Sciences 4, DOI 10.1007/978-1-4614-4316-2,

G

T

MIX
Papier aus verantwortungsvollen Quellen
Paper from responsible sources
FSC® C105338

If you have any concerns about our products,
you can contact us on
ProductSafety@springernature.com

In case Publisher is established outside the EU,
the EU authorized representative is:
Springer Nature Customer Service Center GmbH
Europaplatz 3, 69115 Heidelberg, Germany

Printed by Libri Plureos GmbH
in Hamburg, Germany